THOMAS BEDDOES M.D. 1760–1808

CHEMISTS AND CHEMISTRY

A series of books devoted to the examination of the history and development of chemistry from its early emergence as a separate discipline to the present day. The series will describe the personalities, processes, theoretical and technical advances which have shaped our current understanding of chemical science.

DOROTHY A. STANSFIELD

THOMAS BEDDOES M.D.
1760–1808

Chemist, Physician, Democrat

D. REIDEL PUBLISHING COMPANY

A MEMBER OF THE KLUWER ACADEMIC PUBLISHERS GROUP

DORDRECHT / BOSTON / LANCASTER

Library of Congress Cataloging in Publication Data

Stansfield, Dorothy A., 1914–
    Thomas Beddoes, M.D., 1760–1808.

    (Chemists and chemistry)
    Bibliography: p.
    Includes indexes.
    1. Beddoes, Thomas, 1760–1808.   2.   Chemists – England –
Biography.   3.   Physicians – England – Biography.   I.   Title.
II.   Series.
QD22.B263S83   1984   540'.92'4 [B]   84-8421
ISBN-13:978-94-009-6305-4   e-ISBN-13:978-94-009-6303-0
DOI: 10.1007/978-94-009-6303-0

Published by D. Reidel Publishing Company.
P.O. Box 17, 3300 AA Dordrecht, Holland.

Sold and distributed in the U.S.A. and Canada
by Kluwer Academic Publishers,
190 Old Derby Street, Hingham, MA 02043, U.S.A.

In all other countries, sold and distributed
by Kluwer Academic Publishers Group,
P.O. Box 322, 3300 AH Dordrecht, Holland.

For Ronald

TABLE OF CONTENTS

PREFACE

We meet in Thomas Beddoes an able chemist, engaged in a field where important new discoveries were being made; a good doctor eager to find experimentally sound ways of healing and to make known the principles of maintaining good health; a vigorous, independent man sharing the hope which the ideas of the French Revolution gave so many of his contemporaries. In his life he was a controversial figure and judgement and detached appreciation of his work was often made impossible by anger at his 'revolutionary' political views. It becomes evident that where Beddoes was held in esteem and where he had influence it was not for particular activities but for what he was 'in the round'. With due respect — and with gratitude — to specialist accounts of his achievements as a chemist and of his endeavours to find a cure for pulmonary consumption and his efforts to bring about an understanding of the importance of preventive medicine, I have tried in this account to 'see him whole'. Historians of chemistry and of medicine; educationalists; and those concerned with 'women's studies' will each continue to find particular episodes or parts of Beddoes' life of special interest. At the same time I hope this, the first attempt at a biography — for J. E. Stock's 1811 account is truly named "Memoirs" — will add to our understanding of his varied activities.

This study had originally a quite limited aim; to trace in greater detail the part Thomas Beddoes, a chemist, played in making known to Samuel Taylor Coleridge the work of the German philosopher, Emmanuel Kant and, in general, German ideas and writings. The warm account of Beddoes' personality given by Dr Cartwright in his "English Pioneers of Anaesthesia" led me to widen my scope. During the years 1789–1801 Coleridge was one of the group at the Pneumatic Institute in Clifton where Beddoes was testing his hypothesis that pulmonary tuberculosis might be usefully treated by the use of gases and there he was introduced to new ideas and new experiences. As a result of Humphry Davy's work the experiments focussed on nitrous oxide and Coleridge took part in the serious experiments in the laboratory, not only in light-hearted sessions of breathing this 'laughing gas' just for the sensation. So he came to feel the intellectual excitement of the recent discoveries in the chemistry of gases and the mysterious elation and 'heightened sensibility' that came from a few inhalations of nitrous oxide.

Once we attempt to go beyond this to discover the scientific ideas that made the experience at the Pneumatic Institute more than a short-lived excitement for Coleridge and to endeavour to find out what made Beddoes so important both for the poet and for the young chemist, Humphry Davy, we are in difficulties. For Davy, clearly, Beddoes was important as his mentor in chemistry, but for Coleridge there was much more than a simple introduction to new writers. From the time they met in 1794 a whole cluster of shared ideas and activities linked Beddoes and Coleridge. It was a time of development and change in Coleridge's political and philosophical thinking; he drew away from his older friend, but the excitement of chemistry, the friendship with Beddoes and Davy had important and lasting results. This has been treated in depth by Professor T. H. Levere. I hope the account given here of Coleridge and Beddoes' friendship will be a useful prelude and an addition to our understanding of Coleridge's early work.

The origin of this study itself reminds us that the time was one before specialisation — certainly before the 'two cultures'. Thomas Beddoes himself wrote verse; his older contemporary, Erasmus Darwin, was admired as a poet. A century later, Erasmus' grandson Charles sadly admitted that as his mind had become a machine for "grinding general laws out of a large collection of facts", he had quite lost the power to enjoy poetry. Thomas Beddoes with his many interests and activities can give us some feel of that earlier world.

To do justice to Beddoes' many-sided activities a thematic rather than a strictly chronological treatment has been used. I freely admit that I have written this account of Beddoes' life from a point of view entirely in sympathy with his generally 'progressive' and politically democratic views. The time seems right to correct the over-simplified rejection of Beddoes as a 'revolutionary' and only by standing on his side can we understand what his convictions were and how they penetrated all his work. I am not unmindful of the arguments in support of Pitt and Burke.

My exploration began entirely to satisfy my own curiosity and has continued without being connected with any learned institution. In these circumstances I have been especially grateful to those busy academics who have found time to talk of Beddoes, to challenge and even to encourage. They are so many and over such a long period that some will probably have forgotten and I must make my thanks in general to them all. Among them I most specially thank Dr Hugh Torrens of Keele University, who encouraged me from the beginning, strengthened my flagging will to continue and in the end generously shared with me his large collection of Beddoes material. My sincere thanks are due also to Professor T. H. Levere of the University of

Toronto who has spared time on his visits to this country for helpful and stimulating conversation about the whole area of Beddoes' work, and to Michael Neve of the History of Medicine Unit, University College, London, who has frequently discussed with me the various stages of this work.

Through the kindness of Lord Gibson-Watt, I was able to read the letters of Thomas Beddoes and of James Watt Jr in his private collection of Watt family papers. These letters were invaluable in connection with the help given by James Watt in the making of the apparatus for the breathing of gases and in the setting up of the Pneumatic Institution.

To my friends, Mr and Mrs D. Corser, Miss E. Dudley and Miss B. Lewis, I offer most sincere thanks for their hospitality on many occasions during my search for Beddoes material and for their readiness to enter into my enthusiasm. Miss Dudley's kind and expert help in identifying material in the Bodleian Library has saved me much time and has been invaluable.

At home I have received most kind help from those who have typed my untidy manuscript and who have given me loyal domestic support.

I must put on record the part played by my husband, Ronald, in the making of this book. He has most patiently and kindly undertaken support work, correcting, editing, retyping. Much more: he has been ever ready to live with Beddoes, to make clear his scientific work, to discuss the significance of his activities. Without him, I could only have written another partial account of Beddoes, for, most important of all, it was my husband who first, long ago, led me to appreciate the excitement of science.

ACKNOWLEDGEMENTS

I take this opportunity of thanking the Librarians and Archivists who gave me so much help during the course of my research.

I gratefully acknowledge permission to quote from manuscript collections listed here:

Trustees of the Matthew Boulton Trust	Letters of J. Watt, Junior	Birmingham Public Library
Mrs C. Colvin	Letters of Maria Edgeworth	Bodleian Library, Oxford
Mr C. E. Corbett	Diary of Katherine Plymley	Shropshire Record Office
Mr C. J. Davies-Gilbert	Correspondence of Davies Gilbert	Cornwall Record Office
Mrs T. Fletcher	Papers of Thomas Beddoes	Bodleian Library, Oxford
Lord Gibson-Watt	Watt family papers	Private collection
Mr T. D. G. Sotheron-Estcourt, of Tetbury	Letters of Thomas Beddoes	Gloucester Record Office
Messrs Josiah Wedgwood, Barlaston, Stoke-on-Trent and Keele University	Correspondence of Thomas Wedgwood	Keele University Library

and

Bristol City Record Office	Family records of Dr J. King
Edinburgh University Library	Joseph Black correspondence and papers
Friends House Library	Papers relating to Dr Edward Long Fox
Lichfield Record Office and Shropshire Record Office	Legal documents relating to the Beddoes family
Royal Institution Library	Letters and note book of H. Davy

and from holders of copyright material:

Dr. F. F. Cartwright	*The English Pioneers of Anaesthesia*
Miss M. Maby	*Life and Letters of Dr J. King*
Bristol City Central Library	Contemporary printed records
British Library, Bloomsbury	Extracts from the published works of Dr Thomas Beddoes and from nineteenth-century published sources

THOMAS BEDDOES

1754

                                                1758   W. Reynolds born. d. 1803

1760   Born at Shifnal, Shropshire

1766

                                                1767   D. Giddy born. d. 1839

                                                1772   S. T. Coleridge born. d. 1834

1774

1775

1775

1776   To Oxford, Pembroke College

                                                1778   H. Davy born. d. 1829

1779

1781   To London: studying under Sheldon

1782/3

1784   To Edinburgh

1786   M. D. Oxford December

1787   Working in Oxford

               Visit to France

1788   Chemical Reader

1789

1791

1792   Decision to leave Oxford

1791/2

1792

1793   Leaves Oxford; to Bristol

POLITICAL AND SCIENTIFIC EVENTS

1754   *Black's work on carbon dioxide (fixed air)*

1766   *Cavendish's work on hydrogen (inflammable air)*

1774   *Priestley's work on gases – discovers oxygen (dephlogisticated air) – isolated independently by Scheele 1772. Publication of 'Observations on Different Kinds of Air' 1774–86*

1775   *Lavoisier's work on nitrogen (azote) 1775–6*

1775   War with America – American Revolution/American War of Independence

1779   The Iron bridge erected over the Severn at Coalbrookdale

1782/3 Treaty of Paris – Peace with America

1787   *Publication of 'La Nouvelle Nomenclature Chimique'*

1788   Trial of Warren Hastings (1788–1795)

1789   Fall of the Bastille

1791   Birmingham Riots, Destruction of Priestley's house and laboratory

1792   French clergy required to take oath of allegiance to the state

1791/2 Paine's 'Rights of Man' published

1792   September massacres in Paris. 22nd September Declaration of the French Republic

1793   Jan.: England at War with France; Godwin's 'Political Justice' published

1793/4

1794   Marriage to Anna Maria Edgeworth; Coleridge in Bristol

1796/98
1798–1801   Davy in Bristol: Pneumatic Institute active
1798/9   Coleridge in Germany
1800   Coleridge moves to Keswick
1801   Birth of Anna; Thomas Lovell b. 1803; Henry b. 1805; Mary
            b. 1808
1801   Davy takes up his appointment at the Royal Institution, London
1802
1803   Dr Frank of Vienna visits Beddoes at Clifton, where Beddoes
            had established the Institution for Preventive Medicine
1805

1806   Thinks of moving to London; prevented by serious illness
1807

1808   Death in Clifton, 24th Dec.
1811   Publication of Stock's 'Memoirs'

1793/4 French armies victorious in the Netherlands; Threat of invasion by French armies; Food shortages and riots in England

1794 Trials of Hardy, Thelwall, Horne Tooke for Treason; *Execution of Lavoisier*

1795 The Two Acts (the 'Gagging Acts') against seditious meetings and publications passed

1796/98 French attempts to support rebellion in Ireland

1798 Irish Rebellion

1802/3 Peace of Amiens

1805 *Davy isolates sodium and potassium*

1805 Oct.: Battle of Trafalgar gives England naval supremacy and power to blockade France

1805 Dec.: Battle of Austerlitz. Napoleon in control of Western Europe

1807 Treaty of Tilsit between Napoleon and the Emperor Alexander of Russia consolidates Napoleon's power

1811 War with France continues for another four years

Thomas Beddoes M.D. 1760–1808. (Reproduced by kind permission of the National Portrait Gallery, London.)

CHAPTER 1

INTRODUCTION

Thomas Beddoes died on 24th December, 1808 at the age of forty-eight, a disappointed and lonely man. He had hoped that the new chemistry of gases would provide means of treating disease, in particular pulmonary tuberculosis, an especially tragic condition, since it was both widespread and invariably fatal. With considerable courage Beddoes set himself to make a scientific test of this hope. The Pneumatic Institute which he brought into being was the culmination of his plans. Active only for the three years 1798–1801, it is for this Institute that Beddoes is chiefly remembered; often it is pointed out as the place where Humphry Davy began his career or, less seriously, as the scene of foolish episodes in which nitrous oxide was breathed for the sake of a novel and amusing experience. Yet it has been remembered too as "one of the first of the now familiar research institutes" and as such the seriousness of its work has been recognised. Beddoes' effort ended in failure but it was a humane and reasonable endeavour, supported by serious and practical men. Thomas Wedgwood's contention that the attempt was worthwhile for itself, even to establish a negative result, was right and had it been remembered a much fairer estimate of Beddoes' work might have been made. The circumstances of the time and, it must be admitted, Beddoes' own misjudgements defeated him. Much that was interesting and worthwhile in his work has been forgotten, the failures remembered with scorn. Yet at the time of his death, Beddoes had friends made many years before who continued to hold him in esteem and affection. There were men who might have given an account of his work which would have done him justice and carried weight: Davies Gilbert (Giddy) [1], M. P. for Bodmin and later President of the Royal Society; the poet Samuel Taylor Coleridge; and Humphry Davy, whose scientific achievement was already recognised in 1808. These three had shared Beddoes' hope that in Bristol, where he had worked for the past fifteen years, something might be accomplished to the benefit of men. The friendship with Davies Giddy had begun in Oxford twenty years before Beddoes' death and it was to Giddy that Beddoes had confided, step by step, the progress made during the long period of planning to bring into being his Pneumatic Institution. It was in protesting against the government's repressive measures in the mid-1790s that Beddoes and Coleridge first met and then Coleridge had been caught up in the excitement of the 'new chemistry'; Davy, the young superintendent of the Institute's laboratory, had entered with enthusiasm into Beddoes' plans for the scientific trials of gases. Both were deeply moved at his death. Yet Anna Beddoes was not able to find a biographer for her husband among these intimate friends. She looked for someone who would be able to make a correct record of Beddoes' work and felt that the responsibility should be entrusted to "a judicious medical man".[2] She chose Dr John Edmonds Stock[3] who in 1804 had joined Beddoes' group of doctors at the Institution for Preventive Medicine, the successor to the Pneumatic Institute. Davy, Coleridge and their friend Southey were all aghast. Coleridge, hearing that both Giddy and Davy had felt unable to undertake the work and probably realising that he himself was incapable of the sustained effort it would require, wrote in distress to Davy:

I feel permitted to avow the pain, yea the sense of shame with which I contemplate Dr Stock as a performant. I could not help assenting to Southey's remark, that the proper vignette for the work would be a funeral lamp besides an urn and Dr Stock in the act of placing an extinguisher on it.[4]

Anna Beddoes, anxious that no harm should come to her husband's reputation, judged that Dr Stock would be able to find his way in the confusion of Beddoes' papers, "leaving out what would injure him".[5] In this respect, Stock was a very conscientious biographer and Anna would have had little to complain of when the Memoir was published in 1811. A very large part of the account had been devoted to a summary of Beddoes' publications. Stock collected, and preserved, material that would otherwise have been scattered, but his determination to pass quickly over, or ignore, Beddoes' non-medical activities obscured much that was important and significant in Beddoes' life. In concentrating on Beddoes the doctor, he hid Beddoes the man. This and his inexperience as a writer, brought about the fulfilment of Coleridge's prophecy. Once the obituaries were written, little more appeared in the nineteenth century by way of assessment of Beddoes' work.

In spite of Coleridge's strictures, Stock's biography may not have been the only reason for this silence. At the time of his death, opinions were sharply divided about Beddoes. There were many to admire his originality and intellectual courage and to appreciate his efforts to alleviate suffering. At the same time, his enemies were vehement and remorseless in recalling his sympathy, in its early stages, with the French Revolution and the unscientific exuberance of some of his experiments with nitrous oxide. This criticism was political rather than scientific, though it spared no part of

Beddoes' work. It sprang from the reaction against sympathy with the revolution in France and the ideas associated with it, that gathered force in the 1790s. The underlying mood from which this reaction arose continued to influence people's judgement of the men and ideas of the late eighteenth century and, with nothing but Stock's pedestrian account as a counterweight, stood in the way of an understanding of Beddoes' aims and work. Anna Beddoes, it seems, was well aware of the power of conservatism and the strength of memory: "I am sorry that Tom publishes," she wrote in 1823 of her son's play *The Bride's Tragedy*, "the fooleries of a youth in black and white are sometimes maliciously remembered against him".[6] She could well have been thinking of earlier anxieties connected with her husband's work.

It is possible that she chose Dr Stock for reasons other than his ability to edit Beddoes' medical work.[7] As a biographer, Stock could be trusted to refer discreetly to Beddoes' political activities. In 1808 he was respectably established in Bristol; even his Unitarianism was an acceptable unorthodoxy, but he had once held, and acted upon, radical beliefs, both in religion and in politics. In 1794, when he was a medical student in Edinburgh, he had been implicated in a plot which planned the secret collecting of arms, an attack on the garrison of the Castle, and the seizure of the Lord Provost and of Lord Chief Justice Braxfield. The leader, Robert Watt, had been tried for treason and executed in October 1794; Stock had escaped by fleeing the country. When he was able to return from America, the leaders of the movement for a British Convention who had been condemned to transportation by Braxfield at the Edinburgh Assizes in January 1794 should still have been serving their harsh fourteen year sentences. But of these "Edinburgh martyrs", two had already died and only one, Maurice Margot, was eventually to return to Britain. Though Stock had escaped lightly — he had been able to continue his medical studies in Philadelphia and was pardoned on his return — it is not surprising that he had no wish to recall the time when he, and Thomas Beddoes, had suffered from the fervour of the reactionaries. He fulfilled Anna Beddoes' brief, but his success has hidden from us the unity of Beddoes' work as scientist and doctor and its connection with his political thinking and activities.

It is perhaps understandable that Stock felt that his first responsibility was to put on record an account of Beddoes' medical work, especially as he knew this was the wish of Anna Beddoes. Writing so soon after the doctor's death he must have been conscious of the prejudice that still stood in the way of a balanced estimate of his aims and achievements and may well have

felt that his memoir would disarm rather than encourage Beddoes' enemies.
He was, as it turned out, mistaken. Stock was well aware of the harsh judgements of Beddoes' character as well as the professional criticism levelled
against him. To end his work he attempted a more personal description,
recognising how Beddoes was seen by many and trying to give a balanced
account. He hoped to show the reality behind Beddoes' appearance of meanness, hidden by the bad temper of his writings and the taciturnity of his
manner. Stock at last wrote with more warmth, more 'feel' for his subject
but even so, he revealed almost nothing of what the opposition and disappointments must have meant for Beddoes. The violent language; the irony
and jesting that we see in Beddoes' own writings; the constant trying to find
a new and effective form for his protests, all point to the intensity of his
feelings and even at times to a sensitiveness about the prejudice against him.
These outbursts come to an end but we have no way of telling whether, after
1803, he was happily absorbed in his practice as a doctor and his work for
preventive medicine or whether his constant activity hid regrets and a hope
that he still might achieve something. Beddoes' inability ever to speak about
the Prime Minister, William Pitt, without anger takes us some way to understanding his feelings and so does a cautious comment by Stock about Beddoes'
supposed avariciousness.[8] His carefulness about money at the time when
the Pneumatic Institute was being formed arose, Stock pointed out, because
this was a time when Beddoes was "struggling to emerge from a cloud which
had cast a gloom over his future prospects and dissevered many of his most
valued connections."

At the same time, in concentrating on summarising the writings, Stock
focuses too narrowly and isolates Beddoes' work. It was as an explorer of
new thinking and of insights which were just becoming widely current; in
applying and communicating these ideas, that Beddoes must have been valued
in his own day. It is by virtue of this that he, whose major work was counted
a failure, has interest and importance today. He visited the French chemists
of his day and corresponded with Joseph Black whose pupil he had been in
Edinburgh. He was a friend of Erasmus Darwin with whom he exchanged
letters on geological problems, and on Darwin's theory of perception. Beddoes
was asked to read Darwin's *Zoonomia* before publication and Darwin became
Beddoes' strongest supporter in the trial of gases as a medical treatment.
Darwin wrote publicly in support of Beddoes; tried pneumatic medicine for
his own patients and, by passing on James Watt's design for a breathing
apparatus to Beddoes, made possible the last decisive step to setting up the
Pneumatic Institute. Beddoes moved on terms of equality among the men

connected with the Lunar Society, James Keir, the Wedgwoods, James Watt. Then through his own activities he reached a new group. After 1794, when he came to collect medical reports first on pulmonary tuberculosis, then later on venereal disease and on influenza, he must by his inquiry have had considerable influence on the thinking of those medical men who replied. It is always difficult to assess the importance of someone who influenced rather than achieved; viewed in this way we can perhaps come to a truer estimate of Beddoes and see his influence on Humphry Davy not as almost an accident but as belonging naturally to what Beddoes was doing.

Though there can be no doubt that Beddoes was frustrated by the severe political reaction against anyone who had shown sympathy with the ideas of the French Revolution, so far as we know Beddoes belonged to no subversive association and took part in no illegal activity. But he was hot-headed and incautious. Erasmus Darwin, who also was known as a republican and an atheist and who was aware that he was being spied upon still continued to be respected as the foremost doctor of his time. His reputation was already secure, but Beddoes, thirty years younger, had not established himself and was never able to free himself from suspicion. The effects of political repression in England during the struggle with France are often thought of in political or social terms; possibly, so far as individual lives are concerned, the names that come most easily to mind are Wordsworth, Southey, Coleridge. Among scientists there is always to be remembered Joseph Priestley, the end of his career broken by the impossibility of continuing in England after riots in Birmingham had destroyed his laboratory and library; and the tragedy of Lavoisier, guillotined in 1794 by a revolution that had no need of scientists. Thomas Beddoes gives us another example of how the pressures of the years of revolution and repression affected an individual life.

CHAPTER 2

EARLY LIFE

I. Shropshire Background

Like many others in the west midland counties, Thomas Beddoes' family came from Wales. His grandfather, after whom he was named, came to Shifnal, a small town in the east of Shropshire, at the time of his marriage with Rosamond Phillips and there bought land for a tanyard. Their only son, Richard, married Ann Whitehall from Hardsley, near Ludlow. They had one other child besides Thomas, a daughter called Rosamond after her father's mother.[1]

In his forty-eight years, 1760–1808, Thomas Beddoes lived through a time when great events seemed to rush together and it is not surprising to find him responding to the dramatic changes which took place during his life. He was fifteen when Paul Revere galloped through the night to Lexington and the American colonists turned from reasoned argument to fighting, from protest against unfair taxation to determination to win independence; and twenty-nine when the Paris mob stormed the Bastille, the event which immediately became the symbol of the overthrow of the *ancien régime*. By the time Beddoes was in his mid-thirties these two revolutions had brought into being two nations embodying a philosophy totally antipathetic to the Hanoverian concept of monarchy and even to the idea of a society where authority rested on traditional rights. In the United States of America and in France the new constitution of the state stemmed from the concept of universal rights. Yet ideals of independence and equality were widely held in England too, where there were men who regretted that the dispute with the colonists had come to war and who deplored the brutalities of fighting. There were men who felt strongly enough about these principles to ignore national boundaries. It was Thomas Paine from England who saw that nothing short of independence was the right aim for America and who won the hesitant among the colonists to this view by his pamphlet, 'Common Sense';[2] and similarly it was the Marquis de Lafayette from France who took a decisive command in Washington's army. These were events of Beddoes' student days. Then, as he was settling into his career as a university teacher, there were Englishmen happy to carry congratulatory addresses to the

Assembly of Revolutionary France and to accept French citizenship. Another revolution equally without regard to national boundaries was changing the understanding of the natural world and of man himself. In chemistry, medicine and geology, by experiment and observation new discoveries and new theories were coming into being. Between Black, Priestley, Lavoisier in chemistry, among the medical schools in Leyden, Edinburgh and America, ideas flowed across national boundaries even in the years of revolution and war. Such currents of thought originating among philosophers and theoreticians are in every age kept in being, even given vitality, by men who are willing to risk taking new ideas seriously; to submit them to the test of discussion and to attempt to try them in practical situations. These men are the true eccentrics, not outsiders but respected, well settled in society, yet standing off from the closed circle of established ideas; among them many become the victims of the dialectic of action and reaction. In the England of the 1780s there were such men. Members of the great Whig families, of the landed gentry; some professional men and the leaders of the Dissenting congregations held these ideals, welcomed the evidences of change in America and France and hoped for reform in England. They were linked by a variety of interests, often scientific as well as political, and it is among them that we find Thomas Beddoes.

We know very little of Beddoes' childhood, for Stock is principally concerned with his education and with incidents that reveal his intelligence and determination. The boy Stock describes grew up in the centre of some of the most far-reaching developments of his time and saw around him dramatic changes which made a deep impression. This was no less than the bringing into being of a new industrial society. Shropshire was one of the most important of the provincial centres where, rather than in London, this new life was coming into being. One consequence of this exciting growth, concentrated in place and time, is that the history of Shropshire in the late eighteenth century had been so well described that it is possible to know the changes Beddoes must have seen. Shifnal is today just outside the eastern boundary of Telford New Town, named to honour the engineer Thomas Telford and to keep alive the memory of the ironmasters and canal builders of the eighteenth century. It unites the towns where in Beddoes' lifetime there flourished one of the largest iron manufactories of England and where, if in any one place, it is possible to see the beginning of the industrial society of the next 150 years. In Coalbrookdale, Madeley, Broseley, Ketley Bank and in other parts of the east Shropshire coalfield, the iron-masters were energetically developing new processes of smelting and new methods of mining and were constantly extending the application of steam power and the use that could be made of cast and wrought iron. Transport and travel were revolutionized; the main roads were turnpiked and maintained; new bridges were built and a network of canals developed. The river Severn was the natural means of communication and as a consequence the whole district had traditionally looked towards Wales and the south west. From Shrewsbury to Gloucester and the port of Bristol the river carried the products of the district and from the port brought in groceries, wines and household goods.[3] The new canals as they developed in the later part of the century, at first made it easier to carry goods from the river boats to and from the foundries and coal workings. In the next stage, as the canal network grew, it became easier to transport goods direct from Shropshire to the manufactories round Birmingham and to Merseyside and Lancashire. Yet these changes did not bring about a separation of industry from the countryside. Shropshire remained rural, even in the remoter parts a wild countryside, and Shifnal essentially a country place.

According to a nineteenth century account, Beddoes was born in a house in the High Street and though the date given is mistaken, this may well be correct. Behind the High Street flows the little river Warfe (or Worfe). Further down, its course cuts a deep valley before it joins the Severn at Bridgnorth but at Shifnal it runs through open meadows.[4] This part of Shropshire is a more undulating and gentle countryside than the west of the county, which was wilder and difficult of access in the eighteenth century. But Shropshire roads were notoriously bad and Shifnal might have been isolated if it had had to rely on the old main road which ran just to its north, the Roman road from London to North Wales, the modern A5. But this road was rough and ill-kept westward from Gailey. By contrast the road from Birmingham to Shrewsbury which ran through the town, had been made a turnpike road early in the century and consequently was well maintained. Along it ran regular coaching services to Shrewsbury, and the mail coach to Dublin. This coach route to Birmingham would have been Beddoes' way to Oxford, first as a student and afterwards when he was Chemical Reader in the University. We can look back at eighteenth century life by visiting Weston Park and the Ironbridge Gorge Museum, neither far from Shifnal. On one side was the elegant life of Weston Park where in Beddoes' life-time the paintings of Reynolds and Gainsborough were collected and classical temples built to improve the landscape of the park. On the other was the busy, inventive life of the ironworks, now shown to us in the Industrial Museum.

When the first Thomas Beddoes came to Shropshire he must already have been a man of some substance, for tanning requires capital outlay for land where it is possible to make pits for steeping the hides, and for buildings. Thomas Beddoes[5] bought a house with land behind it running down to a brook and converted this land to a tanyard. On his death in 1769, his son Richard took over the tanyard and prospered sufficiently to buy, when his son Thomas was eleven, 5½ acres of meadowland. This joined the tannery but was on the other side of the brook. Beyond, to the south and west, were unenclosed fields called Haughton Field. In time eight acres of these 'flats' were allotted to Richard Beddoes in exchange for the strip fields which he owned, scattered in the open land around the village. For some years, Richard Beddoes left his fields unmarked but in the end he fenced them.[6] This happened many years later, when Thomas Beddoes was living in Bristol, and it is clear that he must have known Shifnal as a town surrounded by unenclosed strip fields. He would have seen the change that came over the surroundings of the town as the enclosures took place and the purchasers were able to exchange strips and form large fields. Richard Beddoes was prosperous enough to take advantage of this, but he does not seem to have been very active in developing his land. Other[7] farmers in the county were busy with good fencing, planned villages and the building of cottages sometimes designed to harmonise with the landowner's house, sometimes lamentably "castellated or gothicised" and accompanied by "church-like barns" and "fort-like pig styes". Besides buying new land after the death of his father, and taking the opportunity given by an enclosure Commission, Richard Beddoes leased the tannery to a tenant called Duppa who seems also to have rented the house.

Thomas Beddoes and his sister Rosamond must have been familiar with another, rather different farm. This was Hopesay,[8] which belonged to their mother's brother, Richard Whitehall. Unlike Shifnal in the undulating east of the county, Hopesay lies in one of the narrow valleys of the sharp ridges towards the Welsh border, not far from Cheney Longville where there were relations of Richard Beddoes. Rosamond was a frequent visitor at Hopesay and her lively, gossipy letters give glimpses of farming there and at Shifnal: sheep shearing at Hopesay, corn in better supply at her home than with her uncle. One letter, in 1783, mentions so casually that her brother had been expected at Hopesay that it seems reasonable to imagine that at least in his younger days Thomas, too, was a frequent visitor at his uncle's home. This early experience lives in Beddoes' writings about working people, especially country labourers, and in the manner of his advice to them on matters of health. He took care to adapt to the reality of their lives and always wrote with sympathy and understanding.

So intermingled were farming and industrial activities that Beddoes as a boy would have known both and must have seen at least some of the technological changes taking place around his home. There had been coal mining, iron working and small-scale industry in Shropshire since the sixteenth century. Mining was a part of the life of the county; the small colliers also farmed; the owners of large estates left to their agents the management of the mines on their land. An early chapter in a survey of the agriculture of the county is appropriately devoted to coal and mineral working. Even before the period of rapid invention during the mid-century, there was a spirit of innovation; there were, for example, early pumping engines and iron railway wheels in use in the county early in the century. Science here was a matter of practical activity, of the mine owners' search for minerals and the traditional 'know-how' of the men who smelted the lead and the iron ore. There was iron working in and around Shifnal; a forge had long been in operation at the Lizard hill. Further down the valley of the Warfe, from Tong to Bridgnorth where the stream joins the river Severn, there were yet more iron workings. This combination of varied and often rugged natural scenery with the dramatic activity around the mines; the blaze of furnaces and the noise of hammering at the forges; the bustle on the wharves at the river bank and the many and varied vessels sailing downstream or hauled up against the current by teams of men on the banks, fascinated the eighteenth century traveller. By mid-century such sights not only impressed those who passed through on the way to Wales, they were themselves sufficient to draw visitors to the Severn gorge.

During Beddoes' lifetime the best known of these enterprises were the works at Coalbrookdale and nearby, owned and managed by the Darby and Reynolds families. In many respects typical of the energetic and forward-looking men of the district, these particular families and their associates were Quakers. By their membership of the Society of Friends, they formed a close-knit group where it was natural that the control of the various companies should remain in the hands of the same families, by passing from generation to generation. In this way a continuity was established that lasted throughout the eighteenth century. Detached from fashionable society and debarred by religious tests from the English universities with their classical learning, the Quakers, like other Dissenters, had an outlook which valued natural philosophy as an important part of education and which contributed to their success in technical innovation at Coalbrookdale. The decade before

Beddoes' birth was a period of rapid and decisive growth, as a result of technical advances in making good quality wrought iron when coke rather than charcoal was used for the smelting. Though the first Abraham Darby to come to Shropshire had in 1709 succeeded in using coke as a fuel, the iron produced was of limited use. The problems were solved by his son, Abraham Darby II, who first produced forgeable iron from a coke-fired furnace in 1748. His blast furnace at Horsehay came into production in 1757, an event celebrated by a grand dinner for the workpeople. This success, and the processes developed by his associates, put the Coalbrookdale works in the lead for the next twenty years as manufacturers of engines and engine parts. Richard Reynolds, who took over the works in 1763, developed Coalbrookdale in many ways. In the works railways were extended and iron rails introduced; greater use was made of steam engines. Some of this Beddoes must have known of as he grew up. Apart from technical developments within the works, there were changes in the district for all to see. Some sprang directly from Richard Reynolds' own convictions — the walks and viewpoints around his house at Madeley, laid out for his workpeople to use; the Sunday school and — something now taken for granted — the towpath [9] along the Severn, made so that horses could be used to pull barges upstream and men be spared the degrading task of hauling, bent double, as they pulled on the ropes. Most striking of all was the great iron [10] bridge spanning the river at Coalbrookdale, the first iron bridge in the world.

Completed by Abraham Darby in 1779 after four years of planning and of work which involved the solution of quite new technical problems, the bridge immediately caught men's imagination. Beddoes, home from university at Oxford, must have visited it and have shared in the excitement that was generally felt at this great achievement. Contemporary landscape paintings show the springing arch, almost a semi-circle, as an integral part of the landscape grandeur. The bridge, the dark woods of the steep banks, the bright river with its traffic of both working and pleasure boats, were all seen to be in harmony. The earlier paintings show the elegant houses of the iron-masters together with industrial activity and smoky furnaces harmoniously united as part of the country landscape. In later paintings of forges or factories there is a sense of awe and of drama, but men are not dwarfed by this industrial scene; it is in the twentieth century that men become pygmies among the factories. The furnace-men of early nineteenth-century paintings are heroic,[11] and might well share the description John Smeaton wrote in 1798 for civil engineers, "a self created set of men — men who owed their origin not to power or influence but to the best of all protection, the encouragement of a great and powerful nation; a nation become so from the industry and steadiness of its manufacturing workmen and their superior knowledge in practical chemistry, mechanics, natural philosophy and other accomplishments." With such a confident society around him, it is understandable that when he became Chemical Reader at Oxford, Beddoes saw the need to make room in the university curriculum for studies appropriate for young men who were to take part in such developments.

There is nothing to show that Richard Beddoes had any direct part in these industrial activities, or that he knew the Reynolds family who were to play an important part in his son's life. Tanning was an essential industry; leather was needed for clothes, for domestic equipment, for horses' harness and in the construction of vehicles; in industry for pulleys and hinges. The new steam engines relied on leather for valves and seals; and with the use of steam power, bellows grew larger. Richard Beddoes was clearly a successful man; when [12] he died in 1803 he was able to leave land and money to his wife, Ann, and to his son, Thomas, and £5000 to his daughter Rosamond, in addition to the land which he entailed to his grandson, Thomas Lovell Beddoes. The only story about him that has come down to us suggests he was neither well versed in the ways of 'polite society' nor very intellectual. It describes how, when he accompanied the sixteen year old Thomas to Oxford, he took him from college to college to find him a place. It may be an inaccurate tale but it suggests the impression made by the man.

All the same, Richard Beddoes took care to give his son a good education, sending him first to the grammar school at Bridgnorth and then in 1773 placing him for tuition with the Reverend Samuel Dickenson, the Rector of Plymhill in Staffordshire.[13] Richard Beddoes' own education had been completed at Shrewsbury grammar school and he would have been content to give his son the same sort of schooling and then keep him at home. The more ambitious plan, both educationally and socially, was determined on by Thomas' grandfather. Time spent in the house of a Tutor, commonly a clergyman, was an accepted method of education for a gentleman. Mr Dickenson gave Thomas Beddoes a good grounding in the classics and prepared him well for going to Oxford.

His grandfather did not live to see Thomas Beddoes' success. The boy was only nine years old when his grandfather died after being thrown from his horse "upon some timber a few yards from his own door". The violence and suffering of this death and the efforts of the doctors, all of which the young Thomas witnessed, may have contributed to his determination to become a doctor. Quite naturally, the child imitated what he had seen the doctors doing and this led to one of them, Dr Yonge, giving him encouragement and to his play becoming a serious ambition. Dr Yonge was one of the first to subscribe to Beddoes' plan for the project which became the Pneumatic Institute.

On the 15th November 1775 Thomas Beddoes was entered at St John's College, Oxford.[14] An alternative and more credible story about his arrival in Oxford is that he was accompanied by an uncle who did not realise that letters of introduction to the Master of Pembroke applied only to that college and took him first to St John's. The mistake, if it were so, was sorted out and in 1776 Thomas was settled at Pembroke; but such an introduction could hardly have put a sixteen year old at ease or have helped him grow out of his awkwardness. As a boy,

his external appearance was rather uncouth; his manners blunt and not generally prepossessing, and in his intercourse with strangers he manifest a painful degree of . . . shyness and reserve,

characteristics which, according to Stock, were never softened by success or wider experience of society. Fortunately he also had qualities of self-reliance and determination. He could make his own life and concentrate on the studies he chose for himself. Mr Dickenson, his tutor, remembered him as independent but not unsociable and his account — even allowing for bias in a description of a pupil who became famous — gives the impression of an even-tempered boy, calmly determined to achieve his ambitions; a character well suited to help him overcome the difficulties he might meet in Oxford and succeed in his ambitions.

II. Student Days: Oxford and London

Oxford in the eighteenth century was in many respects an uninspiring place for a student. High Anglican, High Tory views prevailed and in many quarters there was remarkable resistance to change. George Horne,[15] President of Magdalen from 1768—1790, a period which covers both Beddoes' undergraduate years and most of the time when he was Chemical Reader, is a striking example and the list of changes to which he was opposed is revealing. He stood against any attempts to make use of classical studies in· solving problems in the Biblical text, even against the substitution in schools of Classical for Hebrew studies; he disapproved of criticism of the place of the psalms in the Anglican liturgy. No wonder he wrote against Hume and Adam

Smith and considered Dr Priestley no more a Christian than 'the Sadducees of Jerusalem'. The University was a clerical society; all the Fellows were in Orders [16] and declaration of belief in the Thirty Nine Articles of the Church of England rigidly enforced. (It was his conscientious objection to making such a declaration that led John Edmonds Stock to leave Oxford without taking his degree.) The University had during this period criticisms of a different nature levelled against it: that the dons were busy with legal wrangles between the Colleges, and that they spent their long evenings after their early dinners enjoying themselves in the town's coffee houses.

This lack of inspiration was also apparent in the chemistry teaching but there was important work being done elsewhere and new ideas were coming under discussion. The discovery of carbon dioxide by Joseph Black and of hydrogen by Cavendish; Priestley's work on airs, and, at just about the time that Beddoes was settling in Oxford, the publication of Scheele's[17] new theory of combustion — the erroneous but in its day impressive phlogiston theory — all these made chemistry an exciting study. Whoever had given Richard Beddoes an introduction for his son at Pembroke College, had chosen well. Pembroke had a tradition of science and medical studies and Dr Adams, the Master, was described by one of the undergraduates as "considerably deep in chemistry".[18] As the teaching at Oxford was, at this time, still in the hands of the Colleges, the right choice was of importance. At the University, Beddoes was no less determined about his studies than he had been at school and he was unusually thorough over his work even at the somewhat perfunctory college lectures. He went to botany and mineralogy courses, as well as to chemistry and pneumatic chemistry and he attended the experimental classes.

For the demonstrations Beddoes must have gone to the chemistry laboratory in the basement of the Ashmolean Museum in Broad Street; now the History of Science Museum, it is known as the *Old* Ashmolean to distinguish it from the present Ashmolean Museum. In 1776 it must still have had the distinction of a new style, of classical elegance in contrast to the older, mediaeval Oxford. The whole building, which had been modelled on a plan by Sir Christopher Wren, was designed to be used for the School of Natural History. When it was opened in 1683, its basement was described as "the lower room, to which there is a descent by a double pair of stairs, [used for] the Laboratory, perchance one of the most beautiful and useful in the world, furnished with all sorts of furnaces and all other necessary materials."[19] In the years between that time and Beddoes' arrival in Oxford, this laboratory had gone through periods of neglect and revival. When he first knew it, it may not have been well equipped, since in 1781 it had had to be completely refurbished [20] for the newly appointed Chemical Reader, Dr Martin Wall. The good furnaces of various sizes and capabilities would not have been much affected for practical purposes, by periods of neglect and the room probably still had some handsome furniture. Grand as the Ashmolean Museum was, it appeared somewhat differently to a "laboratory demonstrator" who was working there in 1787–1788. Thomas Beddoes had returned to Oxford by then, and he probably found the laboratory little changed from his student days:

The chemical elaboratory at Oxford is near six feet lower than the surface of the earth. The walls are constructed with common limestone, and arched over with the same; the floor is also paved with stone .... There is an area adjoining it on a level with the floor, which, though not very large, is sufficient to admit a free circulation of air. The ashes and sweepings of the elaboratory are deposited in it. There is a good sink in the centre of this area, so that no stagnated water can lodge there. The p . . . y which is seldom frequented, is over ground and unconnected with the elaboratory. Notwithstanding all this, the walls of the room afford fresh crops of nitre every three or four months. [21]

Whatever the shortcomings in the Oxford course, Beddoes found it interesting. On holiday at Cheney Longville in the summer of 1780, to give Stock's account,

he amused a few friends ... by giving a history of the discoveries that had been made and exhibiting some of the principal experiments; and his audience were not less delighted with the clearness of perception and explanation of the young philosopher, than by the novelty of the information he communicated. [22]

Beddoes, it is obvious, was very determined and ambitious. He had an extraordinarily accurate and retentive memory and he gained a reputation for his ability to recall all the 'play' of a game of whist. This was his favourite pastime and his dream of happiness was said to be uninterrupted whist playing. His good memory was put to more serious use while he was at University, in teaching himself first French and Italian and then German. He worked without a master, shutting himself in his room with dictionary and grammar. It showed a good insight into what he would need in the future. His school classics was sufficient for scientific texts written in Latin, still the common practice, but the ease with which he could become familiar with the work being done on the continent was greatly increased by this knowledge of modern languages. French and Italian were commonly known, but German was not a fashionable study.

Whist playing and, in the holidays, shooting and collecting mineral specimens, were fairly modest relaxations. Richard Beddoes[23] may not have been over-generous, for a letter written at the end of his first year at college shows Thomas Beddoes feeling that he should justify carefully a request for money which was needed urgently to pay his 'Battels' — his bills for food bought from the College buttery. His total expenses for the year had been "four score pounds" which, he pointed out, could not be considered extravagant when it was remembered that twenty years before a friend had needed as much as £70. In his first year there had been an upholsterer's bill for £19.10 and £8.5.9 spent on books. He budgeted four guineas for lecture fees in the second year, £3.3.0 for natural philosophy and £1.1.0 for mathematics. Like many a student, Beddoes was optimistic that he would spend less another year and reminded his father that the first year is "known to be considerably more expensive". For the next year he anticipated £16.05 for 'taylors' bills and £1.5.0 for 'tea, coffee, candles, Barber, etc.'; even remembering the £6.10.0 spent on 'wine, etc.' in the first year, his accounts bear out Stock's picture of a boy who had not much use for the more extravagant pleasures of many of the Oxford students or even for the sociable breakfasts and the afternoon parties in their college rooms which Stock describes as customary at Pembroke.

Beddoes was five years at Oxford and by the time he left he had equipped himself well to do more than merely continue with professional medical studies. While he was still a student he edited scientific and medical works which he even then realised should be made available in English. Before he qualified as a doctor it became clear both to his fellow students and to his teachers that he was a young man of considerable promise. He might, when he left Oxford, have decided to go straight to Edinburgh where the teaching of medicine at the University and the opportunities for clinical work at the Royal Infirmary were renowned. Instead, after he had taken his B.A. degree in 1781, he chose first to go to London to work under Dr John Sheldon, a former pupil of the outstanding teacher of anatomy, Dr William Hunter.[24]

In 1781 Sheldon[25] was the leading teacher of anatomy in London, for William Hunter was on the point of retirement and in 1782 Sheldon was appointed to succeed him as Professor of Anatomy to the Royal Academy. Beddoes was fortunate in his teacher; Sheldon was a kindly man with wide scientific interests. He had his own well equipped anatomy theatre at his house in Great Windmill St and besides attending anatomy and physiology lectures there, Beddoes made good use of the opportunity to practice dissection. Afterwards, when he was studying in Edinburgh, Beddoes realised how valuable it was for him to have had this time at Great Windmill St. He saw how the students who had to take anatomy and physiology courses at the same time as the rest of their medical studies were rushed and unable to give sufficient attention to this essential part of a doctor's training.

In London, Beddoes had many opportunities to attend a variety[26] of lectures and to join in scientific discussions. At the least formal level, there were the Clubs at the Coffee Houses where he could have met scientists and physicians and there were many distinguished men — William Hunter himself, for example, and Dr Fordyce, one of Professor Cullen's pupils — giving courses of public lectures at a high level. The Irish chemist, Dr Bryan Higgins, graduate of Leyden university, had a laboratory at his home in Greek Street, Soho, where he established a school of chemistry in 1774. This was sufficiently formal for him to have published a syllabus for his course in 1776 and again in 1778. From the title of the second, "Dr Higgins' Course of Philosophical, Pharmaceutical and Technical Chemistry", we can see how much the Greek Street School would have interested Beddoes. William Higgins, nephew of Bryan Higgins, was working there on nitric acid in 1784, possibly as early as 1783, since he demonstrated the reduction of nitric acid to ammonia by means of tin at Sir Joseph Banks' house in 1785. It was William Higgins[27] who gave the vivid description of the basement laboratory at the Ashmolean. There he certainly met Beddoes and worked as his 'operator', and an earlier acquaintance in Greek Street would seem quite possible. Sir Joseph Banks' house in Soho Square was the grandest of all the meeting places, for the fashionable and distinguished and in particular for men of science. Since he held open house each morning and encouraged visitors to his library and museum, there is no reason to suppose that even someone so junior as Beddoes could not have attended. Another distinguished scientist who lived in London from 1777 to 1787 was Richard Kirwan. His house in Newman Street — then almost in the country — was another meeting place for scientists. Kirwan knew many of the Birmingham group who came to be known as the Lunar Society and if he went to Newman Street Beddoes might have seen men whom he would later recognise. Kirwan followed the work that the French chemists were doing on combustion and came, a little later, to feel that it was important for young Englishmen to visit France; in particular, he encouraged them to observe the experiments in Guyton de Morveau's laboratory in Dijon.[28] At the first opportunity after he had left Edinburgh, Beddoes made such a visit and met both Lavoisier and de Morveau.

Beddoes was in London when the news came from Paris of the balloon ascent of Pilâtre de Rozier. This took place on 11th November, 1783, the first free ascent ever made by man. Tremendous interest, both frivolous and serious, followed and determination to make further flights, longer than the five and a half miles Pilâtre de Rozier had travelled from his starting point at the Château de la Muette. Among those whose intellectual curiosity was aroused was Dr Sheldon. A French visitor to England, Faujas de Saint Fond,[29] described how Sheldon

no sooner learned what had been done in Paris than he suspended a part of his anatomical labours to make calculations respecting the weight of the atmosphere. He afterward directed his enquiries to the discovery of the most proper substance for making the covering of balloons, to improving the varnish and to the inventing of the most convenient apparatus for simplifying and perfecting these machines. He visited all the shops and manufactories in London to gain information on these subjects.

Sheldon did more; he involved himself in the making of a balloon which he hoped to use for scientific observations of the effect of the atmosphere on sound and on motion. Through the persuasion of the Duchess of Devonshire, permission was obtained to send this balloon up from Lord Foley's grounds in Portland Place. Two attempts were made, on 16th August and 29th September, 1784; the second was disastrous, the balloon was utterly destroyed by fire and the aeronauts narrowly escaped from the disappointed and angry mob. But on 15th September, when Lunardi made the first manned ascent in England, Sheldon followed on horseback from Moorfields to the balloon's final landing place, the village of Standon, near Ware in Hertfordshire. Sheldon's interest was rewarded a month later on 16th October, for this time he ascended with the aeronaut, Blanchard, and floated westward with him from Little Chelsea as far as Sunbury; Blanchard travelled on in the balloon to Romsey in Hampshire, Sheldon once more following on horseback. By this flight Sheldon became the first Englishman to make a real *voyage* in a balloon, although James Sadler of Oxford could claim he had made the first *flight*, since on 4th October he had ascended from just outside the city and had travelled for thirty minutes as far as Islip. James Tytler, whose attempts had been constantly frustrated, had made two short flights from Comely Gardens in Edinburgh. England took little notice and James Tytler himself felt obliged to refer to each voyage as a 'leap', but the second was witnessed by a very distinguished gathering of city and university dignitaries. Beddoes as a chemist would have appreciated the scientific problems of the gas-filled balloons and the possibilities they presented for scientific experiment. He worked in London under the first Englishmen to realise how observations on the effects of altitude could be made on balloon flights and he later came to employ as his scientific assistant James Sadler, the Oxford aeronaut who had indeed been the first of his countrymen actually to make a balloon ascent.

The physiology course under Sheldon led to Beddoes translating the reports of a series of experiments made over a long period of time by the Italian scientist Spallanzani.[30] These translations made available Spallanzani's work on digestion and, to use Beddoes' terms, into "the generation" and "the artificial fecundation" of certain animals. All Spallanzani's careful and systematic work was of great importance; some of it followed from the possibilities opened up by improved microscopes. It established a fundamentally new understanding of physiological processes in animals and in humans, even though Spallanzani's work on "animalcules", by which he "diligently but in vain fought" the belief in the "wonderful transmutation of vegetables into animals" (a somewhat obscure reference to spontaneous generation) was not fully accepted until much later. Beddoes did considerable editorial work; he added accounts of work by other experiementers – the best known being Dr J. Hunter's study of digestion – as an appendix to each of the two volumes of the translation and, in his own introduction, summarised both Spallanzani's earlier investigations into the circulation of the blood and his study of animalcules. The translation, published in 1784, was a considerable undertaking for a young man in his twenties; he described how he strove to render the meaning of the original exactly and then had to undertake a complete revision because Spallanzani's verbosity became so awkward in the English version. Beddoes presented this translation on the ground that Spallanzani's work was "not very distinctly known, even to some who are employed in this country in pursuits of the same nature". Sheldon approved of Beddoes' work and used it in his lectures for many years, and a copy was bought by Sir Joseph Banks. Beddoes had reason to be pleased with the success of his first publication.

Spallanzani's Dissertations are now only of historic interest yet the translation shows how keenly Beddoes seized on new ideas and how energetically he followed them up. He commented in his preface on the significance of Spallanzani's work which disproved a purely 'hydraulic' explanation of the blood: "the testimony of Spallanzani may be added to that of others, in order to prove that the laws to which inanimate matter is subject cannot be applied to living bodies" [31] And he saw the meaning of the work on spontaneous generation: he thought it probable that "the atmosphere is the mighty magazine in which Nature has reposited her immense stores of the germs of these diminutive beings" (i.e., animalcules). Almost as a final comment on the work, Beddoes wrote, "every new observation in Natural History affords a fresh example of the beneficent care of Providence." [32] Here he was following convention; in Spallanzani, Beddoes found a scientist who held the view that later was to become fundamental in his own thinking:

> If Men would go about to discover things, and not words: if they would investigate Nature, and bring to light some part of her inexhaustible stores, they would render far greater service to the science which they cultivate and deserve better of her republic of letters. [33]

In 1783 Beddoes went back to Oxford to take his M.A. degree and then in the vacation he returned to Shropshire; not, as his sister was expecting, to Hopesay. [34] An outbreak of fever was raging in the countryside and Beddoes was busy working among the country people. He learned what sudden illness meant to ordinary people living in harsh circumstances and unable to afford a doctor. The experience made a deep impression. By the autumn of 1784 when he went to Edinburgh, Beddoes was well equipped for the last stage in his medical education.

CHAPTER 3

EDINBURGH MEDICAL SCHOOL

Beddoes was a student in Edinburgh at a high peak of the intellectual life of the city. The middle years of the eighteenth century were the time when Scotland was pre-eminent in philosophy, in medicine and in science. David Hume spent the last years of his life in Edinburgh; Adam Smith was still lecturing in Glasgow; the professors who taught Beddoes were among the foremost men of the time in medicine and chemistry. It was an elegant and literary city too, at least in the squares and broad streets of the New Town where the well-to-do lived in houses decorated and furnished by Robert Adam. Here fashionable society lionised a young man whose fresh new songs captivated his hearers, Robert Burns, already famous after the publication in 1786 of his early poems. Their popularity brought Burns to Edinburgh while Beddoes, just one year his junior, was still planning his future. Enthusiastic and eager for new ideas, soon to be optimistic sympathisers with revolutionary changes in France, both these young men could easily be seen at this time, as "romantics". But for all his enthusiasm and hopefulness, Beddoes belonged in many ways to the eighteenth century; his ideas related to the philosophy of Locke and Hume and often seem at odds with his impetuous temperament.

The whole system of University life to which Beddoes came in 1784 contrasted sharply with what he had known in Oxford. Edinburgh University had its charter from the Corporation of the City and, in contrast to the English Universities with their ecclesiastical origin, had no religious test and no collegiate system. The students lodged in houses in the old town around the University. Beddoes, remembering his Edinburgh days after some ten years had passed, gave it as his own view that

this dispersion, according to my observation, is not less favourable to diligence and regularity than residence in colleges.[1]

He felt this more normal way of living was an advantage and that mixed company "excited good behaviour." By comparison, college life was less civilising:

Colleges, which after after the example of monasteries seem instituted on purpose to prevent this salutary variety, doubtless give frequent occasions to emulation in those excesses to which young men are particularly prone.[2]

What those excesses were can be deduced from his comment that the sober habits of the Edinburgh lodging house keepers were as good at producing good behaviour as the 'porters' lists' at the colleges. He found that the Edinburgh students were just as hard working and as interested in their studies. Of course the professors were no more obliged to live separated from the town than the students and could be part of Edinburgh society, while in a more formal way public lectures gave the townspeople opportunity to share in the intellectual life of the University.

Edinburgh's medical school [3] attracted students from all parts of the British Isles, from the continent and from America. From its beginning, in the 1720s, it had close links with Leyden where, under Boerhaave, the medical school had become the most famous and influential in Europe. Edinburgh's first professor of medicine, Alexander Monro, had himself been a pupil of Boerhaave and he ensured that this link continued by sending his son, also Alexander — whom he destined to follow him in the Edinburgh professorship — to Holland. Boerhaave, by his stress on clinical experience, had given medical teaching a new direction. He taught the importance of looking for the cause of any condition and encouraged the use of simple remedies. Above all, he was himself a skilful physician and had, to use the language of an older medical system, a 'sanguine' temperament. He inspired confidence in his patients and in his students by his personality and by his bedside teaching. Though his 'Aphorisms' and 'Institutiones' were famous it was rather through his students that his humane and pragmatic tradition passed from Leyden to Edinburgh. From its beginning under Alexander Monro this link continued and in the 1750s clinical lectures modelled on those at Leyden were introduced at Edinburgh. The opportunity of working in the Royal Infirmary still further extended the clinical experience that the Edinburgh medical course gave its students. The wards of the Infirmary could be fearsome, full of risk of infection either directly from the patients or from work at dissection, but time spent there was recognised as a very valuable part of the students' course. In the second half of the century, Edinburgh had succeeded Leyden as the leading place for medical education. It had a style and tradition of its own; these and the outstanding personalities among its teachers had a marked influence on Beddoes' future.

Beddoes found Edinburgh congenial and in retrospect he saw that it had been a successful time for him. While we have to try to reconstruct how he lived in London, we have a lively account of his activities in Edinburgh and of his opinions of his teachers, from the records of student societies and from his own letters.

I have been [he wrote] in truth very much engaged between lectures and societies and fits of idleness. Your opinion of this medical school is I am afraid too just. We are all mad after theory and between the Brunonians and Cullonians Truth and Nature are too much forgotten. For my own part if I was to continue here long I should fall into a state of total apathy for all medical speculation and should become a mere empiric so much have I been disgusted with the wild explanations of the phenomenon of health and discovery I have been obliged to hear.[4]

He was writing without date or address to Charles Brandon Trye, another young doctor. It was probably as fellow students in London that Beddoes and Trye met though they had reached this stage in their medical education by very different routes. His family were too poor to send Trye, who was almost exactly three years older than Beddoes, to a university and he had been apprenticed at fifteen to a Worcester apothecary. When his apprenticeship was over he became House Apothecary to the Gloucester Infirmary. After two years, in 1782, he went to London to an appointment as House Surgeon at the Westminster Hospital. He attended William Hunter's lectures and studied under Sheldon; and, in the words of his biographer, the Rev. Daniel Lyson, "he must have been a fellow student of the celebrated Dr Beddoes, who was at that time one of Sheldon's favourite pupils".[5] Trye assisted in Sheldon's lectures and he seems to have been a good enough anatomist to have hopes of continuing to work with Sheldon. For some reason he felt obliged to return to Gloucester where, in July 1784, he became senior surgeon at the Infirmary. Since Beddoes appreciated the importance of anatomy it is easy to see why the two young men corresponded. But the advantage of the friendship was not all on Trye's side and it would be good to think that Beddoes sent from Edinburgh more than the two letters that survive. Trye had an uneventful medical career, remaining at Gloucester Royal Infirmary until his death in 1811. It is perhaps fortunate that he was a very serious and reliable young man, for Beddoes exhibits the lack of caution and tact which made difficulties for him in later life and continues with a series of comments on the University lectures. We can recognise the style of a confident student pronouncing on the quality of the lectures he attended (or 'cut'):

I do not think you need be anxious about Monro's lectures. You will find most of his opinions, such as they are together with his abuse of Haller and Hewson though in a somewhat more modest form in his late and promised works — and they will cost rather less. Black's course is the best I have ever heard or ever shall hear. But as you cannot see his experiences in a manuscript, you may gain all the information this would give you from Bergman and the

new edition of Macquer's dictionary when it is published. I ought perhaps
to except the doctrine of latent heat but of this I have a copy and you shall
be welcome to transcribe it as soon as I return to England.

Cullen does nothing but read his text book and therefore of late I have not
thought it worthwhile to attend him. Gregory's clinical lectures are good.
I shall have some notes of them, to which you shall likewise be welcome, as
well as the cases.[6]

Beddoes dismisses somewhat briefly professors who were much respected
at the time. Alexander Monro, the second of the dynasty at Edinburgh,
was known for his anatomical work. He disputed Haller's theory of irritability
which gave an explanation of the functioning of the muscles and nerves.
From a student's viewpoint, he was an attractive and lively lecturer, modern
in style and ideas and distinguished by his carefully prepared and organised
lectures. But disputes over theory do not seem to have appealed to Beddoes.
Dr Gregory, who pleased him more, though quite a young man, only seven
years older than Beddoes, was a well respected teacher. At the time he was
preparing his text book of his lectures – in a Latin that became admired for
its elegance – but evidently his descriptions of his cases were more interesting
to Beddoes and more useful to his friend. It is sad that Beddoes found Cullen's
book a sufficient substitute for the lectures and that he made no reference to
Cullen's kindness and hospitality to his students. Professor of Chemistry
from 1755 and of Medicine from 1760, Cullen had for a long time been
known for the excitement and charm of his teaching. He gave his lectures
without notes and in English, one of the first to give up the customary
Latin. Beddoes may have written in a moment of gloom or perhaps before
he had time to appreciate Cullen's work and his custom of encouraging an
equal exchange of ideas between teacher and student. Beddoes wrote notes
for Cullen's translation from Latin of Bergman's 'Physical and Chemical
Essays', and, because the publisher was at such a distance, he also undertook
the editorial work. Beddoes annotated Bergman in the same way as he had
done Spallanzani, referring to new work following the original experiments.
It seems that he was unfortunate in arriving too late in Edinburgh, only six
years before Cullen's retirement and death in 1790, to experience the best of
his lectures and teaching. Those who knew Cullen were aware of "a sensible
decline of that ardour and energy of mind which characterised him at a former
period".[7] Beddoes could surely have forgiven a professor in his late seventies
for reading the book for which he was famous.

Beddoes was dismissive of speculation. He described how the new 'doctrine' of Dr John Brown was gaining ground: "by some strange infatuation

scarce any man except Clegthorn of Dublin and Skeete have come out of his den with the perfect use of their senses"; and the older established and respected theories of William Cullen seemed to him scarcely more likely to be of lasting use. No "exception ought to be made," he asserted, "in favour of those [theories] that descend from the chair. Take away from the sum of Cullen's merit his definitions and descriptions, and what will be the remainder? I am satisfied that as soon as he shall give way to another professor, his hypotheses will go whither all others have gone before."[8] Yet Cullen had developed a system which enabled him to organise and relate observed facts and to give an explanation of life and of disease. Muscle, he taught, was an extension of the nerve and disease related to the irritability of the muscle or to the sensitivity of the nerve. All the functions of the body were seen as under the control of the nervous system. This view of the human body as animated by the sensitivity of the nervous system was one which harmonised with a wider philosophy, David Hume's doctrine that all knowledge and judgement are the product of the association of impressions derived from sensation and experience. This understanding of the nature of behaviour and the consequent importance given to sensitivity found support in the medical explanation developed by Cullen. In later years, when he came to attempt a theoretical explanation of his practice both concerning education and medicine, this was Beddoes' approach.

What Beddoes was seeing when he arrived in Edinburgh was a [9] determined challenge to Cullen's system. During the 1770s Brown taught a different concept: that there was no principle of life inherent in the body, but that external stimuli alone determined the state of disease or health. Excitement from external causes produced conditions of disease which could be classed quite simply as excited or depressed, sthenic or asthenic. Consequently, once the body's state of excitement had been determined, treatment consisted in the adjustment of external conditions. Opium as a depressant and alcohol as a stimulant were the traditional and handiest agents. The debate was heated and made the more unpleasant by Brown's ungenerous behaviour towards Cullen whose pupil he had been and who had treated him with considerable kindness. Beddoes had first met Brown [10] in 1782, while he was practising as a doctor, and had not received a good impression. Despite this, despite his preference for the practical rather than the theoretical, and his insistence that he did not accept Brown's propositions, the 'Brunonian system' and the explanation it gave of disease, though soon to be discredited, proved useful to Beddoes. His own medical philosophy developed along different lines but the notion of excitability became an element in his experiments to find a cure for a group of diseases of which pulmonary consumption is now the best known.

The other theoretical base for his work, Beddoes found in the chemistry he described so enthusiastically in his letter to Trye. By the 1780s Black[11] had given up experimental work to concentrate on his teaching. His lectures were memorable for their carefully prepared demonstrations and he fascinated his audiences by his dexterity. Black had moved easily from the teaching of chemistry to practising as a doctor. He had succeeded Cullen as Professor of Medicine at Glasgow and when Cullen moved from Chemistry to Medicine in Edinburgh Black had again succeeded him. This alternation was natural when chemistry was a part of medicine. What has since come to be seen as chemistry, Black's discovery of carbon dioxide, arose at least in part from his efforts to find a cure for the stone. Between 1757, when Black made his discovery, and 1775, hydrogen, nitrogen and oxygen had all been found. In turn, this understanding of gases led to the investigation of the nature of breathing and to the general realisation that it was analogous to combustion. Once it was known that oxygen was taken from the air and recombined in the body − even though the exact mechanism had not been understood completely − new possibilities for the treatment of disease appeared. Beddoes seized upon this and when he left Edinburgh went straight on to the study of respiration and of gases.

Black did not see chemistry as a study confined to University and medical school. He gave lectures for the townspeople and he founded and was active in the Chemical Society. His discovery of latent heat was crucial for the steam engine as Watt understood. He won Beddoes' admiration and his lectures became Beddoes' model at Oxford. On his side, Black must have noted his ambitious student who, by the time he left Edinburgh, had not only edited Spallanzani but had also produced his own translation of Bergman's *Elective Attractions*.

Beddoes' "fits of idleness" cannot have been many or long. Almost as important at Edinburgh as the University courses were the debates and lectures of the student societies. All we know of Beddoes and the Chemistry Society is that his name appeared in a list of members drawn up by Black, but he took an active part in two other student societies,[12] the Natural History Society and the Royal Medical Society, writing papers for their meetings. The Medical Society had been given its Royal Charter in 1779 and was a highly respected body. With its own meeting rooms and organisation, its papers published and circulated for comment and discussion, it operated as an important and independent part of the medical activities at Edinburgh,

Beddoes must have been both active and respected among the students, for he had the distinction of becoming President of both the Natural History Society and the Royal Medical Society and of being asked to preside over a students' protest meeting. The grievance, which concerned arrangements at the Infirmary, was soon settled.

The societies were good places for meeting foreign students and Beddoes took the opportunity this offered to borrow manuscript lectures of Continental professors – his letters to Trye show that there was an exchange among the students of notes on lectures and of new books. Beddoes offered to get Brown's new book for his friend and was expecting a copy of Hunter's work from him. He was particularly pleased to have the notes of Professor Richter of Göttingen. These could well have been lent to Beddoes by a fellow member of the Royal Medical Society, a member of Göttingen University, Christoph Girtanner.[13] He had introductions to the astronomer Herschel and to James Watt but because of his interest in Brown's theories he chose to spend most of his time in Edinburgh. Girtanner subsequently became one of the most vigorous supporters of Brown's theory and introduced it in Germany – where it had a disastrously divisive effect, to the point of causing riots among the University students. Add that both young men were interested in gases and we can see how it is that they continued in scientific correspondence long after Edinburgh days. Girtanner gives us our only description of Beddoes at this period: "a healthy, strong, rather fat man".

In the summer of 1785 Beddoes, with a party of friends, made a long expedition across Scotland. He wrote on Sunday, September 11th, 1785,[14] to his father describing how they had set out from Edinburgh and by way of Perth, Dunkeld and Blair Athol had gone almost as far as Inverary. Beddoes regretted a little that he could not claim to have made a complete crossing of the country. They paid a "native of the country" 2 shillings a day to carry their small baggage, "out of which", Beddoes told his father, "he maintained himself." Altogether it was an adventure. The weather was often rough, the roads were bad and they could not understand the Gaelic-speaking Highlanders. They were, however, introduced to a useful contrivance:

umbrellas or, as the Highlanders call them screens, by means of which we have always been able to keep our shoulders dry, though it is sometimes difficult to manage them so as to receive benefit from them, it being impossible in high winds to extend them fully.

A storm in June had destroyed bridges and broken up the road and the scene round Loch Lomond was dramatic:

By this time it was nearly dark. The wind was high and some dark clouds threatened us with rain; on our right hand we had, as usual, some rugged hills and on the left the waves of Loch Lomond broke against the shore at our feet and wetted us with spray; the road was uneven, hard and covered with loose stones except in places where it had been mended with loose and unbeaten gravel in which we sunk at every step our own shoes.

When at last they reached Luss they found it "one of the pleasantest spots in the Highlands". Beddoes had a romantic enjoyment of this wild scene and felt none of the 'disgust' that Dr Johnson had expressed twelve years before, when he found on the islets in Loch Lomond, "instead of soft lawns and shady thickets, nothing more than uncultivated ruggedness". Wild scenery had become fashionable since Dr. Johnson and James Boswell made their tour to the Highlands in 1773 and Beddoes found the country was full of travellers, particularly from England. In all these early experiences Beddoes stored up observations which he used later: as a geologist he noted what were to him evidences of volcanic action and as a doctor he was struck by an unexpected cure. Seventeen years later he described to Dr Currie of Liverpool how when he "traversed a part of the Highlands of Scotland on foot" he had found many people ill with fever. In one house where he lodged he had noticed a maid who was delirious. She had

contrived, during the absence of all attendants to get out of bed and crawl naked to the brink of the river, not many paces from the house. When arrived there she perceived a drove of cattle advancing along the road to the bridge. She immediately proceeded into the river and sat, or fell, down with the water near her chin till the drover discovered her from the bridge and gave the alarm. She probably continued five minutes in the water. The result of the immersion was immediate and considerable mitigation of all her symptoms.

Beddoes was led by this incident to prescribe sponging with cold water as a treatment for fever. Such exact memory and power of recall which had served him so well as a whist player enabled Beddoes to hold in mind observations made over many years and in the end to use them when he had developed an explanation which fitted them into a system.

After the walking tour, Beddoes was back in Shifnal, practising as a doctor. He was trying the value of the new drug, digitalis, on which Dr Withering had just published. He told his friend Trye of the cases he was treating, it seems, actually in consultation with Dr Withering himself.

The foxglove has lately performed two great cures here, one of an hydro-thorax, in which I gave it and another of an anasarca of 3 years standing accompanied with every bad symptom and, as was thought by many physicians, with a schirrous liver, in this case it was prescribed by Withering. My patient took it in the form of infusion, zii of the leaves dried and reduced to powder in ilb of water and his [sic] in substance. This plant is mentioned by Kay and Parkinson as very efficacious in scrophula and in a later work, Practical Essays on medical subjects, some cases are related in which it wrought cures by internal and external application. Withering tells me, he is about to publish on its virtues in dropsy.[15]

Dr Withering, at the suggestion of Dr Erasmus Darwin, had moved from Stafford to Birmingham to take over the practice of Dr Small in 1775 and he had, to some extent, taken Small's place as a member of the Lunar Society. In 1776 he had published 'A botanical arrangement of all the vegetables growing in Great Britain'. He and Erasmus Darwin had some disagreements about this book, but later there grew up a more painful dispute between them concerning Withering's claims to have discovered the medical uses of digitalis. The bitterest period was in 1788. There is no hint of this in Beddoes' letter, and both Erasmus Darwin and Withering in time gave their support to Beddoes' Pneumatic Institute.

Beddoes found in Edinburgh just the combination of circumstances he needed to lift his ambition above mere success and to give purpose and direction to his work. Black's lectures and experiments, the debates about new work in chemistry all seized hold of Beddoes' imagination, and when he went to take his M.D. in Oxford in December, 1786 he must have formed a clear plan of research in chemistry. He was already an energetic and ambitious young man in 1784; the Edinburgh years cannot but have added to his self-confidence. His two papers[16] to the Natural History Society, 'On the sexual system of vegetables', read 6 April, 1785, and 'On the chain of being', read 15 December, showed his independent mode of thinking and his wit; and they were well planned to lead to a good discussion. He was already an iconoclast, but thrusting elegantly with a rapier rather than thrashing about with a cudgel as he did later. In his first paper, Beddoes analysed Linnaeus' work, agreeing his achievement in classification but regretting, "his powerful address to the vanity of men, and particularly of rich men, for whom he has provided that *Royal Road* to knowledge, or the shew of knowledge, which Euclid could not prepare for the sovereign of Egypt. ... many ... have been really persuaded that they were acquainted with Nature; whereas they only grasped her shadow, being familiar with a mere barren catalogue." His paper on the Chain of Being amusingly demonstrated the fanciful nature of the concept. The sloth, Beddoes demonstrated using the argument of 'The Chain', could be placed next to humans on account of the similarity of its breasts and its habits. But there was serious work in the papers at the Societies and in his correspondence with other members. He felt he was respected and near the end of his life remembered how he could "never hope to be of so much consequence anywhere else, since the students heaped up on [him] all those distinctions which . . . it is in their power to confer".[17]

From among his own contemporaries he made at least one friend with whom he corresponded for many years, Christoph Girtanner. This personal link with the University of Göttingen led to a valuable exchange of ideas; it was Girtanner who sent Beddoes Kant's *Critique of Pure Reason* and it was largely through him that Beddoes' own ideas were tried out in Germany. From Edinburgh, above all from Joseph Black, Beddoes received support and encouragement. At first it was as a chemist building up a syllabus which would accord with the new teaching of Lavoisier, that Beddoes consulted his former professor. Then, when he was publishing the scientific work that justified his proposed experiments with the medical use of gases, it was Black's approval that he sought. In dedicating his work to Black in 1793, Beddoes found the opportunity to express the admiration which had become all the greater since Black had shown his ability to change to new ideas.

Your late adoption of M. Lavoisier's system has greatly added to the force of this sentiment [he wrote]. And the recollection of so signal a proof, that neither years nor celebrity — the bane of vulgar minds — have had power to blunt your sensibility to truth, affords me greater pleasure than I should otherwise have felt in dedicating to you the following small collection of observations.[18]

Money as well as interest were needed to bring the Pneumatic Institute into being and Edinburgh doctors, as individuals and as a faculty, were generous with their support, an indication of the respect Beddoes gained during his short time, 1784 — December, 1786, at the Medical School and to the impression made by his subsequent scientific work in Oxford and Bristol.

CHAPTER 4

CHEMICAL READER

When he left Edinburgh, clearly with the intention of continuing the work in chemistry he had begun under Black, Beddoes entered into the midst of a vigorous scientific debate. The problems he had to face were quite neatly side-stepped by Stock who passed quickly over Beddoes' "hints, speculations, and experiments", since, in his view, "Many of these it would be useless to record for their value has been destroyed by the changes which have taken place in the whole system of chemical science." In 1811, when Stock's *Life* was published, these debates were yesterday's news; now it is clear that for Beddoes they were issues of immediate importance and involved him in practical decisions, particularly in establishing himself as a teacher, and it was within them that Beddoes had to work. The two questions which were on the point of being decided related to the general problem of establishing a system for the nomenclature to be used in chemistry, and to the more particular issue of the nature of combustion.

The need for a system according to which substances could be named, for long a matter of concern, had become increasingly obvious during the second half of the eighteenth century.[1] Names had originated in a variety of ways, from alchemy or medicine; from popular reference to appearance and even from secret languages. Once the discovery of many new earths and metals accelerated, with a resulting increase in the number of compounds that chemists were investigating, traditional names became inadequate. Besides the need to identify new substances, something was required to replace the lengthy and clumsy phrases which were being invented in the effort to describe new compounds unambiguously. Bergman, the chemist whose work Beddoes had edited and translated, had addressed himself to this problem from the 1770s and, in the two years before his death in 1784, had co-operated with Guyton de Morveau of Dijon. As he continued with the work, Guyton established a set of basic principles which should govern the naming of substances. It was clearly necessary that any such system should not waver between an old and the new theory of combustion which was emerging, and in the last stage before the publication of his *Méthode de nomenclature chimique*, Guyton felt that he had to come to a decision about Lavoisier's new theory of combustion. The meetings between the two chemists at which

Guyton was convinced that Lavoisier was right took place in February and March, 1787 and the *Méthode* was published in the summer. In London and in Edinburgh, two systems of chemical names were already in use, the well-established London *Pharmacopoeia* and a nomenclature developed for his lectures by Dr Charles Webster and in use by 1786. Thus decisions about nomenclature involved questions of national usage and of chemical theory as well as willingness to adopt a new system.

The decision whether or not to adopt the new French nomenclature was complicated for some chemists who were unable to accept the new explanation of combustion. From the early years of the eighteenth century, it had been believed that during combustion, a substance, which had been given the name phlogiston, escaped from the body being burned. The fact that weight was actually gained during burning was not well observed or was ignored. Careful quantitative experiments by Black and Lavoisier's more sophisticated and systematic development of this method showed that burning involved not expulsion, but combination. Lavoisier, in a series of experiments, demonstrated this gain in weight when substances are burned. In October, 1774, Lavoisier learned of Priestley's discovery of dephlogisticated air, which in Priestley's view was air capable of taking in the phlogiston given off when a substance was burned. Lavoisier showed that it was this 'air' that was taken up, causing the gain in weight, and it was he who gave it its modern name, oxygen. These experiments were described by Lavoisier in his *Traité de chimie*, published in 1789. These developments all had to be taken into account as Beddoes thought out his plans for future work.

On 13th December, 1786, Thomas Beddoes returned to Oxford to take his M.D. degree and after Christmas he was back, obviously both reading in the Bodleian Library and working at the laboratory.[2] He would have found opportunities to be active in the University from the outset since a new chemical lecturer was needed to replace William Austin. No new appointment had been made since August, 1786 when Austin left to become physician to St Bartholomew's Hospital, London. Beddoes was able to go straight on with the work in chemistry which had interested him in Edinburgh. In the early part of the year he corresponded with William Goodwyn, the author of work on respiration which had been published in Edinburgh in 1786. Goodwyn was planning a translation of his Latin treatise and sent the work to Beddoes, presumably because he already had scientific translations to his credit. The translation was not published until 1788 and it seems that Beddoes either mislaid it or put it to one side during the summer vacation. He was also working in the Ashmolean laboratory with William Higgins[3] as his 'operator' and, as Higgins later recorded, discussed with him in July work he was doing on calculus and on the action of tin on nitric acid. Higgins had already become an anti-phlogistonist, but Beddoes was to remain undecided between the two theories of combustion for some years yet. At the same time, he was hoping to read the most recent work on respiration and was searching in the library for the 1785 issue of the *Memoirs* of the Paris Society of Medicine[4] where Lavoisier had published an account of his findings. His own reading and translations, followed by the period of study under Joseph Black, had directed him towards new theories and experiments; now his interests and plans were becoming organised.

The summer vacation of 1787 gave him the leisure to visit France and see for himself the important work that was being done by Lavoisier and the French chemists. Although in his *Memoirs* Stock places the French journey in the wrong year, he gives, for him, a lively account of the young Beddoes on the first holiday of his professional life:

He passed some time at Dijon and from thence went to Paris. The particular object which attracted him to Dijon I have not been able to trace. He there became acquainted with Guyton de Morveau, with whom he afterwards corresponded. He spoke, after his return, of the festivities of the vintage with much delight; and seemed fully to participate in the gratification, which most travellers have expressed upon witnessing the harmless merriment of the peasantry at that joyous season.[5]

It is not really surprising that Beddoes chose to go to Dijon. Guyton de Morveau's laboratory at the Academy, equipped with "ingenious inventions for facilitating enquiries in the new philosophy of air", and his experimental work were already famous. Indeed they were so well recognised that in January, 1787, Kirwan had told Guyton that he would advise young people to go to Dijon to study.

Guyton de Morveau,[6] alone among the French chemists who had worked on the new system of nomenclature published in the summer of 1787, was not based in Paris. A holiday journey gave Lavoisier and his friends an opportunity to visit Guyton's laboratory and follow up the decisive scientific discussions of the spring when de Morveau had adopted the antiphlogistic theory. Guyton, it is said, "was pleasingly surprised by the honour of a visit from Lavoisier, Berthollet, and Fourcroy, accompanied by their respective ladies and MM. Monge and Vandermonde." Guyton's biographer continued, "By a very lucky coincidence, Dr Beddoes of Bristol who was travelling through France at the time, happened to pass through Dijon and joined this party, who had assembled to repeat and discuss several experiments explanatory of the new doctrine." In 1787, Bristol cannot have entered Beddoes' mind; his thoughts must all have been on his Oxford course and the opportunity to see the well equipped laboratory, join in the experiments and discuss the new nomenclature which he would need to consider for his teaching, must have been quite as exciting as the festivities of the grape harvest. Besides the experiments in the laboratory, Beddoes visited the Dijon hospital where he saw, and carefully noted, Guyton's system of purifying the air by means of hydrochloric acid. Beddoes certainly was fortunate in joining this scientific party for there were many possible topics other than chemistry to interest him, such as the utilitarian survey Guyton had made of the Dijon region in search of profitable mineral deposits or his balloon ascent of 1784. Guyton and his companion had attempted to steer their hydrogen-filled balloon by means of oars and a rudder, an adventure so hazardous that Kirwan, alarmed at the possible loss to science, had written asking Guyton not to repeat the attempt. Beyond the immediate scientific concerns, a sense of the changes to come in France was in the air. Guyton had already shown where his sympathies lay by criticising the way Voltaire had been denied a funeral service and after 1789 he was to become a loyal servant of the new Republic. This visit to France may have given Beddoes his first experience of 'revolutionary' circles, for Guyton, Lavoisier and their scientist friends were full of lively political talk. Beddoes admired Lavoisier and must have got on well with him and with Madame Lavoisier, for when he left Dijon, he spent some weeks with him in Paris and was able to observe Lavoisier's experiments.

His visit to the Dijon laboratories and to Lavoisier sent Beddoes back to Oxford with a clear understanding of what needed to be done to develop the teaching of chemistry there. He was lecturing at the end of 1787 and by the spring of 1788 became known as Chemical Reader. Black gave his former pupil his support and Beddoes modelled his lectures on those he had so greatly enjoyed in Edinburgh. Unfortunately, no organised record of his lectures has come down to us and they appear never to have been published. There are undated[7] manuscript notes for four lectures in the Bodleian Library, Oxford, which appear to be part of his university teaching. Two of these are notes for geology lectures, and there is also a fragment of a geology lecture conclusion. Two others are early lectures in what was conceived as a chemistry course. The notes seem to have come together fortuitously and are probably the early lectures of two separate courses. The first set of notes for a chemical lecture show Beddoes discussing "the remoter effects of fire". After acknowledging his debt to Professor Black he goes on to what seem to us now very basic matters, expansion by heat and the surprising fact that by contrast living matter contracts when heated; the effects of heat and cold on water and ice; vapour, boiling, and the effect of atmospheric pressure. He shows that heat can be produced by mechanical force. The lecture includes the recent and dramatic work of James Watt, the discovery of latent heat, which was of course of technological importance, and continues with reference to the experiments of "one of our great iron-masters". Joseph Black had been intimately involved in Watt's work and from him as well perhaps as from his own knowledge of the use of steam engines in his home county, Beddoes would have come to appreciate the importance of experiments made in industry. A second set of lecture notes relates to atmospheric electricity. Very naturally it includes an account of the experiments of Franklin which led to the use of lightning conductors — something which Beddoes, if he wished, could probably have amplified by reference to what he had seen in Dijon, for Guyton de Morveau had clashed with the ecclesiastical authorities for impiously setting up a lightning conductor. Beddoes included in the lecture the division of substances into conductors and non-conductors and following this, the way air was electrified on fine days. He was interested in how the intensity of electricity "ebbs and flows" during the day. The study of atmospheric electricity had only begun some thirty-five years earlier and Beddoes' lecture was brought up to date by reference to Volta's contemporary work. Fragmentary as they are, the notes give some impression of the young lecturer introducing current topics into his course.

This absence of full sets of lectures or of notes by students is tantalising. The topic of latent heat and the problem of geological formations, both treated in what we have, show how he was including recent material. His lectures on gases are known to have impressed senior members of the university as well as his students. For the rest we have largely to be content with what we can learn from later writings. Besides the work on gases which, from the outset, he had seen as having revolutionary importance for medicine, Beddoes must have been engaged in analysis to explore the composition of substances including supposed elements. Though perhaps not systematically, he used electricity for this purpose and was interested in its part in muscular/ nervous activity. His observations of the body's reactions to substances was not confined to gases, though the investigations of these came to dominate his work.[8] Before 1793, Beddoes had seen that his projected study of the importance of oxygen in maintaining health or alleviating disease had as its corollary an investigation of the chemical processes of organic matter. The post mortem changes in soft tissues of the body; the production of fat and oil; the processes by which oxygen might be obtained from food, all interested him. He realised these must be complicated and he summed it up, "the chemical composition of the fluids and solids of the living body would influence their properties not less than the properties of dead matter." The connection between diet and freedom from scurvy pointed to the need to understand how plant material reacts to water and air: "Vegetables doubtless decompound water; it appears almost certain that they combine the hydrogen with azote from the atmosphere, to which a certain proportion of oxygen is added." Beddoes realised the importance of heat and light for such synthesis. His observation that plant materials contain combined iron and manganese shows that he had some sense of an organic chemistry. Here is another line of investigation that shows both Beddoes' acuteness and his habit of throwing out ideas and failing to follow them through; through in this instance Beddoes seems to have made a deliberate choice to leave this for the sake of concentrating on the respiration of gases. Characteristically, he could not leave the subject without tossing out a suggestion of practical application. "When, from a more intimate acquaintance with them, we shall be better able to apply the laws of organic bodies to the accommodation as well as the preservation of life, may we not, by regulating the vegetable functions, teach our woods to supply us with butter and tallow." With this work in mind, the seriousness of his later investigation of the toxic properties of tea by means of experiments on frogs becomes clear, though at the time it brought him into ridicule. The practical turn of his mind led Beddoes to see the importance of industrial chemistry and though at Oxford he only wrote two papers in this field, popular lectures at Bristol suggest that he had thought more widely on this matter. Apart from the study of gases, these topics appear only incidentally in Beddoes' later activities.

For the Oxford years the best source of information about his work is his correspondence with Joseph Black.[9] In part, the letters are an exchange of scientific news between equals, though Beddoes is glad to consult his senior and to put before him problems he was encountering in his teaching at Oxford. We can see from these letters how great was Beddoes' esteem for his professor. As soon as he began to lecture Beddoes faced the problem of organising his syllabus. He had to find an order which would be appropriate for the chemical discoveries of Cavendish, Lavoisier, Priestley and Black. He had also to take into account the new work being done on combustion. He quickly planned to write a syllabus of his own because he felt[10] "the want of some tolerable manual of chemistry" at the very time that there was "an increasing ardour for the pursuit of that science at Oxford."

This syllabus was in Beddoes' mind when he returned from France and in

November 1787 he wrote at length about it to Black. Though, not surprisingly, he was inclined to accept Lavoisier's theory, he had hesitations and saw that in the developing state of chemistry, any scheme needed to be flexible. "Till the sciences shall be considered perfect," he wrote, "successive changes of method will be requisite, for as new ss [substances] or new qualities of ss already known are detected, they will require to be arranged somewhere, by which . . . the former order will be more or less disturbed . . . ". Beddoes was satisfied that he had arrived at the right new arrangement by putting gases in the first place, but he found "the system of denomination" a greater difficulty. His comments on the *Nouvelle nomenclature chymique* give some impressions of the objections that would be raised to it. Drawing up a scheme in English was sufficient in itself to present difficulties; in particular, Beddoes found the terminations of the words unpleasant and pointed out that "potasse" was already in use with another meaning.

Black, who accepted the anti-phlogiston theory of combustion and who, in general, approved of the *Nomenclature*, did not agree with Beddoes' doubts though he was happy to discuss the problem. He was clearly glad to correspond with his one-time student and asked him for reports of the experiments on "hot and cold bodys" [Black's spelling] and an account of how experiments[11] on ice were carried out in France. Black was in the midst of a course of lectures and was, "hurried so fast from one subject to another" that, according to his letter, he could "seldom sit down to study new things." Beddoes worked on at the syllabus during the winter, looking at various other schemes, especially German ones. He had it ready for his spring lectures to which he was looking forward and he hoped to publish it in March 1788 in the form of a Letter to Black. He wrote in February to ask permission to put Black's name on the scheme and his request conveyed his tribute to a great teacher:

As I am in no haste to be rich I choose rather to pay the debt of gratitude rather than carry offerings to the vanity of the great.[12]

We can imagine how Black, busy with his demonstrations, was glad to have the account of the work of the French scientists and of their publications which Beddoes said he had "picked up on the continent". Black had no time to read Lavoisier's scheme while he was lecturing but he did enclose a copy[13] of Hutton's *Theory of the Earth*. None of the French scientists, Beddoes had discovered, knew Hutton's work and Beddoes wanted a copy to send to Lavoisier. He also kept in touch with the work going on in Dijon.[14] One letter from Guyton de Morveau dated 19 September, 1788 was carried from Dijon by a Mr Smith, who appears to have been one of Beddoes' students.

Guyton thought well of him: "C'est un jeune homme très aimable et qui apprendra tout ce qu'il voudra, j'ai regret qu'il n'entend pas dans son plan de se livrer à la chymie; mais il a le temps d'en prendre le gôut et vos cours sont propres à le bien inspirer, il a assisté à quelques une des nos séances, et j'ai bien vu qu'il vous avait écouté."[15] Besides this independent tribute to Beddoes' training, the letter gives some indication of the work being done in Oxford and Dijon. Guyton had seen a reference to Beddoes' "expériences sur le froid artificiel par les sels"[16] and was looking forward to having a full account. He imagined Beddoes might be working on fulminating silver and described a dramatic incident in his own laboratory, "il nous est arrivé à ce sujet une chose assez étrange, ayant mis dans un bocal cylindrique de l'ammoniaque sur l'oxide d'argens, pour le faire digérer à froid dans une armoire, il y a une explosion spontanée qui a tout brisé dans l'armoire, il y avait plus de 4 doigts de liqueur sur l'argent, qui est ce qui a pu (déterminer) l'inflammation?"[17] Neither Guyton nor Berthollet whom he consulted, could make it out.

Guyton de Morveau had other news. He had just finished the second part of his first volume of the encyclopaedia where Beddoes would find that the introduction to the article on air would "remet tout ensemble pour la théorie et la nomenclature".[18] Another important book which had just appeared was the translation of Kirwan's essay on phlogiston. Guyton had seen a copy and was taking advantage of Mr Smith's return to England to have one sent to Beddoes: "je charge M. Smith d'une lettre pour M. Berthollet que je prie de lui remettre un exemplaire pour vous".[19] Beddoes' student must be the Mr Smith who appears in letters written by Banks and Blagden, Guyton and Berthollet in the summer of 1787. Unidentified as he is, this young man makes a brief appearance in the history of chemistry and gives us a rare glimpse of Beddoes' work in Oxford.

His chemical lectures met with the success they deserved. Beddoes was able to convince his students that chemistry was neither "one of the black arts"[20] nor merely a subsidiary to medicine and they flocked to his class. "I think," he wrote to Black in February 1788, "my numbers will be greater this than the last course, though I had then the largest class that has ever been seen at Oxford, at least within the memory of man, in any department of knowledge". This was a matter of practical importance as well as gratification, as Beddoes received only fees from his students; his post was unsalaried. Following the example of Joseph Black's skilful demonstrations, Beddoes planned to illustrate his own lectures and consulted his old professor for help in perfecting "some of those simple experiments, which in your hands are so striking and so instructive. I have not yet learned how to show the gradual approach towards saturation by throwing slowly a powdered salt into water . . . . How do you contrive to make that capital experiment which shows the burning of iron in dephl[d] [dephlogisticated] air?" Besides admiring Black's skill, Beddoes was trying to remember details of the apparatus he used. Black was technically inventive and Beddoes asked for his guidance to make a vessel which would stand the heat of the experiment. To carry out his plan of illustrating his own lectures with experiments, Beddoes needed new equipment such as he had seen in Dijon and someone with the skill to design and make such apparatus. He was fortunate in having as his 'operator' just the man he needed, James Sadler, the aeronaut.[21]

After his historic dawn flight on 4th October, 1784, Sadler had remained in Oxford, continuing his balloon experiments. He was unusually skilful in producing the gas needed to raise the balloons and in the actual process of inflating the envelope.

His work interested the university and his last flight in Oxford was made from the gardens of Corpus Christi College, but in the end it was left to the scientists in Oxford to patronise him. He was certainly working for Beddoes in 1790, for on 14th January, 1791 he wrote about an air pump and a barometer he was making.[22] The "very complicated barometer" was nearly finished but obtaining glass for the air pump was proving difficult; he could not get it in London and wanted to know whether he should try Stourbridge. He inquired whether Beddoes wanted anything "got forward in particular" for his lectures – and he needed £15. This same apparatus is part of the laboratory equipment Beddoes described [23] to Black in April 1791. By then he was pleased with his "Elaby"; he had "an improved barometer of which the column of mercury is not altered by temperature . . . an air pump which exhausts perfectly – and is of course constructed on principles totally new – a balance which [I have seen] turn with a 1/100 of a grain when loaded with a pound at each arm." He also had new apparatus to enable him to repeat Lavoisier's experiments and a new 16 inch diameter lens. The apparatus was the work of James Sadler, described by Beddoes as "a pastry cook of this place, a perfect prodigy in mechanics". On one occasion Beddoes even took his experiment out of the lecture theatre and, as he wrote,[24] "astonished this part of England with sending up an air balloon; filled partly with light and partly with heavy inflamable air". Many who, on 9th June, 1790, saw the balloon drifting over the Oxfordshire countryside may have thought it

merely an object of curiosity but for Beddoes himself it had a serious scientific
purpose. He

was desirous to try whether such a mass of inflammable air, burning at a
considerable height in the atmosphere, would produce any imitation of
fiery meteors. It was fired at about three miles distant from the place of
starting, and probably somewhat less than a mile high. This was done by
touch-paper, and it was contrived that the case should be torn and thrown
off as soon as the blaze was communicated to it. We saw this distinctly.
On the demolition of the case the colour of the flame changed, and the
ball of inflamable air, being freed from the load, rose sensibly and suddenly.
It continued compact and burnt away with a lambent flame, 'till it was
diminished to a very small size. It was one of the most perfect and beautiful
experiments I have ever seen. I have never beheld an igneous meteor, and
as descriptions of such objects are very ill calculated to convey accurate
ideas, I cannot pretend to ascertain the resemblance or difference. The
manner of burning, as I conceive, is not unlike.

The design and construction of such a balloon would have needed some
technical skill and, since he was making his apparatus only six months later,
it seems reasonable to imagine Beddoes had help from Sadler. Together
Beddoes and Sadler were gaining considerable and varied experience in the
techniques of handling gases. The collaboration continued after Beddoes
left Oxford and Sadler accompanied him to Bristol.

The experimental work on combustion had its parallel in two publications.
The first, a translation of the Chemical Essays of C. W. Scheele, was published
in 1786. Stock's account is that Beddoes worked with a friend whose English
was imperfect and the most likely collaborator would be Christoph Girtanne.
Though Scheele's experiments led to his false discovery — phlogiston — they
nevertheless made an important step towards the understanding of combustion by showing the existence of two different gases in the atmosphere.
Beddoes did worthwhile work on the edition of Scheele, adding notes
and biographical material. This was the last of Beddoes' translations; his
next publication was an original study of the work of the seventeenth-
century chemist, John Mayow. The work was delayed by difficulty in
obtaining a copy of Mayow's book. Beddoes described with considerable
indignation how he had to borrow the book from a friend. Beddoes' edition
appeared in 1790 as 'Chemical Experiments and opinions extracted from
a work published in the last century', and shows clearly how Mayow's experiments [25] anticipated Lavoisier's work on combustion. In the same manner
as later scientists, Mayow had observed how water rose in a bell jar after a
candle had burned in it, and how a membrane supporting a mouse was drawn

up. Beddoes used his knowledge of the eighteenth-century scientists to help in evaluating the work of the forgotten author, and he pointed to Haller's recognition of Mayow. This was quite startling to those who had failed to recognise the significance of Mayow's researches. Beddoes related with some satisfaction what happened when the book reached Joseph Black:

At the sight of the annexed representation of Mayow's pneumatic apparatus, this sedate philosopher lifted up his hands in complete astonishment.

This insistence on the practical illustration of theory, and in particular the determination to collect new material to support developments in science, represents an attitude to teaching and learning which Beddoes later made some attempt to develop systematically. It is as clear in his approach to the teaching of mineralogy as in his work in chemistry. Early in 1788 Beddoes felt the need for specimens to enable him to explain Hutton's theories clearly to his classes — he hoped Dr Black would be able to send him granite [26] from Dr Hutton's collection and offered Black Cornish wolfram in exchange. Just over a year later, Beddoes felt there were problems in making his views clearly understood because "of the difference between inspection and description, and of another person's report and one's own perceptions", although he was by that time convinced that Dr Hutton's theory of the earth was correct.

He was working on this in 1789 and wrote a paper on the 'Affinity between basaltes and granite' which was read at the Royal Society on January 27, 1791. By then Beddoes had been able to assemble some minerals for, when the paper was sent to Sir Joseph Banks [27] on the coach from Shifnal, Beddoes regretted that he was unable to send the specimens which would have illustrated it. He needed them for his lectures. Beddoes referred to the "two opposite hypotheses which at present divide the observers of fossils", and by describing in detail the evidences of a common origin for basalt and granite argued for the igneous origin of the great chains of hills and mountains of Europe. He concluded with a firm dismissal of the claims of the Neptunists:

It seems impossible to attribute the disorderly deviation, which is so general in the mountains of slate, etc. from that position which all sediments from water assume, to anything but a force lifting up from below, and sometimes bursting through.
    It is moreover certain, that all these lifting masses, from granite to acknowledged lava, are found squeezed up through fissures found in the strata by their own expansion. This, and not the infiltration of water . . . appears to be the true origin of such veins of lava. [28]

The lecture is widely illustrated from various accounts of rocks and of mountain formations but is most vivid when Beddoes draws directly on his own experience. He turned from travellers' descriptions of the rock formations in Europe to discuss the rocks in these islands, the granite, schists, limestone and basalts, and their relations to one another. This part of the lecture traced the geological lines running from Dolgelly through to Welshpool; from Welshpool eastwards to the Long Mountain and to the Shropshire hills on the road to London, and south to Malvern. The Wrekin and Lilleshall Hill Beddoes saw as centres from which lines radiate north through the whinstone to Scotland and south through the Brown and Titterstone Clee Hills to Malvern. All this would have been familiar to him journeying from Shifnal to the farm at Hopesay, for whatever route he took he would have passed by Wenlock Edge and the Clee Hills to the foot of the Long Mynd. It would have been easy for him to have explored westwards from his home into Wales, as he seems to have done in 1791 when Joseph Reynolds was waiting for him to come home "to go into Wales". His knowledge of the Scottish mountains gained on his tour from Edinburgh completed this survey. In commenting on the nature of the rocks themselves Beddoes had for reference his own collection of mineral specimens and could compare those he had himself found to those from farther afield.

Though he was careful to reject for geology a complete reliance on chemical analysis, Beddoes drew on his experience as a chemist of the growth of crystals. He refuted the suggestion made by de Saussure in *Voyages dans les Alpes* about the granite found in the vertical strata lying at the foot of a mountain: "This naturalist supposes them to have filled up gradually by the rain water dissolving particles of granite, carrying them down and depositing them in the form of granite again". He continued that these "two operations one may safely deny to rain water the power of performing." The large size of the crystals shows the falsity of the view, for rain water evaporated would always deposit further small crystals and not enlarge those already there. But it was above all his knowledge of the working of a furnace that led him to a better understanding of igneous rocks. Referring to the making of porcelain he observed how, "the very same mass, according as the heat is conducted,[29] may without any alteration of its chemical constitution be successively exhibited any number of times as glass, or a stony matter with a broken grain," and from his own personal experience, he explained the variety in the forms of igneous rocks;

In the slag of the iron furnaces, the same piece generally exhibits both these appearances; the upper surface cools fast and is glass; what lies deeper, loses its heat more gradually, and is allowed time to take on the crystaline [sic] arrangement peculiar to its nature .... it is easy to conceive how, under variations of heat and mixture a melted mass may coagulate into quartz, feldspath and shoerl, or mica.

Even the term 'shoerl' — presumably 'schorl' — an old miners' term for black tourmaline and hornblende, recalls Beddoes' home in a mining district.

In June, 1791, Beddoes lectured [30] on 'The natural history of the strata, rocks and mountains of the earth' and the two manuscript lectures in the Bodleian Library could belong to this series or to the earlier one in 1789. He faced the problem of putting forward a controversial account:

I have adopted a more circuitous plan than would have been necessary than if I had been to explain a theory familiar from frequent repetition and sanctioned by a general reception.

Where you have to weigh the evidence in the first instance without any bias of authority, there you are full of scruples of hesitation. A new theory may be regarded as a fort in danger every moment of being attacked; and those who are interested in its defence must survey every part with great care, replace it where it is weak and new model it where it is ill-constructed.

This weighing of evidence was applied to 'The theory of the Earth'. In the second lecture, Beddoes was at pains to make clear that he was not concerned with the study of individual fossils but with the way they "exemplify those forces which seem universally exerted in bringing this earth to the form in which we find it." He gave examples of the dispersal of fossils, both of plants and of animals, in "latitudes where the present climate was such that they could not live". This is typical of Beddoes' attitude; he habitually searched for connections and a system, not for mere collection and classification. Typical too was his exploration of German and Russian writings. At the same time, he illustrated his account of the "struggle" between water and land vegetation by the example of the Solway Firth. Though we have only the one published paper and these fragments, we can trace Beddoes' steady interest in geology/mineralogy in letters over several years, from the references to Dr Hutton's work in the letters to Black, through the discussion of the strata beneath which coal was found with Erasmus Darwin, to plans for a Geological Survey of Cornwall which appear in letters to Davies Giddy. Always he was keen to exchange and collect specimens, from the Reynolds' mines, from Erasmus Darwin, who offered him about a hundredweight of fossils [31] and, from the time of these lectures, with Davies Giddy at Tredrea in Cornwall. Looking forward somewhat, we see how important geology was for Beddoes. Throughout the 1790s geological matters appear in his letters to Davies Giddy and when in 1798 Beddoes, by then in Bristol, planned a public lecture on geology, he wrote to Giddy for a specimen of copper ore to replace the one he had left at Oxford, 'pro bono publico'. Pneumatic Chemistry led to Pneumatic Medicine and Beddoes had to relinquish geology but it was an important subject for him, closely linked with chemistry.

At the end of his Royal Society paper, Beddoes expressed the wish to see for himself "Whether the basaltes proceeds southward ... till it join the Elvin or whinstone, and granite of Devon and Cornwall, where I imagine they may be found incorporated, I wish for an opportunity to examine." He found this opportunity soon after his paper was written, and in the summer of 1791 made a tour to the West Country. The visit was all the more enjoyable for being made in company with Davies Giddy. Seven years younger than Beddoes, Giddy was also a member of Pembroke College, and had attended Beddoes' lectures. Although they had met as teacher and student, the two men had common interests and formed a real friendship. Giddy went to Shropshire to meet Beddoes and was shown a blast furnace, possibly as an illustration of the lectures he had attended as well as for general interest. Beddoes, staying at Tredrea, shared with Giddy more than the intellectual interest of the geological survey, for at this time they thought alike about the promise of reforms in society that the French Revolution seemed to give. In 1791, far from any town mob, they could wear the revolutionary cockade and celebrate the anniversay of Bastille day with undisturbed cheerfulness. Though politically they were soon to part company, their friendship lasted until Beddoes' death and was to prove important for him in many quite unforeseeable ways. To see for himself the rocky mass of Devon and Cornwall where, in his opinion, "Dartmoor should be regarded as the centre of lifting up",[32] and to fit it into his view that "the whole western side of our island has probably been raised by the basaltes on which the superficial strata now rest, though from particular circumstances the suffused mass has now and then crystallized into granite", must have given Beddoes immense satisfaction. He and Giddy planned to produce a geological survey of Cornwall; though Beddoes had to put this aside for the sake of other ambitions, Giddy was still inquiring about it in 1795. It was on his return from this journey that Beddoes was amused to overhear himself described, by a young man who did not recognise him: "excepting what he may know about fossils and such out of the way things he is perfectly stupid and incurably heterodox. Besides he is so short and fat he might almost do for a show."[33] Beddoes' views "about fossils and such things" did, in themselves, if the young man had but thought, put him among the heterodox. To assert a volcanic origin for the earth in face of the traditional view that the land emerged as the waters receded, was to deny the whole religious view of creation and its time scale. Safer to be a Neptunist. But neither his views about the earth's origin nor his political enthusiasm, especially in the company of a respected Cornish landowner, had yet put Beddoes 'beyond the pale'. It was he who gave Giddy an introduction to Sir Joseph Banks and supported him when he later, successfully, applied to become a Fellow of the Royal Society; and Beddoes' paper 'On the affinity between basaltes and granite' was presented to the Royal Society by Banks.

The years at Oxford until almost the end must have been happy ones for Beddoes. Though in 1791 he wrote to Black about the ecclesiastical dominance of the university which would prevent for ever the free development of the scientific spirit, yet in the same letter he was describing enthusiastically the growth of his laboratory and the discovery of James Sadler. His lectures went well and he was, even if disliked in some quarters for his 'revolutionary' views, in good standing among his fellow dons. He was welcome in the Senior Common Room at Christ Church where, it seems, he was introduced by no less a figure than the Dean, Dr Cyril Jackson. The friendship of Joseph Black may have given Beddoes support. In 1788 when he came to visit Watt in Birmingham, Black found time to meet his former pupil in Shropshire.[34] From Birmingham, Black went to Oxford where he stayed with Dr Jackson in Christ Church; in the opinion of the Edinburgh professor, his host lived very soft.[35] Understandably, Jackson had links with the informal group of scientists meeting in Birmingham and it was he who introduced Beddoes to their meetings. Two of their number, Dr Withering and Dr Erasmus Darwin, Beddoes had already met in Shifnal. So, even from conservative Oxford it was still possible to keep up with progressive ideas and enjoy the company of others interested in new developments.

Yet in one important respect Beddoes was frustrated by a conservatism which he was unable to change. He needed, and expected to find in Oxford, a well administered and well stocked library. He soon discovered that the Bodleian's librarian was grossly negligent and that the book stock in what should have been a great institution was seriously deficient. Beddoes was indignant, even angry, not solely because his own work suffered but because to him a university library was "a department which the common interest requires above all others to be well administered". The difficulties which he had himself experienced, and which his informal protests had failed to remove, led him to write a forceful and well documented complaint. This he drew up formally as,

A Memorial concerning the state of the BODLEIAN LIBRARY and the conduct of the PRINCIPAL LIBRARIAN addressed to the Curators of that Library by the Chemical Reader.[36]

The Memorial had at the foot, "Thomas Beddoes, Pembroke College May 31st." The copy now in the British Library, which appears to have been Beddoes' own, has the year, 1787, added in ink below the month. This, together with his description of how he failed in March, four months after its publication, to find the autumn 1786 number of the *Journal de Physique* makes clear that he wrote his account of the deplorable state of the library very soon after his arrival in Oxford. When a student in Edinburgh, he had been prepared to lead a protest against unsatisfactory lecture arrangements. In 1787 it was with student-like boldness that, though only a newly-appointed lecturer, he rushed to criticise what had long been tolerated. In the preamble to his report he rebuked those who had acquiesced in abuses:

Freedom of inquiry into the state of the university seems to have been too much discountenanced. Had such discussions been encouraged, the close of the eighteenth century had never found us with so many wants, and so patiently acquiescing under them (p. 4).

Beddoes was clear about the possible consequences of his action, realising that either he or the librarian would be publicly condemned as a result of the 'Memorial'. But he took the risk so that the university might benefit and wrote with characteristic bluntness and independence. His memorial opened forthrightly:

I offer no apology for calling your attention to the state of that institution which chiefly distinguishes a seat of learning from other places, and to the conduct and abilities of that public officer, to whom the charge of this institution is committed (p. 2).

More than once, he admitted, he had been discouraged on the grounds that he would appear to be making a personal attack on the librarian. A marginal note reveals that it was the Vice Chancellor who warned him that he was taking a measure that would hurt himself and not the librarian. What he had himself discovered at the Bodleian and the contrast with other libraries that he knew, had shocked Beddoes and he felt that the need for reform must be made known. He wrote on behalf of many, having found that it was,

a complaint of very long standing, among frequenters of the Bodleian Library, that they cannot derive from it such advantages as might be expected from so ample a repository of books, and that the Librarian has sometimes thrown obstructions in their way (p. 2).

The 'Memorial' described in detail both slackness in the day to day administration of the Bodleian and failure to keep the stock of books up-to-date. For the first, the librarian was responsible but for the second, there were others who shared the blame. In Beddoes' view,

those to whom the care of buying books is entrusted have not sufficient knowledge or do not take sufficient pains; and . . . the sum which is understood to be raised for this express purpose is not expended (pp. 4—5).

Even in the very basic matter of being present at the right times, the librarian was remiss. The Bodleian was often shut long after eight o'clock in the morning, the time laid down for it to open. In addition, the librarian was absent without permission for more than the seven days allowed in each quarter of the year; and even these should not have been taken without the Vice Chancellor's permission. Still more irritating to Beddoes seems to have been the fact that he could not use the library on Saturdays and Mondays, the days that fitted best with his arrangements for the chemical lectures. This difficulty arose from the librarian's holding a curacy some eleven miles from Oxford, where he would go to take the services on Sunday. He probably also found it a comfortable change from life in an Oxford College. This holding of a curacy was in direct breach of the rules laid down by statute. Beddoes recalled the rule,

that the librarian be not encumbered with a beneficium curatum (but) should be at hand everyday (p. 5).

He quoted at length the very detailed statutes governing the library from its foundation in 1602 by Thomas Bodley. These in themselves were already sufficient to prevent the laxity which Beddoes noted. He made a practical suggestion of his own, that "It would be much more convenient if the Library were open in the afternoon from four to six or seven, during the Summer half-year. As three is the hour of dining at almost every college, it may as well be shut from three to four" (p. 7).

The most important tasks of the librarian were neglected. The new books were not catalogued and the library was not arranged so that the readers could easily find what they needed. Beddoes repeated the story about the Rector of Lincoln College who was encouraged to keep a fine copy of James Cook's *Last Voyage* in his college rooms. The accounts of Cook's great voyages in the southern hemisphere, the last of which was described in *A Voyage to the Pacific Ocean in 1776—1780*, published in 1784, were valuable for their scientific importance as well as for their descriptions of unknown seas and lands. Nevertheless, the newly presented copy remained in the Rector's rooms, "to save the librarian being plagued about it" (p. 13). A less well-known example of the librarian's resistance to being disturbed had been observed by Beddoes himself. He described how a "gentleman who had as much as possible avoided applications to the Librarian" had asked for help and had been curtly dismissed with, " 'Sir, you give me more trouble than all the rest of the University.' If the statutes give the librarians some discretionary powers over the younger part of those who are admitted into the library," Beddoes continued, "they do not, I think, direct it to be used in this way" (p. 9).

Beddoes was particularly aware of the failure to purchase books regularly and systematically. He drew attention to serious deficiencies. Even among English writers, there were in the library no complete works of John Locke or of Addison. Of Swift's works, the Library had only *Polite Conversations*. [37] It is hard to imagine that, of all Swift's satires, from *A Tale of a Tub.* to *Gulliver's Travels*, only this "collection of genteel and ingenious conversations according to the most polite mode and method now used at court and in the best companies of England" by 'Simon Wagstaff' had been purchased by the Bodleian. So extraordinary a statement raises doubts about whether Beddoes could have searched thoroughly. The librarian's choice of Sir John Hill's *Vegetable System* was an example of the way his inadequate knowledge led to the purchase of less than authoritative works. To make matters worse, the buying of Hill's "useless writings" continued and a whole collection took space and used up money that could have been better spent.

In place of Hill's writings and of Reynault's *Botanique* which had cost the considerable sum of £37, Beddoes would have preferred works by Bloch, Jacquin, Pallas, Schreiber and Schmiedel. Such a list is a reminder of Beddoes' familiarity with continental, and in particular German, works. He was especially disturbed to find that the writings of the Swiss medical scientist, Albrecht von Haller (1708–1777), professor of anatomy, surgery and medicine at Göttingen from 1736 to 1753, were incomplete. Though his *Elementa Physiologiae* had been bought in 1784, it was an imperfect set and there was no possibility of getting the three missing volumes. Two series of Haller's anatomical writings had parts missing and the library did not have his *Bib. Med. Practice.* It is understandable that the incomplete holding of the works of "this great author" was a cause of serious annoyance, for they were all of great importance and relevance to Beddoes. Haller had made advances in the understanding of the muscular activity of the heart; he had worked on respiration and had his own theory of irritability and sensibility. Beddoes' interest in and his knowledge of German authors was not limited to scientists. He knew the German writers whose works ought to be bought for anyone wishing to make a serious contribution to current theological debate — Jerusalem, Doederlein, Michaelis, Reimarus, Mendelssohn and Lessing. As he considered that German authors might "justly contest the palm of Science and Literature with those of any other nation", it was natural that he felt the need for adequate dictionaries and was critical of the librarian's foolish purchase in the current year of a German and French dictionary printed in 1739. Beddoes considered foreign books should be published in the original language. To counter the argument that few people knew German he wrote:

But in a place of education, I think it is rather to be expected, that the means of making literary acquisitions should be provided, than that persons should come already furnished with them. And who can doubt, but that if the poets, divines, philosophers, and lawyers of Germany were within our reach, we should be tempted to study them as we are tempted to do other things, by the solicitation of opportunity˄ We cannot surely be afraid, lest the labour of acquiring the language be thrown away unless we can suppose that the powers of Haller, Heyne, Meiners, and Michaelis, desert them, when they write in their mother tongue (p. 16).

The catalogue of his scientific and medical library when it was sold after his death shows that Beddoes was an avid and systematic purchaser of learned journals, both from English societies and those published in France and the German States. At this early stage in his career they were essential for his work and it was the librarian's obvious ignorance of the importance of periodical literature that Beddoes criticised most vigorously; to him, "No other species of neglect (was) so ruinous to a library" (p. 9). He found uncut copies of foreign periodicals, evidence of the librarian's failure to read as he ought. Parts of important continental journals were not up-to-date, ones from Berlin ended at 1779; the *Memoirs* of the French Society of Medicine and the Petersburgh Transactions too, were imperfect. It was in the *Memoirs* of the Paris Society of Medicine, 1785, that Lavoisier had published his 'Observations on the alteration produced in the air of places where a great number of persons are assembled'. Beddoes recognised this investigation of the composition of air as a key paper, "intimately connected with the subject of diseases that may be cured, or relieved by breathing different airs". Later, in the collection of letters relative to asthma and consumption, which he published in 1793, Beddoes gave a very full and careful summary of Lavoisier's paper, explaining that he did so because the account was not easily accessible in England. The risk of missing such important accounts of scientific work was good reason for Beddoes stressing the need for up-to-date issues of foreign journals, especially those from Germany which he considered were among the best. Those journals that were purchased were not accessible, for the librarian kept them in his own college rooms. Even the *Philosophical Transactions of the Royal Society of London* were not at hand, as Davies Giddy had found.

Beddoes descended to such minutiæ as the failure to examine books as they were bought, so that defective copies found their way on to the shelves and money was wasted. His exasperation at such carelessness provoked a description of his own search for Volume 29 of Rozier's *Journal* and his discovery instead of a duplicate copy of Volume 28. Money was also wasted by inefficient buying; Beddoes had bought Margraaf's *Opuscules* for 3/6 and Baume's *Chymie* for 14/- while the library copies had cost 7/- and a guinea, and books that he had requested, among them a copy of Anderson's *History of Commerce*, had not been bought. He ended his devastating catalogue of the librarian's inefficiencies,

my own reflections can suggest but one excuse. Being unacquainted with modern advances in science and improvements of language, he has taken no survey of the extent of a librarian's duty and is therefore not aware of his own insufficiency. According to this, the most favourable view of the case, there are but two steps a conscientious man can take; either to resign his office, or to acquire, at whatever labour or inconvenience, that information which is essential to the discharge of his duty (p. 18).

Towards the end of the 'Memorial', Beddoes suggested the first steps in reform. The responsibility for the purchase of books should be taken over by the university departments and not left solely to the librarian. And the departments should take more note of the Request Book. There should be separate funds for book purchases and for the maintenance and fitting up of the library's rooms. Beddoes' solution to financial problems was a direct one — if the library had insufficient funds both for books and for an adequate salary for the librarian, it should apply to parliament. He pointed to Edinburgh's £700 and Göttingen's £1100 and could "discover to reason why an English should be inferior to a Scotch or an Hanoverian University in any respect" (p. 19).

The 'Memorial' was the work of an impetuous young man, eager to get on with his work and conscious that he was breaking new ground. He brought forward sound evidence that reform was needed but he conspicuously lacked any sense of decorum. Beddoes had insight to see what needed changing; energy to press for change when others would not stir themselves and a brashness which defeated his best intentions. The combination is seen for the first of many times in the 'Memorial on the State of the Bodleian Library'. It was inevitable that the Curators passed over his address; perhaps it was even best for Beddoes' career in Oxford that his outburst was ignored. But in time the changes he suggested were made; and it may well be that in the reform of the Bodleian Library Beddoes made his greatest contribution to the intellectual life of Oxford.

When he was drawing attention to the apathy that he found in Oxford, Beddoes slipped in a very pertinent observation on the university. He regretted that it had,

no institution for instructing the youth of a great commercial state in the principles of commerce and manufactures (p. 4).

The comment is curiously at odds with the prevailing tone of Oxford at the time, with its ecclesiastical conservatism and High Tory outlook. Beddoes, it would seem, had noted the absence of any appreciation of the social and industrial changes that surrounded him in his home on the edge of the Coalbrookdale works and the intellectual activity that accompanied them. It was a forward-looking, independent criticism and some parts of Beddoes' work while he was at Oxford show his own interest in the application of chemistry to industry. He also knew from experience the connections between mineralogy and the industrial search for ores. But his greatest success was in his exposition of the current debates and experimental work in chemistry, his knowledge of pneumatic chemistry and his interest in its relevance to medicine. Unhappily, numbers at Oxford declined[38] and by 1792 Beddoes faced the possibility that he would have no chemical class. As he felt he would not be justified in keeping possession of the 'Elaby' if he did not lecture, he saw the Vice Chancellor and offered to arrange his resignation in whatever manner was most convenient for the appointment of his successor. This the Vice Chancellor recognised as generous. He then made a new point,

as it is much to be desired that some salary would be annexed to the Chemical Chair and we cannot ask it in the name of any person so eminent as yourself, I wish you would draw up a memorial to be sent to the Secretary of State, and keep the lectureship till an answer be given to that memorial.

At that time his standing in the university must have been high indeed. Beddoes was being invited to help bring about an important reform of the teaching of chemistry and this time his recommendations had every hope of being put into action. He did draw up the memorial. It was sent to the Secretary of State and no reply was ever received. At the end of the year

Beddoes, realising that he had become *persona non grata* in official circles, planned to leave Oxford. His personal life was changed and any hope that his influence might direct the course of chemistry teaching in Oxford vanished.

The collapse of the plans for a Regius Professorship was not in any way due to a change in academic thinking in the university; it was as desirable as ever to give chemistry the same status as medicine and botany. In Beddoes, the university certainly had a teacher well qualified to impress the Secretary of State and to introduce the kind of chemistry teaching established by Black. He was incomparably more suitable than the Cambridge professor, Richard Watson,[39] later Bishop of Llandaff. Watson had conducted an energetic campaign for an official stipend for a chemistry professor and succeeded in 1766. Yet he once had to send to Paris for an 'operator'. Even though he had decided for his own reasons to leave the university, there is no indication that Beddoes became impatient and ceased to support the 'Memorial'. The reason the plans were abandoned is to be found elsewhere, in the fears engendered by the events in France.

On his visit to Dijon and Paris in 1787, Beddoes was in the company of men whose scientific work would lead them to hope for an open, liberal society. Intellectual inquiry needed freedom alike from rigid religious restraints and from the cramping of opportunity in a society dominated by an unyielding aristocracy. Such a liberalising movement had been growing in France in the eighteenth century[40] among the bourgeoisie and had gained adherents even among some of the nobles. It had its origins in the criticisms of the *ancien régime* made by Voltaire and Montesquieu, and in Rousseau's concepts of the worth of the individual and of contract as the foundation of society. Such ideals need the spur of practical grievances before they lead to action; the revolution in America was a clear example, and in France there were practical difficulties enough.

The bankrupt state of the country's finances, brought to a crisis by the support given to the American colonists, had forced the King into a position where he had inevitably to accept change. By 1787 it was clear that radical alteration of the system of taxation which favoured the nobility and clergy, was essential; but once Louis XVI attempted consultation he had opened the way for the expression of all the grievances of the time. In the first stages, hopes grew among the aristocracy that the powers the King had assumed, which had denied the nobles any effective part in government, might be curtailed and that they might re-establish their traditional rights. In the provinces it was the feudal powers of the nobles themselves which had been steadily increasing, that men found oppressive and wished to see abolished. The bourgeoisie suffered from the tax system which exempted the aristocracy and the clergy from contributing to the revenue: the poor were hungry. It seemed that everyone had reason to look for change, and there was equally reason to believe that change was justified and possible. On the one hand, the ideas of the *philosophes* and of those, on both sides of the Atlantic, who supported the arguments of the American colonists, upheld the movements against absolutism and for social justice. In practice, changes were in fact taking place even in Europe where such benevolent rulers as the Emperor Joseph of Austria abolished the harsher features of absolute rule. Most powerful example of all, in America men had successfully protested against unjust taxation; freed themselves from the autocratic rule of George III and brought into being a new state with a declaration of the Rights of Man as an integral part of its Constitution.

All this was keenly observed in Britain where the American Revolution gave new life to discussion of the nature of the state and the rights of the individual. The 'Glorious Revolution of 1688'[41] had established the sovereignty of parliament, and a certain degree of religious freedom and personal liberty. The claims of the colonists were a reminder of reform still needed, in England as across the Atlantic. It was not only the Americans who were taxed but not represented or who suffered from the autocratic attitude of George III. More deeply, the Declaration of Independence led to debate about the principles of government and even to the possibility of the right to rebel. The discussion ran deepest in the circles where scientists and Dissenters met and continued long after the war had ended. Their work directly or indirectly influenced Beddoes. Among the members of this circle, known from Benjamin Franklin's description, as "The Club of Honest Whigs", were men such as Joseph Priestley, highly respected and influential in the scientific circles where Beddoes made friends after 1788; and Richard Price, the leading Dissenting minister and political thinker whose *Observations on the Nature of Political Liberty* was published in 1776, the year Beddoes first went to Oxford. Linked with them, too, was the editor of the *Monthly Review*, to which Beddoes was to contribute later in his life. More generally, their thinking went far beyond concern with the American problem. They formulated ideas which had wide application; for example, Joseph Priestley's assertion that the purpose of government was to ensure the well-being of its people was influential both in France and in England. Many in both countries agreed with his argument that neither the holding of overseas possessions nor preparation for war was part of the duty of the state. Their own disadvantaged position led the Dissenting community to work for the ending of limits to religious toleration. They contributed to the circulation of such ideas both in this country and in America and France. Once he was introduced to the midland scientists and industrialists, Beddoes would have met men who held and discussed such views; who sympathised with the principles of the French Revolution or simply welcomed the abolition of oppressive laws and customs. They did not fear that the overthrow of the *ancien régime* would bring anarchy but looked forward to the establishment of a stable society which would have learned from England's constitutional government and which would in turn set an example for further reform. Beddoes' earliest writings on contemporary affairs show how enthusiastically he had made these ideas his own. There was yet another focus of liberal thought, which the Marquis of Lansdowne had formed at his seat, Bowood, near Calne in Wiltshire. No longer active in politics, Lansdowne had invited to his household on one pretext or another men whose work could be expected to influence public life. Priestley found the duties of librarian left him time to carry out his major chemical experiments; Jeremy Bentham worked at Bowood on his plans for a rational judicial system, which he published in France. The house was open to refugees from pre-Revolutionary France and it was there that Mirabeau met the man who was in 1790 to become the writer of his revolutionary speeches, the Protestant Etienne Dumont, himself a refugee from Geneva. Through these contacts, Lord Lansdowne sought to influence the planning of the future constitution of France, believing that he might support the adoption of a system modelled on the English two chamber parliament. This hope, widespread in England, is clearly seen in the activities directed from Bowood. The English aristocracy, having themselves restricted the autocratic powers of the crown, could conceive no better future for France than the adoption of a similar system. The Dissenters and those in England who were pressing for constitutional reform were not alone in this complacency about developments in France. In his welcome for the Revolution; in his confidence that the way had been opened for political change and for intellectual freedom; in his distress when the threat of counter-revolution grew strong, Beddoes was at one with many thoughtful and influential men in England. But he was more vulnerable than many in face of the disappointments that were to come from developments in France and from reaction in England.

As well as ideas, events in France had their counterpart in England. Inevitably, Beddoes with his lively mind and interest in the changes in society which he saw taking place around him, responded to these fluctuations. He must have felt the enthusiasm of men whom he knew and admired: Erasmus

Darwin [42] who greeted "the dawn of universal liberty" and who felt himself "becoming all French both in chemistry and politics"; Joseph Priestley who urged the young to "partake of the glorious enthusiasm" [43] and who accepted honorary French citizenship. The London Revolution Society and the many reform societies in the provinces, corresponded with clubs in France [44] and sent their members on fraternal visits. Even as late as April 1792, after the revolution had become more menacing, James Watt junior and Thomas Cooper visited the Jacobin Club in Paris on behalf of the Constitutional Society of Manchester [45] carrying greetings and assurances that "il y a des hommes partout (mêmes parmi des peuples que les intrigues des rois et des courtisans ont trop souvent fait paroitre ennemis) qui prennent un vif intérêt à vôtre cause, la cause non seulement des Français, mais du genre humain." There were widespread and enthusiastic celebrations on the anniversary of the fall of the Bastille. One occasion in particular, in 1790, the great 14th July dinner held at the Crown and Anchor Tavern in the Strand, showed the Dissenting congregations, the Whig Clubs and the reform societies united [46] in their admiration for the new France and in their determination to achieve similar progress in England. These demonstrations of solidarity were repeated in 1791, sometimes with care to display both symbols of loyalty to the English crown and the emblems of revolutionary France, sometimes with less caution. When Pétion, [47] who was soon to become mayor of Paris, visited England in the late autumn of 1791 he was guest of honour at a dinner held on 4th November, the London Revolution Society's traditional anniversary remembrance of the 1688 revolution. Though Pétion had come ostensibly to escort Mme de Genlis and two daughters of the duc d'Orléans to Bath, he~used the visit to make many contacts with English friends of France. At the dinner, Pétion was moved by the warmth of feeling for France; by the flags of the two countries twined together with garlands and by his meeting with two English "friends of liberty":

Payne cet ami de la liberté qui a tant contribué à ses développements en Amerique, vint à moi; nous nous serâmes la main, signe d'amitié entre les Anglois. Je fus voir le docteur Priestley, qui était à quelques pas de moi, et nous nous dîmes des choses obligeantes. J'admirai sa bonté et la candeur de sa phisionomie.

The whole company joined in singing 'Ça-ira', the song of the Revolution. Whether or not as early as 1790 Beddoes had already been caught up in this revolutionary feasting, he was wearing the tricolour on 14th July, 1791 and singing, even writing, revolutionary songs. On that day in Birmingham a

Revolutionary dinner was not saved from disaster by a careful display of loyalty to George III and by July 1792 it was no longer possible to celebrate this anniversary.

Public debate in England, from the autumn of 1789 to summer 1792; outbursts of violence in France, more terrifying on each new occasion; and increasing reaction here, both in the country and in government, combined to bring about this change. At the end of 1789 there was hope that there would be no need for further violent action in France.[48] The fall of the Bastille, symbol of the overthrow of tyranny, had been followed by the abolition of feudal privileges and, on 26th August, by the Declaration of the Rights of Man and the Citizen; the King had assented to the drawing up of a constitution. A new era seemed to have begun and this was the theme of the sermon [49] which on 4th November, 1789 preceded the London Revolution Society's annual celebration of the English Revolution. Recalling the achievements of 1688, Dr Price described true patriotism and liberty as transcending national frontiers and looked forward to the day when the freedoms won in France would spread throughout Europe, with the

ardour for liberty catching and spreading; a general amendment beginning in human affairs; the dominion of kings changed for the dominion of laws, and the dominion of priests giving way to the dominion of reason and conscience.

He turned to those who had waited for this day:

Be encouraged, all ye friends of freedom and writers in the defence! The times are auspicious. Your labours have not been in vain. Behold Kingdoms administered by you, starting from sleep, breaking their fetters, and claiming justice from their oppressors! Behold the light you have struck out, after setting America free, reflected to France, and there kindled into a blaze that lays despotism in ashes, and warms and illuminates EUROPE!.

He was answered by Edmund Burke's *Reflections on the French Revolution*, still remembered though the sermon is forgotten, and the two speeches represent opposed views of the changes in France. To Burke, there was danger in any attack on the edifice of government; anarchy must follow. In part a defence of tradition and precedent, the *Reflections* were also intended to expose the danger threatened by the activities of Lord Lansdowne and his friends. Burke feared that they might play the same part as the liberal French nobles and that their theories and their intrigues would topple the structure of society. Burke's argument gained strength from its emotional language and from his descriptions of Louis XVI and Marie Antoinette forced from their elegant life in Versailles and escorted back to Paris by the rabble of the city, "a swinish multitude" as Burke called them. This never-to-be-forgotten phrase was quickly turned into an all-embracing description of the common people. The only possible answer to Burke was to attack the edifice which he sought to preserve and such an attack came quickly in Thomas Paine's *Rights of Man*. The first part appeared in 1791 and gave an account of the French declaration of August 26th and of the way this had come into being. At the outset Paine utterly rejected Burke's appeal to antiquity and precedent:

A thousand such authorities may be produced, successively contradicting each other. But if we proceed on, we shall at last come out right; we shall come to the time when man came from the hand of his Maker. What was he then? Man. Man was his high and only title, and a higher cannot be given him.[50]

Paine's clear and straightforward account was followed in the next year by the second part of the work, a description of the social reforms needed to bring into being a just and 'rational' state. Simple and direct in style, *Rights of Man* was dangerous in the way no sermon could be. In plain language it made clear the rights that even the poorest could claim and described the system of government which would reform the wretchedness of their daily lives. It cost very little and could be easily bought as well as easily understood; it was acclaimed and execrated. Copies of the book and effigies of Paine were burned in public and finally in May, 1792 the Royal Proclamation against seditious writings laid its author and all who printed or sold it open to prosecution. Thomas Paine left England; he was outlawed in December, 1792 and lived until 1802 in Paris, a Citizen of the French Republic. These great documents of revolution and reaction dominated the years when Beddoes was working in Oxford and have echoes in his writings.

The violent feelings roused by *Rights of Man* and the government action which it provoked were only part of the mood of reaction. There was long-standing opposition to Dissenters[51] and to their efforts to secure removal of the civil disabilities which were so irksome to them. This readily combined with fears of revolution directed from France or simply with horror at outbreaks of violence both in Paris and in the French provinces, where the peasants rose to demand food or to search the châteaux for the documents which legalised their feudal servitude.[52] There were refugees in plenty to spread rumours, from the aristocrats who were early arrivals in England to quite humble parish priests who in 1792 found themselves no longer able in conscience to live under the new régime. The contacts which the radicals maintained with France, the humiliations of Louis XVI and the Queen were probably better known to most Englishmen than the intrigues of the French monarchy with its foreign supporters. There came into being a concerted movement of alarm[53] which led to a general willingness to inform against people suspected of revolutionary sympathies. This took an organised form in November, 1792 with the founding of the Association for the Preservation of Liberty and Property against Republicans and Levellers. Such an association of settled, prosperous men strengthened the government in its stand against the radicals and identified those who would help in uncovering disaffection and drawing attention to dangerous publications.

In Oxford, Beddoes saw these alarms reported in Jackson's *Oxford Journal* where there were accounts of the planting of subversive agents, of horrors in France and of the raising of the militia at home. A very carefully worded advertisement for the London Friends of Liberty's dinner on 14th July, 1791, with the caveat that no "cockade or other Badge of Distinction" was to be worn, led to the publication of a letter of protest against any plan to celebrate this anniversary. It came from Dr Edward Tatham, Rector of Lincoln College; he must have congratulated himself when ten days later he read the *Journal*'s account of the riots that followed just such a dinner in Birmingham. As he was leaving Oxford, Beddoes could have seen the alarmists organising their supporters when "The Oxford Association for the *support* of our happy *Constitution* called A general meeting of the inhabitants of the City and precincts of Oxford ... for the purpose of entering into an Association for preserving Liberty Peace and Property against Republicans and Levellers of every Description".

From the time his student days ended and he entered into his professional life, Beddoes lived in this political turmoil. By temperament he welcomed new ideas, and by temperament too he seems to have hated violence. He realised how hard it was to find accurate reporting[54] of events at at time when the provincial press was manipulated, even though by bribery and subsidy rather than by direct censorship. His determination to be well informed on scientific matters was paralleled by his wide reading on political topics. But it was not until reaction was strengthening and sympathisers with France were being shaken by attacks on the monarchy, and finally by the September massacres in 1792, that Beddoes himself wrote on political matters.

He must have felt that he was living in two worlds. The new friends he was making were men who had had their education at the Scottish universities, the Dissenting Academies, or in industry at chemical or iron works. Quaker, Unitarian, atheist, liberal Anglicans, they were advocates of toleration, opponents of war and of slavery. As industrialists and entrepreneurs as well as scientists, their interests were wide and not confined to this country and in their company rather than in Oxford Beddoes would have been able to discuss new theories in geology and chemistry, new applications of science and the political events and movements of the day.

CHAPTER 5

THE MIDLAND CIRCLE

The group in the Midlands to which Dr Jackson and Dr Erasmus Darwin belonged was, of course, the famous Lunar Society. This company of scientists, doctors and industrialists met for discussion each month and, as they gathered from the surroundings of Birmingham, it was a sensible arrangement to choose the time of the full moon for their journeys. Beddoes had various friends among them and though never one of the central members of the Society he was to some degree connected with the group. The first introductions came from Beddoes' university life but there were soon other links. Shortly after his appointment in Oxford, Beddoes in 1787 or 1788 met William Reynolds of Ketley and Dr Erasmus Darwin[1] and, in 1791, we find him writing of his first meeting with the chemist, James Keir, an older man and a central figure in the Lunar Society. The young Chemical Reader was being accepted into a very distinguished and progressive circle and his pleasure is obvious in a letter written to Reynolds in 1791, where he described how he had met "interesting and welcoming people in Birmingham".

William Reynolds' father, Richard, had introduced many innovations at Coalbrookdale, both in smelting processes and in the use of railways for wagons. Even more important for future developments in the industry was the encouragement that he gave his son to pursue his scientific interests. William had opportunity to carry out experiments at home and then went to Edinburgh University to study chemistry under Joseph Black, becoming, in a family where there was a tradition of practical experiment and innovation, the first to have any academic training. When in his turn William Reynolds[2] became active at Coalbrookdale, he too was the author and encourager of new processes and inventions. He introduced new methods of handling; steam engines; rail roads with iron rails; he developed canals and the famous inclined plane where barges were raised from the Severn up to the works, was his invention. Reynolds' surveys of the district and his carefully kept records were extensively used by Archdeacon Plymley[3] in reports on Shropshire made for the Board of Agriculture. Joseph Plymley acknowledged the contributions, confident that Reynolds' name would "add interest and value . . . in the opinion of all those who have the pleasure of knowing him." Over and above his surveys "in the field", William Reynolds had a well-equipped laboratory at his home at Ketley and he kept himself informed of scientific developments both in England and on the continent. He has come to be recognised as "the outstanding figure among the iron-master-scientists". A friendship which went beyond these common interests in geology and chemistry soon grew between the two men and until his death in 1803 William Reynolds was Beddoes' faithful friend and supporter.

During the 1780s and 1790s Reynolds was searching for a scientifically based means of producing a steel which would compete with the imports of Swedish hard steel; it is in connection with this that we have direct evidence of Beddoes' interest in the 'applied chemistry' needed by his manufacturing friends.

Reynolds gave Beddoes the opportunity to observe in working conditions the processes of iron smelting. He saw the heaving and swelling, the blue lambent flame which in time becomes pale and dies away, the "comminution" of the ore to the size of gravel. An experienced workman was assigned to explain the work and answer questions. Beddoes gave an account of these observations and suggested their causes in a paper for the Royal Society in 1791: "An account of some appearances attending the conversion of cast into malleable iron".[4] Although the paper is a formal account, it is clear that Beddoes' scientific imagination was captured by the dramatic industrial processes. He described the changes in the molten ore with great exactness and noted, to the minute, the time taken by each stage. It was a paper that gave him real satisfaction; Beddoes wrote to Black promising to send him a copy as soon as it was printed. Crell[5] published it in his *Chemical Journal* and noted the second paper in 1793, commenting on the importance of the work. The new process was suggested to Reynolds by Bergman's analysis of Swedish iron, which showed how large a percentage of manganese it contained. Since Bergman was the Swedish chemist whose writings Beddoes had translated earlier, Beddoes' work, and presumably discussions of it with Reynolds, could well have contributed to the scientific understanding of the steelmaking process. William Reynolds did in time succeed in patenting a process for manganese steel which anticipated later developments. Beddoes was also interested in William Reynolds' trials of steam engines for various purposes in the Coalbrookdale works. Beddoes' assistant in Oxford, James Sadler, as well as experimenting with balloons and making Beddoes' new laboratory apparatus, had designed a steam engine with the technical improvement that it did "not condense in the cylinder." Sadler described this engine in the letter he wrote to Beddoes (about laboratory apparatus) in 1791 — he was in particular concerned with a dispute with Boulton and Watt over the patent.[6] Beddoes appears to have tried to further Sadler's interest with Reynolds. He later sent Reynolds sketches of engines by Sadler; these appear in Reynolds' sketch books [7] with a note that they were sent by Dr Beddoes from London. There are in Beddoes' later letters somewhat anxious hopes that Sadler should settle down with Reynolds. Sadler does indeed appear to have had some success with his engines at Coalbrookdale and to have got on well there . . . at least it was he who was sent by Reynolds to fetch arms from Birmingham when it was realised that there was danger of an "attack" by 2000 men on Reynolds' home, The Bank.[8]

James Keir, whom Beddoes met in 1791, was another important figure in the developing science-based industry of the Midlands. He and Erasmus Darwin were contemporaries and friends from their Medical School days in Edinburgh; like his fellow countryman James Watt he had moved to Birmingham from Edinburgh and had quickly become a valued member of the Lunar Society. For a short period he managed the Soho works for Matthew Boulton and, as a chemist, he was able to give Watt advice on the composition of the special ink that it was necessary to use with the copying-machine Watt had invented. Keir's most important achievement was his great chemical manufactory at Tipton. There he developed new processes which, with his work on potash, led him to manufacture soap. His business prospered so well that Erasmus Darwin, sending news of the Philosophers to Richard Lovell Edgeworth in February 1788, could write, "Mr Keir amuses his vacant hours by mixing oil and alcaline salts together, to preserve His Majesty's subjects clean and sweet — and pays 1000 guineas every six weeks to an animal called an Exciseman; who would otherwise trouble this diversion of his."[9] Beddoes visited Keir on the way back from the 1791 geologising expedition in Cornwall and gave Davies Giddy a lively description of what happened.[10] He was well received and evidently the two men were quickly at ease with one another, since Keir felt sufficiently sure of his visitor's scientific interest to show him round the soap manufactory — a mark of confidence, for industrial espionage was rife. William Reynolds had given Beddoes opportunity to apply his chemistry at the very heart of the working of the industry he had known from childhood and at Tipton Beddoes was able to observe still other applications of chemistry. What he saw must have strengthened his views about the inadequacy of the courses offered at Oxford. He seemed to have realised that *all* the secrets of the manufacture were not revealed to him but he admired what he saw and was impressed by Keir's canny business acumen. Before they parted Keir made a suggestion which surprised and must have gratified the younger chemist. He asked

Beddoes to collaborate with him in some chemical writings he was planning. We know from Erasmus Darwin that Keir was planning "a new improved edition" of his chemical dictionary. At first the invitation was declined; Beddoes feared that difficulties would arise from their differences over the theory of combustion since Keir still held to the old views and Beddoes inclined to the new. In the end, Beddoes agreed to write a limited amount.

Beddoes was happy to make friends among this group of scientists and to feel that he was keeping up to date with developments in chemistry. Though the *Dictionary of Chemistry* Keir had published in 1777 and 1779 held to the old theory of combustion, he had also written a pamphlet on gas which had earned Priestley's commendation. Beddoes wrote to Black in 1791 that Priestley had abandoned the phlogiston theory — the last to do so. (Priestley was not so settled on this opinion as Beddoes thought, for his later writings suggest he never whole-heartedly adopted the new theory.) There were opportunities too for discussions of problems in geology. Possibly Beddoes called on Keir to talk over his geological findings in Devon and Cornwall and his recent paper on 'Basaltes and granite', for Keir had been the first to note the possible likeness between some rock formations and the crystallization resulting from cooling. As early as 1776, he had suggested that the basaltic formations of the Giants' Causeway were formed by the crystallization of molten lava. So in Keir Beddoes found another vulcanist who later in 1798 wrote arguing the volcanic origin of strata. Earlier in this period geological topics had already appeared in Beddoes' correspondence with Erasmus Darwin.[11] They differed vigorously in their opinion of the origin of coal, Beddoes getting the worst of the argument, though Darwin forgave him and invited him to visit.[12] It was, on the other hand, for practical reasons that William Reynolds consulted Beddoes on geological matters, valuing his help in the search for coal and iron ore.[13] All this must have been valuable to him as a scientist and teacher. Beddoes for his part did not escape observation. The impression he made was such that Matthew Boulton planned to send his son to study under him in Oxford.

Keir and Beddoes also talked of Thomas Day who was well known to the older members of the Lunar Society. The visit came at the right moment for conversation about this most attractive member of the Lunar Society. Day had died, aged 41, two years before and Keir had just finished a biography of his friend, the *Account of the Life and Writings of Thomas Day Esq*. Keir's personal reminiscences and his wise judgement were well worth hearing, especially as this was just the time when Beddoes himself was beginning to turn from a concentration on scientific matters to wider concerns. Perhaps he felt himself sufficiently established as a chemist and geologist and as a university teacher to be able now to express himself on social and political matters. We know he was impatient of the conservatism of Oxford and among the friends he found in these contrasting circles he would have found encouragement to try out his ideas. From his writings in the 1790s it is clear why Beddoes must have valued a first-hand account of a man he admired but had never met. Oddly enough, this admiration for Thomas Day played quite an important part in Thomas Beddoes' later life, for when Beddoes settled in Bristol, Keir gave him an introduction to Richard Lovell Edgworth, knowing that an admirer of Day would be welcome to one who had been his close friend. Day's robust character, where firmness and integrity combined with 'sensibility' and with a shyness which was by Keir's account most marked with women, made him warm friendships in his lifetime. It still excites our sympathy, yet it is hard not to smile at some of his oddities. He courted first Honora and then Elizabeth Sneyd and agreed to make an effort to polish away his awkwardness of manner. For this purpose, he went to France, accompanied by Edgeworth, in a serious effort to improve. Despite all his efforts, which even extended to lessons in dancing and deportment, he was rejected again on his return. Not surprisingly perhaps, for he had embarked on a remarkably unconventional experiment in education. At this time Day was endeavouring to bring up two adopted young girls according to Rousseau's principles of education and as can be imagined, he brought numerous difficulties upon himself. In the end he had the honesty to acknowledge that the experiment was a failure and to make more orthodox provision for Sabrina and for Lucretia. Day did eventually marry and settle down to farming, first at Abridge, near Epping, and then near Chertsey. His wife was a wealthy heiress, Miss Esther Milnes, but Day refused to touch her fortune. He carried his eccentricities into his farming methods and it was an unschooled horse which he had refused to 'break in' that caused his death. Day's quixotic behaviour makes him the most eccentric among his friends but in his ideas he was typical — perhaps even might be taken as representative — of the Lunar Society group. He wrote against slavery and in support of the American Colonists in their struggle for freedom. After the end of the war with America, he campaigned vigorously for reform, not merely of the constitution but of the corruption which was doing so much harm to government. Day's belief was that "public abuses never correct themselves", that reform is only brought about by positive effort; but though he worked selflessly for all these purposes it is for the book he published in 1783 that he is still known. Keir describes how Day wrote, "with a view to guard the rising generation against the infection of the ostentatious luxury and effeminacy, which, amid many excellent qualities, characterise the present age." The book was *The History of Sandford and Merton*, a story of two boys, intended as a companion to Edgeworth's *Harry and Lucy*. It far outlasted Edgeworth's book in popular esteem, being still in print (acceptable as a school prize) a century and more later. Day and Edgeworth had both discovered from experiment the impracticality of Rousseau's system. Their aim was to put forward new methods of teaching, consistent with their ideas of social reform, but based on better principles than Rousseau's. Thomas Beddoes in his turn came to devote much thought to the principles and practice of education and to give attention to this for the rest of his life; and from his writings and activities in the next years we can see how closely Beddoes came to identify with this cluster of interests which make up the matter of Thomas Day's writings.

These topics first appear in 1792 in a handsome book, illustrated with a number of woodcuts and printed on fine paper, entitled *Alexander's expedition down the Hydaspes and the Indus to the Indian Ocean*.[14] At a first glance it gave the impression of having been for sale by John Murray, 32 Fleet St and James Phillips, George Yard Lombard St, but this was simply a 'blind'. The printer was in fact J. Edmunds of Madeley in Shropshire; the engraver of the woodcuts was the Madeley Parish Clerk, Edward Dyas, and the expense of printing was met by William Reynolds. For good measure, be it added that the type was set by Edmunds' daughter. How Reynolds came to provide — and provide generously — for this strange book is made clear in a letter from William Anstice of Madeley Wood to Dr Samuel Parr. William Anstice would have had the responsibility of deciding what to do with the library of his uncle, William Reynolds, since he was the executor of his will and Trustee for the children. He had come to Madeley as assistant to Reynolds in 1796, when he was fifteen years old, and had grown into an intimate member of the family. He decision to offer Beddoes' "literary fragment" to Samuel Parr was very understandable since Parr was well known to Lunar Society circles and his library,[15] at his home in Hatton, Derbyshire, was an important collection which included in addition to classical writing the scientific works of his contemporaries. Parr, for scholarship and political vigour, was the Whig counterpart of Dr Johnson. Written in 1819, the letter to Dr Parr accompanied a copy of the book:

The late Dr Beddoes must be known to Dr Parr, by character if not personally. In any case, the literary fragment which accompanies this cannot prove uninteresting. It owes its origin to a conversation which took place at the table of the late Mr Wm. Reynolds, a noteable Madeley ironmaster, in which some men of taste and genius contended that the poetic effusions of Erasmus Darwin were inimitable. Dr Beddoes maintained a contrary opinion, and to try the point produced a short time afterwards the manuscript of the present piece as from his friend Darwin, and sent to him for his inspection previous to publication. The advocates for Darwin's style were deceived, and the Doctor triumphed. Mr Reynolds had it printed at his own expense, but for obvious reasons it was not published and therefore may never have met Dr Parr's eyes.

"The obvious reasons" for not publishing are worth investigating. On reflection, Dr Beddoes may well have felt that much of the verse was sorry stuff and would do him little credit in print, however much it might entertain his — and Dr Darwin's — friends. And it was poor compliment to Darwin that it had ever deceived Beddoes' readers. There are indeed lines which give hope for something at least tolerable in the heroic couplet style:

> Impetuous rivers clash with headlong force —
> Dire seethes the foam, and loud the surges roar;
> The deafened Bands suspend the uplifted Oar;   (Ll. 36—38.)

but our hopes, like Alexander's boats, founder at

> Back reels the flood — devouring eddies curl —
> And foundering keels revolve with dizzy whirl.   (Ll. 39—40.)

The best passages were chosen by Southey for a collection of contemporary verse which he published in 1799 as *The Annual Anthology. Alexander's Expedition* was in the press while Darwin was still working at his *Oeconomy of Vegetation* and though this frees Beddoes from the charge of slavish copying, it made publication, with the risk of embarrassing Erasmus Darwin, even more inappropriate. Besides, it is unlikely that any group of friends would want such an intrusion into their private parties as publication would have involved.

A closer reading soon reveals that there were other reasons for not publishing. Far from being simply a friendly literary jest, the work is a well sketched out contribution to the discussion of topics which were currently interesting liberal-minded men. Beddoes widened the scope of his writing by appending a number of essays to the poem itself. This was a method used by Darwin in the third edition of *The Botanic Garden* and Beddoes was at pains to point out that his work was in the press before this edition was published. In spite of this disclaimer, the essays add to the resemblance between Darwin's work and Beddoes', but in subject matter *Alexander's Expedition* was more like a little-known poem written in 1776 — the year of the American Declaration of Independence — by another member of the Lunar Society, Thomas Day whom Beddoes so admired. *The Devoted Legions*, Day's heroic-couplet narrative, told the story of Crassus, the corrupt Roman leader who, against the will of his people, made war for greed and who was finally put to death by his own soldiers. This catastrophe was foretold by one of the Tribunes who warned against a war which he knew to be unjust and his impassioned speech formed the whole of Day's poem. Day chose his subject skilfully to make a parallel to the situation he intended to expose. Crassus' unjustifiable adventure; the opposition it aroused and the cruelties on both sides in the campaign against the tribes of Scythian bowmen, would be seen to stand both for the American war itself and for the horrors which followed after the English commander brought in Indian tribes to attack the settlers' villages. More particularly, Day intended the Tribune's speech to turn the reader's thoughts to Burke's great oration on behalf of the colonists and in support of the constitutional rights of free men. Beddoes too took from ancient history a famous expedition but he turned the situation round and, as the subject of his poem, chose a hero who was by tradition noble and admirable, a contrast to any contemporary ruler who might come to his reader's mind. With Alexander as hero and the expedition to India as subject, Beddoes made opportunity to treat two problems of his own time — the tyrant and the exploitation of a less powerful people by an aggressive and greedy nation — and to arouse the reader's sympathy for the civilisation of India. The notes which accompany the poem, he explained, "were chiefly written with a view to diffuse more widely a knowledge of old and new Hindoo literature, which although sufficiently familiar to the learned, is but just reaching the circle of ordinary readers".[16] But Beddoes was aware even so early as this that his opinions would be "warmly disapproved".

William Anstice appears not to have been quite accurate in believing that *Alexander's Expedition* "owes its *origin*" to the conversation at Reynolds' home. It seems more likely that the friendly discussion encouraged Beddoes to bring forward work he was already preparing. He must have been reading widely and systematically on the subject of India, its civilisation, climate and topography and keeping himself informed about the effects of the operations of the East India Company. The notes to the poem and the essays which follow it name the sources which he consulted. These range from classical authors — Strabo, Aristotle — to French and German writers on contemporary conditions. He went to Blumenbach's *de Generis humani varietate nativa* for support for his contention that Indians were not different in kind from Europeans, and to Townshend's *Journey Through Spain* for a description of the heat of Catalonia. Here he found evidence about the effects of a hot climate. He used a contemporary military atlas and memoir for his descriptions of scenery and his macabre lines on famine in Calcutta were based on the Annual Register for 1771. At a deeper level he had been reading with obvious enjoyment a translation of the *Bhagavad Gita.* It is hard to imagine that he collected all this material on the spur of the moment. We cannot know whether he intended to write a pamphlet on India or whether a poem following Day's example was part of his original plan. On the other hand, in the eighteenth century it would have been quite within the capabilities of an educated man, especially if he had Beddoes' lively mind, to turn out in response to a challenge a heroic couplet narrative the length of *Alexander's Expedition.* Perhaps this was what Beddoes did, finding when he knew Reynolds was arranging for it to be printed, the appropriate illustrations for Edmund Dyas to engrave. *Alexander's Expedition* would then be the first, and happiest, example of Beddoes' habit of letting his sense of humour enter into the treatment of a subject on which he felt deeply.

The poem presents Alexander as an explorer who does not enslave the people whom he meets. "He stands honorably distinguished among conquerors by his eager thirst as well as liberal encouragement of science and in his character the romantic traveller is blended with the adventurous soldier." Alexander, in Beddoes' view, had not been hardened by the "brutal Institutions of Sparta" and did not share the "Athenian disdain of barbarians and tolerance of slavery" — this comment, culled from a scholarly article which Beddoes had noted in a Göttingen journal, is versified when Alexander is described:

> Large was thy thought, and liberal was thy soul
> Nor stooped thy glance beneath bright Honour's goal;
> Beyond the sage's amplest glance thy mind
> Embraced the mighty mass of human kind,
> And spurned, with firm disdain, the barbarous rule
> Framed by the Founder of the subtle school.   (Ll. 231—236.)

So, as he comes to the end of his journey, Alexander can hope for a time

> When every clime shall see my flag unfurled
> And boundless Commerce mix a cultured world
> From mad mis-rule reclaimed and brutal strife,
> Trained to the soft civilities of life.   (L. 200.)

If Alexander's dominion could have continued the rule of peace would have been established and

Had *General Concord* from her finished fane
Shed her pure light, and breathed her strains humane
Man's varied race, from far dissevered lands
Her courts had thronged and pledged discoloured hands; [!]
Her shrines had witnessed varying voices blend
The vow, and in the stranger hail the friend   (L. 258.)

At this point the poem continues with a prophecy of India's grim future. Here was Beddoes' opportunity to castigate the activities of the British in India and the support given the East India Company by the Government:

Mourn, India, mourn — the womb of future Time
Teems with the fruit of each portenteous crime.
The Crescent onward leads consuming hosts,
And Carnage dogs the Cross along thy coasts;
From Christian strands, the Rage accursed of gain
Wafts all the Furies in her baleful train:
Their eye-ball strained, impatient of the way
They snuff, with nostril broad, the distant prey.
— And now, the Rout pollutes the hallowed shore,
That nursed young Art, and infant Science bore.   (Ll. 275–284.)

Six short essays [17] follow the epic. Four are on general topics concerning India, and two are scientific curiosities. The first of these, Chapter III, 'Conjectures on explosive compositions', concerns the unknown nature of "Greek fire" and the second, Chapter V, 'On the complexion of the natives of hot countries', includes an account of an extraordinary experiment. This appears to have had as its purpose to show that skin colour is in truth only "skin deep" and hence to assert that all men are one species. The other four essays are on Hindu religion and culture, on Hindu manufactures, and on the British presence in India. These made a more general contribution to the current debate on India and the East India Company. As he developed his attack, Beddoes must have known that his outlook was shared by the group of friends for whom he was writing. Hatred of tyranny and exploitation; of corruption and selfish luxury; of superstition and ignorance; belief in the power of education and a sense of the bond between lovers of freedom which over-rode national divisions, these were feelings common to the men who were likely to have gathered at Reynolds' home. It was in this apparent *jeu d'esprit* that Beddoes first found an opportunity to express other than scientific ideas.

In the first essay, Beddoes answered those who saw in the Hindu religion nothing but superstition and practices of self-torture repugnant to Western

Christians. He agreed in condemning the folly and inappropriateness of ritualised and dramatised religious observation but in assenting to this criticism of Hindu practices the reader had to face the fact that he had committed himself to an equal condemnation of such practices in the Christian religion. The asceticism of friars and monks and the masochism of such holy men as Simon Stylites have to be rejected along with the Hindu fakirs. Nor would Beddoes allow that the fatalism and indolence of Eastern races is an inevitable result of the tropical climate. It was in writing his fourth essay on 'The antiquity of the Hindoos' that Beddoes found so valuable the work of the St Petersburg scientist, Dr Pyotr Pallas, whose "travels it is a reproach to our language not to possess". Memories of the Bodleian Library's deficiencies still vexed him. This fourth essay emphasised the antiquity and variety of the Eastern nations whose cultures are to be respected. Beddoes went so far as to stress that the Hindus could not be converted to Christianity because they found satisfaction in the beauty and quality of their own religion. He supported this with a long quotation from a translation of the *Bhagavad Gita* and urged that we should learn to know and understand other religions for this, Beddoes believed, would "diffuse peace and goodwill".[18] Knowledge and understanding would lead to a more peaceful and productive relationship between the English and the people of India. Such sympathy for other civilizations would be easier to attain if we could come to understand the great age of the earth as proved by the new theories of geology: "the system of subterranean Nature, which is beginning to be understood and which exhibits as well as the system of the Heavens, an arrangement highly worthy of admiration, proves the earth to have existed for millions of years, perhaps of ages". This open-minded view led Beddoes in the second essay to attack a defence of the caste system he had found and wished to overturn. Beddoes considered, for example, that it is wrong to suppose that the beauty and skill of Indian work arises from the rigid division of crafts; he put forward the view that freed from the caste system the Indians would "still have fabricated delicate wares in equal abundance, and by virtue of a freer exertion of genius, their manufactures would have extended to a thousand elegant and useful articles besides". Equally mistaken was the attempt to justify the English presence in India as preferable to the tyranny of native rulers such as Tippoo Sahib. Beddoes, with noteworthy clarity of judgement, pointed out that all the reports reaching England had as their purpose the blackening of Tippoo's character and that in any case, there was nothing in his activities to exceed the cruelty of the British soldiers. A subject people, he stressed, cannot be fairly estimated, for, "A people under a foreign commercial tyranny can least of all people, attain an erect and independent mind, that base of all excellence. It is no more possible for them to advance in science or in virtue than for the brutes who draw our ploughs and carts to become rational".[19]

Practical considerations too, were against England's position in India. Those who hoped for commercial benefit from India had only to look to Britain's experience with America. After much costly effort nothing, in the end, had been gained by the subjection of the colonies. Indeed, Beddoes pointed to the decline of Spain, Portugal and Holland attributable to their foreign possessions and prophesied that perpetual wars would follow from our pursuit of wealth in India. The warfare would only increase our taxes, just as the taste for easily acquired wealth would corrupt our society.

This corruption was, at the very time that Beddoes was writing, being exposed to public view. Astounding wealth had been won by the servants of the East India Company, clearly by the most dishonest and shameful means, and almost unbounded power by its officers and soldiers. Beddoes was twelve when the hero, Clive, narrowly escaped censure in Parliament for the vast riches he had amassed during his Indian career – glorious though he might appear as the establisher of British ascendancy in the continent by his military victories over the French in the 1750s. As he grew up, Beddoes must have been familiar with the opulent "Nabob" returned from his Indian service, with wealth that enabled him to join the ranks of the country gentlemen or even the aristocracy. Anxiety about the extent of corruption and extortion mounted steadily among liberal-thinking and responsible people and reached its height in the years after 1786 when Burke called for the impeachment of Clive's successor, Warren Hastings. His trial by Parliament had begun in 1788 and was still in progress when Beddoes was writing *Alexander's Expedition*. Warren Hastings was not acquitted until 1795, by which time the evils of the Indian situation had become clear. Fox, Sheridan and above all Edmund Burke attacked Hastings and called for reform in impassioned oratory and it is in this context that *Alexander* can be properly understood. Beddoes saw that direct control of the East India Company by the government had resulted in a form of despotism and described it harshly:

> unchained by Hate
> Commissioned Murder moves in guilty state,
> And strews, with impious arm, the human wreck
> O'er heaven loved realms, which Peace and Plenty deck.
> With courtier glance, meanwhile, a fawning ring
> Of Priests and nobles eyes the vengeful King,
> Lists the shrill horn proclaim the spreading ill
> And hymns to Flattery's harp, his SOVEREIGN WILL.
> Secure the Coward, on his distant throne
> Smiles as the smitten sink, the tortured groan.   (Ll. 371–380.)

The verse might veil the attack on the King, but when Tippoo Sahib, the autocratic ruler of Mysore, was described in the last of the Essays, 'On the possessions of the British in Hindoostan', as no more a tyrant than any monarch, the reference became dangerously obvious. Beddoes followed this with an appeal to his readers to "join in execrating despotism in all its forms and degrees whether mercantile or monarchical". Yet Beddoes was not alone in his own time in wishing for a system giving the Indians greater respect; Burke also had made a plea for a moral basis for the relations between peoples. Nevertheless, through all the years of the British Raj, Beddoes' strictures would have been unacceptable to the English rulers. If, with that phase of our Indian adventure now closed, we turn to the writing of the twentieth-century Indian statesman Jawaharlal Nehru,[20] we find that his account of the East India Company is in many points curiously like Thomas Beddoes'. But representative of liberal opinion as it was, the tone of Beddoes' work and its strongly radical elements indeed support the comment that there were "obvious reasons" for not publishing it.

Efforts had already been made to restrain the English abuse of power in India with a system of dual control by the British Government and the East India Company, yet it was not by any such means that Beddoes saw change coming about. In his view, a freer and more liberal society would in time be established in India as in England, not by Government but by the efforts of the people themselves. The people of England, he believed, would expose the exploitation of the Indians as they had opposed slavery and he was confident that if ever a Clarkson should "arise in behalf of the Hindoos his appeal will not be in vain". For this reason, Beddoes was optimistic. "One reflection consoles us, while we contemplate the past or present calamities of Asia. The posterity of the oppressed will at last receive from the posterity of their oppressors the doctrines and the spirit of freedom."[21]

Such a spirit of hope and confidence in the future was the immediate result of the revolution in France and Beddoes was only one among many who felt that a new age was coming into being. "The genius of Freedom is only now growing and stretching itself", wrote a young lady in Shropshire in her diary in April 1792.[22] She was Katherine Plymley, the sister of the highly-respected Archdeacon Joseph Plymley, friend of Clarkson and of the Reynolds family. Two months earlier, she had noted that Clarkson had suggested holding a collection for the French Assembly and Plymley had not rebuffed him but had consulted Reynolds, who in his turn had been anxious only lest they might appear to wish to stir up revolution in England and so provoke a counter-collection. These were hardly the circles of extremists, let alone revolutionaries; indeed Beddoes himself, writing to Davies Giddy, refers to Archdeacon Plymley somewhat lukewarmly as "a democratic or semi-democratic"[23] parson. Such warmth of feeling towards the Grand Experiment in France was openly expressed in the first period of the Revolution and of course the favourite occasion was on the anniversary of the Fall of the Bastille. Groups of well-wishers up and down the country held celebratory dinners, the most famous of all being the one held in Birmingham on 14 July, 1791. The gentlemen who met for dinner to celebrate the new régime in France had no seditious or violent intentions; indeed, in place of honour at their table was a portrait of King George III; but they were associated in the minds of ordinary people with the Dissenters and this in itself appeared to provoke the hostility of the Birmingham mob. Not content with attacking the hotel where the dinner was to be held, they rioted through the streets, wrecking and burning on the pretext of "defending Church and King". The best-known victim of this anarchy was Dr Joseph Priestley, who saw his Meeting House burned and his home and irreplaceable laboratory destroyed. Priestley himself narrowly escaped with his life and had to take refuge in the house of the Wedgwoods. Not only did the militia fail to protect the citizens, it was subsequently widely believed that the mob had been whipped up by government *agents provocateurs*. Priestley was not the only one of Beddoes' friends to be at risk in the three days' rioting. Boulton and Watt had to convince their workpeople at Soho that it would be folly to join the loyal friends of the King (i.e., the mob), and they then set about to organise them to defend the works. Dr Withering and Dr Parr also narrowly escaped. Parr was specially at risk as he had been visiting Dr Priestley, and Mrs Parr,[24] fearing that the house would be burned, energetically hurried off the books of his great library to safety; fortunately the mob did not get as far as Hatton. We do not know where Beddoes was on 14 July, 1791, possibly he was already with Davies Giddy; what we do know is that he was present some time that summer at a happy, possibly even slightly romantic, dinner in Cornwall where a young lady — "Rosalind of Cornwall's Bowers", as a light-hearted verse calls her — wore a French Revolutionary favour. The song about her tricolour cockade must have been quite popular among Beddoes' friends, for he reported to Giddy that it had become a hymn.

The aftermath of the Birmingham riots was followed with increasing concern by Beddoes.[25] He noted the growing strength of the feeling against Dissenters, especially among the clergy, and described to Giddy the depth of popular prejudice: "I conversed with several friends of Church and King and what was no trivial penance I heard their political sentiments. Their ideas exactly resemble a mass of felt. It would be certain loss of labour to disentangle and put them straight". Beddoes, by contrast, had endeavoured to keep his own ideas straight by making a detour through Exeter on his journey home from Cornwall to seek out a copy of the French gazette. This, as he feared, showed how events in France were being represented in the English accounts which had included only the bad points about the French Assembly. As the months passed, there came stronger indications of government manipulation. Beddoes felt in December 1791 that he had evidence to show that the judges were so prejudiced that they would drop charges against the rioters in Birmingham. By the spring he had no doubts about such prejudice, for Keir, who had attended the trial, wrote describing the failure to allow fair compensation to the victims of the riots. Beddoes sent Keir's letter on to Davies Giddy and in a sinister and mysterious note wrote that he had information "respecting the apprehensions of our government and the measures they are taking to continue the people in their prejudices". These were not exaggerated fears on his part. Even the cautious Samuel Parr,[26] advocate of waiting for "those gradual changes which the Spirit of Freedom will most assuredly produce", when he wrote in the summer of 1792 to dissuade the Birmingham group from holding yet another 14th July dinner, admitted the bias of the government in its use of "professional stirrers up". Later in the same year Beddoes was writing anxiously to Reynolds about the activities of the reactionaries and hoping that Reynolds,[27] seeking the "annihilation" of ecclesiastical influence, would agree with him. He appears to have had personal as well as general reasons for concern, and perhaps felt that government informers had been at work, for in the same letter he claimed that a young man had been told not to associate with him.

Not all Beddoes' actions at this time were of a nature to excite suspicion. When, in the early months of 1792, he took an active part in the Shropshire efforts to support the Bill for the abolition of the slave trade, he was in very respectable company.[28] It was probably in this connection that Beddoes met Archdeacon Plymley — the "democratic or semi-democratic parson" of the letter to Giddy — for he was the leader of the 'Society for effecting an abolition of the slave trade in Shropshire'. His sister Katherine was an eager supporter of the cause and her diary gives a lively account of its progress in and around Shrewsbury. She recorded a meeting on 13 July, 1791, obviously called to launch a petition from Shropshire. Beddoes, as we know, was not there, but the list of the local gentlemen who sponsored the petition included his friends William and Joshua Reynolds, their highly respected father, Richard Reynolds, and Dr Darwin, presumably Dr Robert Darwin of

Shrewsbury. Beddoes himself appears in the diary in February, 1792, when he sought Archdeacon Plymley's permission to organise a petition from Shifnal. After some hesitation on the grounds that Shifnal was such a small place, this was agreed, Beddoes pressing for it because "it would help to shew the universal voice of the nation". In the event, he collected 150 names from Wellington and Shifnal to add to the nine and a half foot long petition that was sent up from the County.

The failure of the Bill for the abolition of the slave trade; popular violence and government reaction at home; growing reaction in France, all combined to make this a despairing time for those who had hoped so much. Beddoes, who had to contend with the ecclesiastical spirit of Oxford, must have been particularly vulnerable, the more so because of his confidence that the new spirit of liberty would bring about change in so many spheres. He had hoped that there would be a fruitful interchange between new knowledge and greater political freedom; that science would be freed in the society he saw coming into being and that, in turn, new discoveries would lend it their support. Education too would continue to progress. In its new forms it would rouse the imagination of young people just when it is at its liveliest and the choice of the right material for study would lead men away from war and from such brutal civil violence as had been witnessed in Birmingham. Beddoes contrived to squeeze all these ideas into *Alexander's Expedition*! In fact his one criticism of his hero is that unfortunately Homer's poems had led him to admire the warrior Achilles. As opinion in England hardened against the French experiment and as events there led on to the imprisonment of the Royal Family and to war which it seemed would involve England, small wonder that Beddoes made anxious endeavours to sort out his ideas.[29] "These untoward circumstances and the constant fluctuation of my own opinion respecting French affairs", he wrote in February 1792, "have contributed to render my feelings very uncomfortable this afternoon. I wish you [Davies Giddy] or anyone else could give me a fixed principle by which to direct my judgement." By August the excesses of the Terror had led to the despairing cry, "I flattered myself that the tree of despotism was decaying at its roots. But this infernal Club of Jacobins with its mad mob will water it with innocent blood; and it will take root and put forth new branches and cover the whole earth with its blasting shadow." Stock wrote that this was Beddoes' last word on the French Revolution; but it was not quite the last.

Among the many Frenchmen who sought refuge in England during the time of the Terror there was one distinct group which aroused especial sympathy. These were the emigré clergy, whose numbers had increased

dramatically ever since 10th August, 1792, when Louis XVI was taken
prisoner and the monarchy overthrown. The oath of allegiance to the constitution of the kingdom, required by the Civil Constitution of the Church
which the French Assembly had enforced in November, 1790, had not been
expected to create great difficulties for the clergy. There was no attack on
the doctrine of the church and reforms in organisation were on the whole
acceptable. But when in April, 1791, the Pope condemned the Civil Constitution, the clergy were faced with an absolute choice of either denying their
church or, by refusing the oath, being seen as opponents of the Revolution
itself. It was inevitable that they became involved in counter-revolutionary
activities of all kinds. The Annual Register [30] published a pathetic account
of groups of these clergy walking the road from Dover towards London and
set up "a general subscription to prevent them from perishing in our streets.
Those unfortunate helpless men are here", it continued, "under the sacred
protection of our hospitality". There was a highly emotional response to the
plight of these particular emigrés and eventually in the summer of 1793
their cause even received royal support when George III ordered that appeals on their behalf should be made from pulpits throughout the country.
In Beddoes' home county the appeal, according to Katherine Plymley's
account,[31] was not popular, "not through new principles, but [by] the
old rancour of the farmers and lower tradesmen against the French". But
Beddoes in Oxford would have seen with considerable indignation the "whole
subscription from the University and the several Colleges and Halls amount
to eleven Hundred and Twenty three Pounds Sixteen Shillings" and this as
early as December, 1792. Oxford must have been a painful place for Beddoes
throughout that year, for, according to Jackson's Journal,[32] town, university
and clergy were all, in separate addresses to the King, proclaiming their
loyalty. "Horrible Acts in France" were reported and the French were
accused of having "sent their emissaries over to London and to all great towns
to distribute seditious publications and to excite dissentions among us."
It all seems to have been hard for Beddoes to bear. In the autumn he was
at home in Shropshire and on 9th October he published at Shifnal a two-page
"Fly Sheet" entitled 'Reasons for believing the friends of liberty in France
not to be the authors or abettors of the crimes committed in that country".
Though occasioned by a local appeal for the refugees, the paper is in fact an
impassioned defence of the Revolutionary government in France. It is also
quite openly an attack on the English government for whipping up enmity
against France by means of false information — "fabricated lies" as Beddoes
rashly calls the official accounts.

A sympathetic reading could find much that was sense in Beddoes' "reasons". He pointed to the achievements of the revolution; to the corrupt and mercenary nature of the clergy in France when considered as a body and to the possibility that there had been barbarous violence on both sides. Nor, he reminded his readers, are the French unique in their violence: "the design to roast Dr Priestley alive ... betrays as sanguinary a spirit as that of the Paris mob". Just as we would not want all Englishmen to be condemned for the excess of some, he argued, so in the same spirit we should guard against condemning all Frenchmen. The one paragraph that was directly concerned with the plight of the refugee clergy did not deny their need for help but emphasised that they should be relieved as "distressed *Men*". Despite his antipathy to priests, Beddoes asserts that "the most vicious ... should not be left to die of hunger", a test of charitable feelings that still remains valid.

Beddoes must have known quite well that his appeal would not, in fact, be read by sympathisers but he made no attempt to persuade or to moderate his language. The phrase about Dr Priestley is typical of the violence of his tone. He claimed that he had proof of his accusations that the clergy were also guilty of violence, and of his contention that the King was planning a counter-revolution with their aid (a point he also made in a private letter [33] to Davies Giddy) but he never brought this proof forward. There were in fact grounds for his fears. Instead he accused those who signed the appeal, including the M.P. for Bridgnorth, of ignorance and of making a "foul mouthed charge". If only he could have urged his case moderately he might have had some effect, or at least have escaped official notice. In his indignation against violent prejudice and misinformation he became as intemperate as those he attacked. He was not alone in his sense of the inappropriateness of appealing specially for the clergy; we find Wedgwood [34] a year later making a similar reply when approached for support. He, by contrast, confined his argument to the main issue, that *all* the emigrés were in need of help – though he allowed himself to comment, "it is too true that the prevailing party" will always abuse its power. "We need not name St Bartholomew, or the revocation of the edict of Nants; for melancholy instances without number will occur on all sides." Beddoes had, it seems, no such skill in making his point nor any self-discipline in controlling his feelings. On this occasion Reynolds did not save him from unwise publication, nor did he heed the advice Dr Parr had given his friends in July, 1792, to keep discussion of revolutionary ideas to private gatherings.

It cannot surprise us that a month later a secretary in the Office of Lord

Dundas, wrote from Whitehall[35] on 1st November, 1792 to Isaac Hawkins Browne, Member of Parliament for Bridgnorth,

Sir, It was yesterday intimated to me, that Doctor Beddowes has lately been very active in sowing sedition in your neighbourhood, particularly by the distribution of Pamphlets of a very mischievous and inflamatory tendency, and such as ought to be publickly noticed.

Neither of the Pamphlets alluded to, have yet been in the possession of Government, nor is the Conduct of Dr Beddowes more fully described than I have now stated to you. — But as it is understood that you have been a witness to some of his proceedings, and are likely to be in possession of some of the pamphlets, Lord Grenville (in the absence of Mr Secretary Dundas) has desired me to apply to you, and to request the favour of any information upon the subject which you may be possessed of, with your opinion how far the particular matters, of which Dr Beddowes may stand accused, can be authenticated upon oath.

I shall be obliged to you for an early reply to this letter and for any of the pamphlets or handbills in question, which you can conveniently spare. I have the honor to be Etc[a].

Evan Nepean.

Here we have the missing piece of the jigsaw which explains why Beddoes' 'Memorial to urge the establishment of a salary for the Chemical Reader at Oxford', was "forgotten or destroyed", even though it had been asked for by the Vice Chancellor of the university and received "very favourably" by Lord Dundas. Beddoes, in the account of the circumstances of his leaving Oxford which he gave to Davies Giddy, merely wrote, "I went into the country, became eminently and much beyond my importance, odious to Pitt and his gang as I knew from a hundred curious facts",[36] without more detail of what happened or of his own feelings. The phrase suggests that Beddoes may not have imagined that his writings would be noticed and viewed as seditious. This was typically incautious; he was well aware of the Royal Proclamation against Seditious Writings and Publications of 21st May, 1792. He referred to it in correspondence with Davies Giddy[37] who was somewhat embarrassed by finding himself obliged to support a loyal address from Cornwall in which the King was thanked for the care he showed for his people by issuing the Proclamation. The realisation of the possible consequences of his impetuosity must have been alarming for Beddoes. If followed up, the charge could well have led to severe penalties and there was no guarantee, considering the Government's state of alarm at revolutionary activities in England, that the incident would not lead to prosecution. Much less was sufficient for a spy to be sent to Somerset to tail Wordsworth; merely the habit of taking long country walks and a household which his neighbours could not quite understand. In fact, the manner in which he was investigated suggests that Beddoes was more discreetly treated than might have been the case. Isaac Hawkins Browne [38] was a landed gentleman who had a large estate at Dawley, bordering on Shifnal. He was one of the few Shropshire landowners to be active in the development of the iron workings on his land, rather than leaving them to the care of his agent. In the early 1790s his large iron works at Old Park were entering a period of rapid growth. Interested in the provision of canals and in the turnpiking of the road adjacent to his estate, he shared the concerns of the Darby and Reynolds families and could have known Thomas Beddoes as a friend of William Reynolds as well as remembering him as the son of the prosperous Beddoes family at Shifnal. That Isaac Hawkins Browne should be asked to report on his activities suggests that Beddoes and his family had a certain social standing and merited being dealt with more tactfully than by being reported on by some paid observer. No official action followed whatever report was sent to Evan Nepean, but Beddoes needed opportunity to plan his future and perhaps to live inconspicuously. He stayed for the next few months with William Reynolds at his home at Ketley.[39] There he was able to discuss his medical plans with Reynolds and with Dr Yonge of Shifnal, friends who understood his aims and who could offer practical advice. This time spent with William Reynolds enabled Beddoes to recover himself, as the gradual change in the tone of his writings shows. The publishing of the Fly Sheet was not the cause of his leaving the University; but it meant that instead of resigning with the prestige of having helped set up a Chair of Chemistry, Beddoes left Oxford in April, 1793, marked as a revolutionary, a serious handicap at a time when reaction was hardening.

CHAPTER 6

REVOLUTIONARY AND EDUCATIONALIST

The Flysheet attacking the appeal for the French emigré clergy was an outburst of violently expressed feeling but this alone did not satisfy Beddoes. He was full of enthusiasm to make some practical contribution to bringing into being a new and more equitable society. Like Wordsworth, he felt the "bliss" of being alive, "When Reason seemed the most to assert her rights"; he shared the more sober mood that followed and knew that all

> Were called upon to exercise their skill
> Not in Utopia, subterranean fields,
> Or some secreted island, Heaven knows where!
> But in this very world, which is the world
> Of all of us, — the place where in the end
> We find our happiness, or not at all.[1]

This is the determined optimism of the last of the essays which follow *Alexander's Expedition*. Beddoes extended his consideration of Indian affairs to the wider hope that reform of society would be brought about by the people themselves. He was firm in his belief that such a movement could only come from conviction — people were nowhere changed by punishment. It is clear that he also saw the need for new principles in education if the mass of people were ever to become capable of developing to this point. The Advertisement to his poem gave some general indication of the kind of education which he considered would help produce such a society. In his view, it was essential to capture the mental energy of young people and to take care to encourage what we today would call their creative powers — a view by no means so easily accepted then as it has become since. For Beddoes, "It is excited Fancy that has worked so many miracles in art and science and one may lament, both for the sake of knowledge and humanity, that some attention is not paid to this truth in education".[2] He cited as proof the early achievements of many great men, with Alexander, Newton, Locke, Boerhaave, Linnaeus and Lavoisier as examples. Furthermore, Beddoes believed that ideas conceived in youth might bring fruit later. In politics too, he asserted, great changes have been the work of young men. Imagination, to Beddoes, was no trivial power: "A mind vigorous in imagining", he wrote, "is also vigorous in judging"; its power is not confined to one kind of activity: "Probably in the most abstruse researches of science as much imagination is exerted as in the highest flights of poetry; and in the latter we judge and compare as much as in the former. It seems too, perfectly indifferent to the power by which we combine ideas, what sort of ideas it has to combine." Some young people receive, Beddoes allowed, an education which encourages these powers and which gives the discipline of reflecting on ideas. They are the fortunate ones and humanity loses by such teaching not being widely available. Beddoes felt confident he knew the reason: "if this advantage is at present confined to the few, where does the fault lie, but in those institutions which, by every direct and indirect means, counteract the designs of creative wisdom, and check the improvement of the individual, and by consequence of the species."

These passages in the work he produced in response to the challenge at William Reynold's dinner table, deal on the whole in generalisations, but Beddoes had in fact already made an attempt before this to write about the more practical details of education. He had set himself to work out a scheme for a school for children of the poor, such as might be organised by some philanthropist. He must have known of such schools and of the difficulties they encountered. William Reynolds' father, Richard, had "built a school house near his own residence at Ketley; but at that time",[3] as his grand-daughter later described,

such was the ignorance of the population by which he was surrounded, that not only were the children unmanageable and most unwilling to learn, but the parents would only send them on condition of being paid for their attendance. His exertions at Coalbrook Dale were more successful. The ministry of the excellent Mr Fletcher of Madely had produced a beneficial and lasting effect on the people in that part of the country and, so far from being indifferent to the value of good education for their children, they eagerly availed themselves of efforts made on their behalf

— the efforts produced a weekday school and two Sunday schools. The Quaker Reynolds, and the Methodist John Fletcher, had their counterparts among the more liberal Anglican clergy who set up parish schools and to some extent trained the teachers for them. So far as private individuals are concerned, possibly the best known example is Elizabeth Gurney Fry[4] and her school in the laundry at Earlham. Beddoes' plan was in the form of a letter dated 25th January, 1792. It was published in that year as, 'An extract of a Letter to a Lady on Early Instruction, particularly that of the Poor'.[5] In the circumstances of the time, Beddoes might well have had a particular project in mind. The 'lady' of his essay is not however identified nor are there any circumstantial details to suggest a real school, so we must regard her and her school, at least for the present, as a fiction invented by Beddoes to give him a plausible reason for setting out his educational ideas. Far removed as elementary instruction and the teaching of reading would seem to be from the experience of a university lecturer in chemistry, Beddoes had strong reasons for choosing to write on this topic, though his purpose is not apparent until his essay is well under way.

From the opening of the 'Letter' and from various later passages, we can see that Beddoes presented a clear 'reading method'. The child should begin with letters and then pass on to words and sentences. The phrases he is given to begin with should relate to his own life and he should then progress to read stories which tell of natural things. Beddoes suggested specific material to the 'lady': Mrs Barbauld's work; two stories by Mrs Trimmer; and Day's *Sandford and Merton*. The choice of Mrs Trimmer is a little surprising, for she was very pious and narrow; but Thomas Day and Mrs Barbauld were teachers with whom Beddoes could agree on many points. *Sandford and Merton* is full of lively factual incidents and the friendship which grew up between the two boys of the story points a good egalitarian lesson. Mrs Barbauld, who was the daughter of Dr Aikin of the Dissenting Academy at Warrington, designed her lessons to arouse her pupil's interest in the varied life of the world around him and, where religion was concerned, to connect it with the child's experience of that world. Her conversation lessons range in subject matter from a study of the oak tree and its usefulness to a discussion of great men which she introduces with a description of Mr Brindley directing work on the Duke of Bridgewater's canal. Her attitude to religion was exactly what Beddoes could approve, for she was clear that men "have a right to worship God as they please. It is their own business and concerns no one but themselves". Both Thomas Day and Mrs Barbauld find opportunity to condemn slavery. They taught that however its cruelty was mitigated, it remained an evil since "all men are brethren and equals". These two authors were putting into practice the educational principles Beddoes had in mind.

Beddoes went further than describing appropriate material. He suggested that reading and writing should be taught at the same time. This was unusual, for children were commonly taught to read without being given instruction in writing. Beddoes had a reason which has since proved well founded; he considered that "the action of imitating the letters ... would prove a source of amusement and very often confer the powers of distinguishing them quickly". (*See Note 5*.) He attributed the idea of letting writing and reading reinforce one another to "a very ingenious French writer"; he may have had in mind Mme de Genlis,[6] governess to the children of the Duc d'Orléans, for she certainly made great use of instructive toys and models and a variety of direct experiences. Beddoes' comment that the physical effort of making the letters would help the child, rests on his theoretical ideas about the imagination and the value he placed on what we should call kinetic experience — something he developed later in connection with the use of models. In the tradition of Hume, he believed that "the soul of a child . . . essentially resides in his senses". This is the reason for selecting material for teaching reading from the child's own experience and Beddoes asserted that it is injudicious to force into the child's head any idea which does not arise from the senses; indeed, "the greater number he acquires of such ideas (i.e, ideas arising from the senses) the better, provided you can engage him to dwell long enough upon each to fix it firmly and distinctly in his mind".[7] Coupled with this was another objection to the consequences of learning from material which has no real meaning for the child; "will not children long used to insignificant sounds become careless of the meaning of words and will they not acquire an indistinctness of conception and dimness of understanding . . .?"[8]

This stress on the importance of imagination as it arises from the senses is fundamental to Beddoes' thinking and at first in the 'Letter' it, and the choice of material it dictates, appears to be innocuous in application. It then leads Beddoes to a criticism of the use of religious maxims as teaching material. Even this point is initially made in a fairly restrained and reasonable manner but Beddoes quickly developed it into a strong attack on the teaching of dogma. Here the Established Church is as harshly criticised as Presbyterianism or Papacy. Beddoes advocated a thoroughly rational approach; religion was to be solely the concern of the individual who must in turn show tolerance to others. Indeed, Beddoes went so far as to praise Hinduism for seeing the different religions as varying approaches to god. If this were not unconventional enough, as the pamphlet got into its swing it moved over to a positive attack on the Church, asserting that its authoritarian teaching was harmful both to the individual and to the community. The miseries which Dr Johnson suffered as he forced himself to assent to orthodox statements of belief are cited as an example of the harmfulness of dogmatic religion. Beddoes suggested as an alternative to an authoritarian church that it is only necessary to accept what is common to all religions, which is the belief in an omnipotent god. Each individual should then be free to build on this his own pyramid of beliefs. He saw that religious tenets need to be experienced to be believed and warned of the danger that learning by rote would lead to intolerance.

To Beddoes, "the temporal welfare of mankind ... requires charity in preference to faith, and philanthropy rather than zeal; [nor could he believe]
that a quiet and humane inhabitant of the Earth is on this account disqualified
from becoming a denizen of heaven".[9] Another lesson should show that a
man's religion was a product of his social environment and children should
know that religious boundaries are co-extensive with geographical and political
ones. He went into this in some detail, drawing attention to the religious
and political boundaries on the Continent. In general, Beddoes argued that
religious beliefs were projected from earthly systems of government:

The material world easily supplied the notion of power; and our propensity
to sit in judgement on our neighbour's conduct and our own — the invisible
court of conscience — soon supplied an analogy, strengthened however by
visible tribunals, which led to the idea of remunerative justice — and that
with scarce an exception, in another life. Thus man created a God, an Heaven
and an Hell.

He confessed his own conviction: he "considered Deism very well in its
place". Such an outrageous attitude created difficulties later for him in
Bristol.[10]

He followed his very sound suggestion that stories for children should
come from natural history with the advice that natural objects themselves
should be studied. Although to us this seems obvious, Beddoes felt it necessary to reassure the 'lady' that she and any teacher she might employ, would
in time overcome the difficulty of arranging such lessons. Knowledge of
nature would have the beneficial result of making children less prone to
cruel behaviour to animals. Such cruelty, in his view, often springs from a
child's need for physical activity and then the tormenting of animals leads
to callousness and inhumanity. In this same sensible vein he urged that
opportunity should be made positively to encourage 'benevolence', and that
teaching should be reinforced by action. If we cherish "this tendency to
general benevolence we shall hinder malice from taking possession of the
breast ... while we exercise the minds of children in the contemplation
of their own sympathetic feelings we must be cautious to prevent them
degenerating into the grimace of affectation and ... be careful to lead on
their attention to beneficent actions".[11] It follows that stories and sayings
which glorify war should be excluded. So far, so good: this hardly seems
objectionable. It was not only by his religious opinions that Beddoes put
himself at risk. He allowed himself to be led into a similar unguarded position
in political affairs and these thoughts let loose in him a torrent of indignation,

at times amounting to rage, concerning the violence of the mob. He cited as his examples the uprisings of slaves, particularly the violent rebellion in St Domingo, when more than 2000 people had been killed and over 1000 plantations destroyed. He came in the end to the Birmingham mob who, in his view, would not have been so vicious had they not been so suppressed and had not the militia been called in. Such outbursts of violence, whether at home or abroad, Beddoes saw as nothing but reaction to tyranny, coupled with provocation from authority. There is deep feeling in his plea, "I beg your attention for a few moments to the urgent necessity for humanizing the minds of the people. Upon this the welfare of civil society immediately depends for a savage spirit in the people and tyranny in the possessors of power are to one another cause and effect".[12] His language became more and more excited and he abandoned any attempt at relevance. Memories of the Birmingham riots led him to dash off an attack on the government's provocative action in placing soldiers in barracks near towns, something which many thought had contributed to the violence of the rioting. New barracks were needed in the 1790s for the growing number of men under arms and such a standing army was by tradition generally distrusted. Beddoes' indignation led him to thoughts of the way the army was raised. It was a system well adapted both to corruption and to the needs of an autocratic ruler, for the Crown contracted with individual commanders to raise their regiment and paid a bounty for each recruit. Here Beddoes found a cue to bring in examples of tyrants in history and to explain how violence accelerated inevitably as tyrant and slave reacted one against the other.

Beddoes quite lost the original subject of his essay, but he finally succeeded in recovering it. At first he reintroduced his more philosophical point concerning the effect of education on society, emphasising that progress will come when we concentrate on "improvements at home instead of destruction abroad". To his mind, "a well digested system of public instruction will secure the peace of society more efficiently than the gallows and the bayonet".[13] He was convinced that mankind is capable of improvement and that not to believe so is "calumny of the human race". To meet this challenge he returned once again to his plan for the school. He conceived that it should be a place which would, over and above teaching reading and writing and inculcating sound standards of behaviour, "confer the habit and teach some art of industry". He then came to practical details. In the school of his imagination children would be occupied in simple crafts: he suggested sewing, spinning, making bird cages and baskets. He had thought this out thoroughly:

Suppose we have sixty scholars, a schoolroom, a working room, a small garden adjacent, and a school mistress as well as a master . . . . Ten of the scholars should apply to reading and writing, while the others, who should succeed in their turn are employed in sewing . . . . While at work, the mistress should encourage and support a conversation, calculated to fix the favourable impressions we suppose [the children] to derive from books. Such conversations might powerfully stimulate the less advanced and habituate them to that sort of reflection which applies itself to the ordinary affairs of life and constitutes plain and practical sense . . . . It would certainly be worth a trial if it promised no more than to render schools less like gaols.[14]

Beddoes pointed out that this was no dream school. Such a scheme was already successfully practised in Germany and some of these ideas were, he said, carried out in the Poor House in Shrewsbury, though with "some sacrifices to those established absurdities I have endeavoured to expose". Beddoes did not name his source of information about the German schools, but he clearly could have heard talk of these experiments from his Lunar Society friends. The midland scientists and industrialists were, of all groups in England, the most interested in developments in Germany and maintained active contacts for trade and manufacturing purposes. The way in which the picaresque R. E. Raspe [15] was accepted into these circles when he arrived in London from Hesse, is a good example. His was an enforced visit, to escape the consequences of 'borrowing' from the coin collection which, as librarian, he was cataloguing for the Landgrave. Even though he was such a doubtful character, he was useful as a link with scientists in Sweden and Germany and Matthew Boulton himself employed Raspe in various capacities – as industrial spy and, later, as assay master in the Cornish tin mines. It was Raspe who made the set piece of George III's head as a portrait medallion for the 1791 July 14 banquet in Birmingham. Given the interest in education in these circles, it seems not unlikely that contacts such as this would be made the opportunity for discussion of interesting educational institutions in Germany. Beddoes followed his account of the German schools with best wishes for the "little institution" and it would seem his 'Letter' was at an end. Not quite: he felt he must warn that constant effort was needed to keep any ground won. What a pity he was not as succinct as his contemporary [16] who warned that "The condition upon which God hath given liberty to man is eternal vigilance" – for Beddoes felt so, and bitterly observed that we need to persist in our efforts because men ever will feel, "Horror at new discovered truth". Were it not for the practical details in the 'Letter', it would be possible to believe that the whole thing was merely a cover for a political pamphlet of a subversive nature.

It was a very unguarded piece of writing to appear at that time, when opinion was hardening against liberal ideas and when anyone sympathetic to the changes in France was seen as a potential traitor to England. It appeared at the same time as the second part of Paine's *Rights of Man* which, with the London Revolution Society's correspondence with the French National Assembly, greatly alarmed the government. When the Flysheet brought about the decision to investigate Beddoes' activities, Evan Nepean asked Isaac Hawkins Browne to be on the look out for "Pamphlets of a very mischievous and inflamatory tendency" and said he had seen "Neither of the Pamphlets alluded to". This suggests that there was more in question than the one publication. If the solid gentlemen who later organised themselves into the Association for the Preservation of Liberty and Property had seen the 'Letter to a lady' they would have found its contents and tone very alarming, even before the outburst occasioned by the appeal for the French emigré clergy. It too would have been sufficient to cause a watch to be set on the author. Although Beddoes' open advocacy of political reform declined, and his hope for a speedy amelioration of men and society faded, his concern for education remained. This continued as his main interest in 1792–1793. Among all that he wrote at this period, only one work can really be allowed intrinsic merit, but all have interest in revealing Beddoes' ideas and his ardent, vehement character.

In the two other pieces published in 1792 Beddoes, rather than exhorting others, took upon himself the task of education. He is distinctly better as a practitioner than as a preacher. His little tale, *Isaac Jenkins and Sarah his wife*,[17] was designed as a lesson against drunkenness and, like *Alexander's Expedition*, was printed in Madeley by J. Edmunds. The printer and Stock's account combine to suggest that it was written while Beddoes was staying with William Reynolds. Although there are passages looking forward to reform of society, the story is free from a harsh spirit of controversy. It has indeed a relaxed air and a charm which set it apart among Beddoes' writings and which suggest that William Reynolds' friendship and his pleasant home, where plans for the future could be discussed, helped Beddoes regain a calmer mood. Coleridge admired it greatly and wrote later of "the genius of him who wrote *Isaac Jenkins*" — indeed he hoped that parsons seeing preaching and influencing individuals as distinct parts of their duty "would preach as Luther and would converse as Mr Langford to Isaac". Part of the attraction of *Isaac Jenkins* lies in its romantic descriptions of the country-side, descriptions which may owe something to Beddoes' reading of German novels. In spite of this, Beddoes' aim was not to impress the literary but to capture the imagination of working people, for his story belongs among the improving tracts which were so freely produced at the end of the eighteenth century and in the early part of the nineteenth. Its aim was to stress the evils of drink and its form immediately reveals the readers for whom it was designed. It is a tiny octavo book, easy to hold in the hand, let alone slip into the pocket, in fact a chapbook to be sold by a pedlar at the lowest possible price. It had a great success with this public, running into several editions and being popular in Ireland as well as in England. James Atkinson's *Medical Bibliography* of 1834 lists a 1796 edition with a note, "This had a great sale". It is the only one among Beddoes' books to have such a comment. In 1819 it appeared in a little book published in Edinburgh with two other short stories, one a translation of a tale by Mme de Genlis, where it stands up well in comparison with the other two. It last appeared in 1860 in a series of tracts.

The success of *Isaac Jenkins* lies in its freedom from condescension and from exhortation. It set before its original readers a scene which they could not fail to recognize and, showing how the misery and hardship of poverty are intensified by the debts and the moroseness of a drunken father, left them to draw their own conclusions. The cottage where Isaac and his family live is given a precise location, at Titterstone under Clee Hill, some twenty miles south-west of Madeley — not so far away as to be quite unknown to the Madeley workpeople for whom the tale was probably written in the first place and quite near to Hopesay, the home of Beddoes' uncle and aunt. The narrator tells how Sarah Jenkins, in desperation when her children were gravely ill of 'spotted fever', waylaid the doctor who was on a visit to someone fortunate enough to be able to afford him. She heard him go up the hill to Hopton to attend the parson whose arm had been shattered when he "was out a cock shooting [and] his gun barrel had burst in his hand". Sarah listened anxiously for his return. When she heard a horse, even though it was impossible it could be the doctor's, "she bolted her head from the door and looked wistfully towards the hill".[18] In this way Beddoes introduced the doctor into the poverty-stricken and lowly household in quite a credible manner and at the same time turned the action so that he could draw on his own experience. Isaac's drinking too, is given a credible origin and described in a sympathetic way. He gave the time of the story as, "A terrible time, the latter end of 1783", just the time when Beddoes himself had worked in the Shropshire countryside during an epidemic. He must surely be expressing what he had sometimes felt at that time, when he described the thoughts and feelings of the doctor in the story, Dr Langford. After he had attended the

children, the doctor discovered the father's drunkenness and, "so thought he
to himself, 'I have done these creatures little service at the last'." By skilfully
questioning Sarah before he remonstrates with her husband, Dr Langford
discovers that it was in the time of grief after the death of his son that Isaac's
behaviour changed. The boy died after a shocking accident that happened
when he and his father were returning with a team of horses from work in the
fields. Some 'collier boys' let off a cracker and startled the horses and

The lad, whom Isaac had put to ride the fore horse because he was tired, was
very soon thrown down headlong; and the horses ran over him. It was in vain
that Isaac hurried and bawled after the horses. They had trod and crushed
the child underfoot before he could come up; and all he could do was to
carry him home, bleeding and mangled with his face all one wound. And he
actually died the next day after suffering terrible pain. [To add to his bitterness,] [t]he collier lads however got safe into the pits and he could find no
trace of them.[19]

Beddoes might have been remembering his own grandfather's death, but the
scene must have been recognisable to many among his readers. Such pits,
clearly drift mines with tunnels leading in from the hillside, were common on
the Clee Hills. It was understandable that at such a moment Isaac could easily
be led to drink heavily at the alehouse by a group of men who had come
from London to cut and "saw the hollies into thin boards for fineering"
(veneering). By introducing such an episode Beddoes avoided harsh judgement and the assumption that drunkenness is a vice of the poor.

The account which follows is full of lively touches and some humour.
Sarah was sent to get wine or ale from the alehouse. She took her black
crock which she had washed and rinsed "briskly at the spout of water close
by the garden hedge".[20] At first she was rebuffed on account of her husband's
score behind the door. Then the alewife who, fortunately, shared something
of the spirit of the Wife of Bath, relented. She remembered how respectfully
Sarah had always behaved to her:

Now I bethink me, Sarah, you always stood back when we were at the church
door together; and when we meet in the lane, you always stop short and look
down upon the ground, and make a curtsey and say 'Your servant, Mrs
Pritchard' and, Sarah, did I not always make answer, 'Sarah! How be you,
Sarah!' For I always loved to be lowly and humble in spirit. So wait here a bit
and I'll bring you the drink directly.[21]

Mr Langford went roundabout to his remonstrance with Isaac, describing
how hungry the beasts are in the frozen fields before he drew a parallel with
hungry children and pointed out that pain and hunger will make them vicious

just as it does animals: "A blind man may see the way to make children wicked is to use them ill".[22] He saw that Isaac was ashamed and the little boy too sensed something amiss. Puzzled by the look on his father's face he can only suppose that he is unwell and offers him his potatoes and milk with the words, "Here, Daddy, eat a mouthful, do now, and dunno be sick". The doctor enabled Isaac to establish his self-respect by letting him take up the talk about farm animals, show off his knowledge and describe how the

kine have come to the gate in the lane and hung their heads over and lowed so 'twas pitiful to hear them. For all they were dumb creatures I knew their meaning, as well as if they had said 'give us a mouthful of dry food, for we have more snow than grass to eat and our bellies are aching with hunger and cold'.

Then, Mr Langford said, "I felt him abruptly . . . an attempt on my part to humble him by extorting a full confession of his guilt might only set him upon devising some excuse or defense".[23] To give him time to recover himself, Mr Langford distracted the child by taking out his watch and swinging it before him; then he motioned Sarah to silence, slipped some money into her hand and went quietly away.

The description of Isaac's reformed state is less lively and perhaps a little more moralising. There are two memorable episodes. The first is a grim, professionally detailed description of the dropsy of a wealthy drunkard. Isaac was called in to help with the heavy nursing and as he lay dying, the sick man sang over and over again the familiar psalms he had not heeded early enough. Beddoes was clearly giving a hint that formal religion had been of little real use to the dying man. The second is by contrast a humorous episode. The Londoners come again to cut holly and, confident that he would not refuse, invite Isaac to join them in another drinking bout. They are well put down by their former drinking companion who has lost all interest in "Three Horseshoes". This part of the tale takes the form of a letter from Dr Langford telling of the ways in which Isaac's new attitude had led him to repair his cottage and behave in a more kindly and responsible way to his wife and his children. It lacks the dramatic qualities of the early episodes and there are no more descriptions as romantic as the opening when

nothing was to be heard in the dusk of the evening but the church bells tolling for funerals, unless it was the howling of the wind or the hooting of the screech owls. It was melancholy for a man to hear as he walked home across the fields from his work towards the fall of night.

There is scarcely a hint of Beddoes' revolutionary feelings in this story but his more affluent readers were surely intended to be impressed by the vivid descriptions of the tumbledown cottage and the poor diet of the family.

REVOLUTIONARY AND EDUCATIONALIST 91

"The house was as cold and open as a barn; the plaster was fallen by patches from the walls and you could see through the bare wicker work; the wind whistled in at the chinks in the door. The floor was damp, dirty and ill smelling withal."[24] Beddoes also made it clear that the improvements Isaac was able to make when he gave up drinking were only marginal. He did not hide his view that what is needed is not preaching, but a change in circumstances: "there would be little wickedness in the world if there was no distress — Vice almost always begins among the poor from misery, and among the rich from idleness".[25] In 1792 he was still hoping for change: "I wish with all my heart things could be so ordered that every poor family should be comfortably clad and plentifully fed, and have besides the wherewithal to make decently merry at times; and I hope this may come to be the case," he told Sarah and Isaac. It was a very different attitude from that expressed in, for example, the tracts of Mrs Trimmer.

Those well intentioned people who, at the end of the eighteenth century, set 'out to provide improving reading for the humble soon found that the matter was not quite so simple as they had supposed. First of all, there was the problem of persuading 'the poor' to accept the tracts. Katherine Plymley[26] saw quite clearly that servants might resent having material written down to them and pointing morals. Her diary for spring 1795 describes a scheme originating with Mrs Hannah More — a leading spirit in the activity — to have tracts printed by subscription. These were then sold to pedlars to sell cheap in the hope that this would be more likely to get them read than giving them direct. "It is hoped," Miss Plymley wrote, "they may in time subvert the mischief arising from the improper ballads and irreligious stories at present circulated". She was very doubtful whether the scheme would succeed, partly on account of the quality of the material: "The tracts seem very unequal and I could have wished some had not been printed", she commented. In fact, she thought the most good they might do could possibly be to the *donors* who would read them and so themselves receive moral instruction! That her fears were justified she soon learned from her brother. Mr Eddoes, a Shropshire bookseller, had told the Archdeacon that "he has sold numbers of the tracts to those who give away such things but could not dispose of any to the Hawkers". We learn that she was happier about *Isaac Jenkins* from her description of an unusual way in which it was sometimes distributed:

Mr Houlbrooke took a very good way to dispose of a publication of this kind of Dr Beddoes a very good book the history of Isaac and Sarah Jenkins he used to carry some of them in his pocket when he rode, and drop them here and there on the highway this was likely to excite attention and prevent

the jealousy I suspect they [i.e., servants and workpeople] are apt to feel
when these things are given from a idea that you have and in endeavouring
to infuse into their minds ideas that you do not think necessary for yourself.
. . . Mr W. Reynolds in a letter to my Brother encourages us to hope that he
had given numbers of the last mentioned book among his miners and with
much success [K. P.'s punctuation].

She was probably all the more appreciative of *Isaac Jenkins* having the right
touch as she had to admit to a disastrous attempt to influence her own
servants and to finding such tracts — Mrs Trimmer's *Servant's Friend* in
particular — were disliked and rejected.

The success of *Isaac Jenkins* contributed to the popularity of the other
little chapbook of this year: 'A Guide for Self Preservation and Parental
Affection or Plain Directions for keeping themselves and their children free
from several disorders'. The same basic ideas for success in capturing children's
attention recur here but in the main the pamphlet is just as its title describes
— practical advice of a medical nature; for this reason it more suitably belongs with the medical writings of the next period. It is interesting to see
that Beddoes had been at pains to adapt this work, as he did *Isaac Jenkins*,
to its readers. It is only 24 pages long and cost 3 pence; Beddoes wrote of
it to Dr Darwin, "I published just before I went into Ireland, a little book of
directions for people respecting their health . . . . I thought it trifling and
did not expect the sale it has had. Perhaps the plain style recommended it,
perhaps the fame of *Isaac Jenkins*." Dr Darwin considered the pamphlet a
great public benefit.[27]

In the next work, *Of the Nature of Demonstrative Evidence*, published in
1793, we are back in the world of theory, but as he sets out his aim we
realise that in this essay Beddoes was attempting a more ambitious development of his ideas. He explained

It was partly in order to strengthen those arguments that have been urged in
favour of a plan of education which shall pay some attention to the senses
and understanding . . . , partly to take away from the revivers of exploded
absurdities, that support which they have been desirous to gain by forcing
into an unnatural alliance with their cause, so respectable a science as mathematics; and partly to show what false measures of objects are taken by those
who have no better rule than ancient metaphysics, that these remarks are
offered to the public.

His concern was still to convince his readers that ideas and also the power of
abstraction rest on the activity of the senses. Language too, he argued, is built
on sense experiences. Obviously, his authorities are Hume and Locke and for

the argument concerning language he referred at length to Horne Tooke. The educational conservatives who wished to maintain the importance of authority and to uphold the contention that abstraction is a distinct power had taken their last stand upon the ground of mathematics. Beddoes vigorously contested this ground, maintaining that as much as any other science the "mathematical sciences are sciences of experiment"[28] — and reminding his readers that Euclid himself began with experiments. He was as practical in support of this contention as he was in advice about teaching reading, for much of the early part of the essay is concerned with the way young children could best learn geometry. Not surprisingly, he urged that children should have something to handle and to move around and that this would lead on to an understanding of the properties of geometrical figures: "What I want is the sensible evidence of the thing . . . a point is first the end of anything sharp".[29] The habit of observation which would be fostered by such a method of teaching would be valuable in many aspects of mathematics. He instances how the idea of proportion can be based on sense experience — illustrating by reference to "fatness" (an appropriate choice in his case!). Beddoes denied that his method would impair the power of abstraction; "I answer, that this method will confer the power of abstraction in a superior degree. The senses will deliver more distinct ideas to the memory; the more firmly the memory holds ideas the more easily will the imagination and judgement be enabled to perform their functions".[30] It seems a little odd that in going on to discuss the relationship between observation and experiment Beddoes does not make reference to verification or to number. It might have been even more appropriate here than in the *Advice to Parents* where his plan of health care was supported by appropriate statistics, set out in tabular form. Above everything, he was convinced this way of teaching did away with the "revulsive sensation" which comes to all who have been condemned to rote learning and to the unexplained acceptance of proofs.

To a considerable degree, though it avoids the political excesses of the earlier work, 'Of . . . Demonstrative Evidence' suffers from the same defects as the 'Letter' — an undisciplined use of any material which comes to hand to support his ideas. He was out of his own field and appears rather as the enthusiast, not as the scientist. Yet two somewhat extraneous aspects of the essay give it a special interest, providing an insight into Beddoes' own life and even into developments in eighteenth century thought.

In the first place, we have in Beddoes' essay probably the first account in English of Kant's *Critique of Pure Reason*.[31] Beddoes abandoned his initial decision to refrain from quoting or discussing the opinions of other authors because he recognised the originality and importance of Kant's work and because, in spite of the growth of his reputation on the continent, Kant's writings were not accessible in England. Beddoes was happy to see in Germany growth of freedom to express ideas and felt that those in England who were interested in new thinking would "wish that an explicit account of Kant's doctrines were published in English, which the terminology would render a difficult undertaking". There follows a lengthy extract from the *Critique*; Beddoes chose the assertion "that we are in possession of knowledge a priori" and in particular the passage discussing necessity and universality as ideas a priori. In the end Beddoes held to his view "To observation and to induction alone, whatever Mr Kant may imagine, it is easy to see that we owe our knowledge of the absolute necessity or strict universality of geometrical truths", but he is led by this challenge to pay attention to the problem arising from the limited powers of our senses to make observations. Beddoes referred to the Göttingen journals as the source of his information about Kant's reputation; he was, of course, still in correspondence with his Edinburgh friend, Dr Girtanner. Beddoes was already knowledgeable about Kant before he and Coleridge met in Bristol; it is reasonable to assume that at least some of Kant's works which were in his library at the end of his life were bought in the years following 1793 and that Coleridge could have known them.

The other matter of note is the special interest in mathematics. The essay was dedicated to Davies Giddy, himself a mathematician, and it is the first glimpse of another project which may even then have been forming in Beddoes' mind. This was the idea of making apparatus to be used in teaching mathematics. The scheme is discussed more fully in a letter written to Thomas Wedgwood sometime after 1796.[32] In it he hoped that his models will "smooth the difficulties which render the study of geometry so repulsive to young people." Thomas Wedgwood was both patient and patron of Beddoes and as such had a part to play in Beddoes' work in Bristol. He was also a scientist in his own right, and as a rich man, anxious to use his wealth to further human progress. He wished to do more than relieve distress – his approach was more that of a Nuffield or a Rockefeller – and he saw in education a means of reforming society. Ultimately, he settled on the idea of helping the 'formation' of a genius who would in turn lead mankind – Godwin came to his mind, but (as is well known) he decided in the end on Coleridge whom he had opportunity to know well when both were part of Beddoes' circle in Bristol. Thomas, like his brother Josiah, did considerable thinking about education and Beddoes exchanged ideas with both; Josiah

Wedgwood certainly included Beddoes among those he consulted in drawing up his plans for the education of his own children.

In writing on educational topics, Beddoes was taking part in a debate that was of widespread interest at the time. He was unusual in developing his ideas at two very different levels. 'Of the Nature of Demonstrative Evidence' belongs to the world of the Wedgwoods with their theoretical interests and their ambitious schemes. Beddoes clearly had in mind the child of an affluent family taught at home by a tutor or governess, as were Josiah Wedgwood's children, and he possibly imagined a parent interested in educational ideas. In 'Of the Nature of Demonstrative Evidence' Beddoes does not develop practical suggestions very far, but when he returned to this topic later he had come to believe that a small group, gathered together from neighbouring families, would be the best arrangement for the education of young children. Mrs Barbauld, in the introduction to 'Evenings at Home', described something approaching such a group in a house where there were often visiting children. Beddoes was just as concerned that the children of the poor should be taught well. We can see that the underlying purpose of the 'Letter to a Lady' is a determination to argue for something more truly education than the instruction devised by writers who concerned themselves only with schools for the 'labouring classes'. At the same time he understood the practical needs of humble people; he might wish to free them from dogmatic religion, but he knew that they needed to learn skills. Direct observation of working people, particularly as a doctor, gave Beddoes an unsentimental sympathy for those whose lives were a constant struggle with poverty, and especially with the women and children. He was at his best when writing directly for them.

These writings, from *Alexander's Expedition* to 'Of the Nature of Demonstrative Evidence', were written or prepared while Beddoes was at the height of his career as a university lecturer. They may appear unrelated to his teaching of geology and chemistry but in fact this is not so. As a scientist he had to argue for himself, and for his students, new theories of the formation of the earth; new explanations of the process of chemical reactions; and new attempts to explain rather than merely describe the great variety of forms of life. Engaged in such speculations it is not surprising that Beddoes looked for new social conceptions of society. The argument with which he begins 'Of the Nature of Demonstrative Evidence' and the opening of his first medical work, 'Observations on the cure of calculus, etc.' show that Beddoes himself made this connection. He saw "chemistry daily unfolding the profoundest secrets of nature". That men should unite in working for liberty regardless of national boundaries; that all men, of whatever skin colour, should be entitled to human dignity; that the satisfying of physical needs was a pre-condition of a harmonious society, these were ideas to which Beddoes could subscribe for scientific as well as humanitarian reasons. He was working at a system of coherent thought, an endeavour we associate with a poet or a philosopher more easily than with a chemist. The attack on him, when it came at the end of 1792, must have been deeply felt. There is a tone of desperation in some of his utterances which raises the question whether a sense of persecution did not in fact intensify the irascible side of his nature, possibly even permanently. The essays of 1792–1793, for all their lack of restraint and balance, still make it clear that when he came to Bristol and to practice as a doctor, Beddoes was already a man with interests and contacts outside medicine.

CHAPTER 7

BRISTOL: REVIEWING FOR *THE MONTHLY REVIEW*

During the months Beddoes spent with William Reynolds a more detailed plan came into being for testing the efficacy of gases in the treatment of disease. William Reynolds, his younger half-brother Joseph, and Dr Yonge, the Shifnal doctor who had encouraged Beddoes' early decision to study medicine, each agreed to contribute [1] £200 towards a 'Pneumatic Establishment'. Beddoes himself was to find a fourth £200. The agreement was that a house should be rented where a laboratory could be set up and where there was suitable accommodation for patients. It seems that the first plan did not envisage a hospital but simply provision for occasionally receiving patients in the house. The whole undertaking was to be under Beddoes' supervision and he was to find suitable assistants. Beddoes left to explore possibilities of setting up the Establishment in London; he was accompanied by Dr Yonge and James Sadler his "laboratory genius". Together these two would have been able to assess the suitability of any premises for both the medical and the scientific work. Sadler found time to work on his steam engine plans and Beddoes continued to interest himself in them, for it was on this visit to London that Reynolds received the drawing that he marked in his note book, "sent while he was in London with Dr Beddoes". The London plan came to nothing and at Beddoes' suggestion, they settled on the Hotwells instead. It was there, on the outskirts of Bristol, that Dr Thomas Beddoes set up in practice and with the help of Sadler organised his laboratory.

Bristol had much to attract a newcomer. It was an important town in the West Country and a prosperous port, though it must be admitted that it owed much of this prosperity to its large share in that slave trade which Beddoes and his friends hoped to see abolished. The city had close links with the Midlands and its situation on the route from Shropshire to the sea and to Wales led to personal ties between the two communities. The developing iron and coal fields had for many years attracted ambitious young men from the city, among them members of the Reynolds and Darby families. Richard Reynolds, father of Beddoes' friend William, after his years managing the Coalbrookdale mines and forges, returned in his retirement to Bristol and played a considerable part in the cultural development of the city. William Anstice, who passed on to Dr Parr the copy of *Alexander's Expedition*, also came from Bristol. He belonged to a family which Beddoes very probably met, since it was connected by marriage with Thomas Poole of Nether Stowey who certainly belonged to the group which took part in Beddoes' experiments in breathing nitrous oxide. Yet it was more likely to have been the established spa at the Hotwells that influenced Beddoes decision to come to Bristol rather than the city's links with Coalbrookdale.

A warm spring rising just to the west of Bristol gave the village its name of Hotwells. This spring rose at the foot of the high cliff bordering the river Avon just downstream of the gorge which is today crossed by Brunel's suspension bridge. Originally the village lay between the steep, high cliff and the busy river. By the late eighteenth century the sadder purposes for which people came to Hotwells were to some extent disguised and an account[2] published in 1803 described how,

After quaffing the salutary beverage, those who are inclined have the advantage, during rainy or cold weather of walking under a colonnade, in a crescent form with a range of shops. There is likewise a fine gravelled parade about 600 feet long, by the side of the river, shaded with trees; and here during the heat of the day, the company may retire, and be amused with the ever-varying scene of ships passing and repassing. Little excursions are frequently taken down the river in boats, sometimes accompanied with music, which, re-echoed by the rocks, has a delightful effect. Companies sometimes sail as far as Portishead, where they land and dine in the cool and shady woods, and, from different stations in the vicinity, enjoy delightful views of Bristol Channel, the steep and flat Holmes islands, and the opposite Welch mountains.

The writer was at pains to stress that the medicinal waters themselves were pleasant

Bristol Hotwell water, when received into a glass from the spring, appears sparkling and full of air bubbles which adhere to the sides, as if it were in a state of fermentation. It is without smell, pleasing and grateful to the stomach.

This same account passes over the difficult time of uncertainty just before the war with France and describes how much had been done in the previous twenty years to improve the village, a new bath for bathing had been added, the spa extended and a new colonnade built; "Many handsome piles, mansions and houses of free-stone" had been lately erected, among them two squares associated with Beddoes, Hope Square and Dowry Square. These new houses were built back from the river at the foot of the cliff and up its steep sides. Climbing up to some of the squares must have been hard work for invalids and even for Beddoes, who as early as 1794 was complaining of shortness of breath. A busy main road now cuts the Hotwells in two but in 1803 the writer clearly saw uninterrupted walks from the hotels and lodgings to the riverside and the public meeting places. There the dream was to develop a life similar to the polished society of Bath, with the same rules and style. It does not seem that Hotwells ever quite succeeded and Beddoes certainly scorned its pretentions.

Above the Hotwells, on the high downs, the village of Clifton was develop-ing into an elegant place to live. Sweeping rows of buildings were built along the contour lines, hoping to achieve the same grandeur as the crescents in Bath. Above on the downs were the squares and terraces; sometimes the plans were over ambitious and the contractors had to break off the work unfinished. When he married, Beddoes' home in Rodney Place was in the heart of this fashionable area:

The beautiful village of Clifton, which for the purity and salubrity of its air, has been denominated the Montpelier of England, from its elevated situation furnishes the most charming views over the western part of Bristol, and of the Avon for a considerable way, with its moving scene of ships. It stands on a hill, which rises by a gradual ascent from the river, and is, in a great measure, covered with villas, and elegant piles of building.

Many private families of opulence and respectability, make this their principal residence.

This grandeur was some years ahead for Beddoes. When he arrived in 1793 besides experiencing the mood of depression which had halted expansion, he was soon to see what a turbulent place the city could be. At the end of the year the corporation's failure to end the tolls paid on the new bridge provoked such opposition that there was rioting for several days and before the city could be pacified people had been killed. Apart from local difficulties in Bristol the national situation made 1793, the year of the outbreak of war with France, an unfortunate time to be a newcomer in any city. Everywhere the country was unsettled; there were protests about shortages and wages; fears of seditious publications and activities. In Bristol such fears, added to anxieties about the war, were a check to growth. The optimistic building of new houses, of elegant squares and houses, ceased. Buildings were left empty, even unfinished. At such a difficult time, it cannot have been easy for Beddoes to make a start in the city. The whole country was beset by almost hysteric fears of revolution. The Proclamation against seditious publications issued in May, 1792 had begun a period of increasing tension. Though John Thelwall, a leader among the London radicals, protested that the Proclamation had not the force of law, the sympathisers with Revolutionary France and those who wished for even modest reform in England, now felt a growing sense of urgency. His pamphlets of 1792 show how Beddoes was carried along with this movement and throughout 1793, while he was settling in Bristol, he would have watched the continuing efforts of the radicals.

They sought some means of expression that would be seen as both peaceable and widely representative, since the Westminster Parliament, corrupt and unreformed, manipulated by the King's patronage fund, provided no forum for the expression of their hopes. At first the hope was that Parliament might be influenced to pass a moderate measure of reform in representation by petitions from the towns and counties; but the organising of signatures and the presenting of such petitions proved too difficult a task. A very moderate measure for reform was defeated in Parliament and from that point the only instrument left for the advocates of reform was a national convention.[3] In November 1793 such a convention was actually held in Edinburgh and attended by delegates from England. The resulting trials for sedition and the harsh sentences of transportation imposed by Lord Chief Justice Braxfield shocked even parliamentary leaders. For the members of the reform movement the trial and the judgements were ominous; if the English courts followed the lead of Scotland, many could face sentences of transportation, even of death.

With the fear, the sense of urgency grew among the radicals in England; they first expressed their abhorrence of the Scottish trials and then began to plan to hold a National Convention in England. Protests against the building of barracks in towns, such as Beddoes himself had made in his 'Letter to a lady', now turned into fears that there were plans to use such barracks to quarter Hessian mercenary soldiers. People remembered how Hessian troops were used against the American colonists and feared that they might now be employed to suppress popular protest at home. It was beyond doubt that Volunteer regiments were being raised and armed, ostensibly for defence in time of invasion, but perhaps for the same purpose. There was the possibility that soon Parliament would move to make a Convention illegal. On its side, the government received accounts describing the protest meetings held by the London Societies and reports from its spies; these gave details of offers to provide arms for the London radicals, and accounts of the audiences at the various meetings held by John Thelwall in London. Mrs Thelwall,[4] in her account of her husband, insisted that his revolutionary beliefs rested on "principles of reason and humanity", but his speeches in 1794 were unrestrained and violent in their attack on the government and on the Ministers. Though William Godwin[5] endeavoured to restrain him, Thelwall persisted.

The story of his determination to find ever new means of holding his meetings witnesses to the government's equal determination to close them and to prevent publication of his articles. Pressure on landlords to refuse to hire him rooms; libel action against his editor; and, when Thelwall had rented a house where he could live but which had a room large enough for a lecture, arranging that he should be charged with "nuisance"; all show the sort of harassment he, and others, had to face.

In the spring of 1794, Beddoes must have followed the course of these events, and of similar activities in the provinces, with alarm. Apart from the optimistic expressions of hope for a new society in 'Demonstrative evidence' and 'Observations on Calculus', incautiously expressed but not overtly anti-government writings, so far as is known, he took no further part in this movement of protest. He busied himself getting established as a doctor in Bristol and looked for patients in Wales. There must have been mixed feelings about him. On the one hand he soon felt confident of earning sufficient, but he was still regarded with suspicion; the Bristol Library Committee even doubted whether he was a fit person to be entrusted with books. Apart from his reputation, it would not be in character for him to hide his opinions and feelings.

In mid-May came the arrest of the leaders in London; first Thelwall then Joyce, Lord Lansdowne's secretary, and on May 16th the cautious Horne Tooke whose works Beddoes had used to support his argument in the essay, 'Of the Nature of Demonstrative Evidence'. On 22nd May, *habeas corpus* was suspended. There followed one of the most dramatic trials [6] in English history. The twelve leaders were accused not, as might have been expected, of sedition but of high treason. As newly defined in a panic pronouncement by Lord Chief Justice Eyre, treason could be any course of action which might result in the death of the King without there being any such intention by the accused; or any attempt to reform the system of government except by the decision of Parliament. The twelve accused had little hope of escaping the death penalty in its most barbarous form; a verdict of guilty was almost inevitable after the alarming disclosures of plotting made in an 'Address to the Crown' passed by Parliament between the arrests and the opening of the trial. On October 20th, only five days before the trial was to begin, the *Morning Chronicle* appeared with an article which brought about a dramatic reversal of the legal situation. Entitled 'Cursory Strictures on the charge delivered by Lord Justice Eyre', it made clear the unconstitutional nature of Lord Eyre's ruling. By careful reasoning and reference to legal precedent it exposed the argument for what everyone knew it to be: a device to trap the accused. The way the article came to be written was as dramatic as its effect. The last of the accused, Thomas Holcroft, had not been arrested but had surrendered himself to join his friends; to share the appalling conditions of their imprisonment and to stand trial with them. Before he gave himself up he was able to get a message to his friend William Godwin, author of *Political Justice*, who had only just gone into the country, to Warwickshire, to stay with his friend, Dr Parr. Godwin himself was at risk since he had been active for the defence in recent political trials where there had been fears of heavy sentences. He immediately returned to London; visited Holcroft in Newgate Prison and then, without going to the Tower where his friend Thelwall was confined, shut himself up and devoted himself for two days to the writing of 'Cursory Strictures'. Though he may have seemed cold hearted in not visiting Thelwall, and may even have acted in accordance with his theories rather than as a friend, Godwin probably felt the visit to Holcroft was risk enough. He must remain free to make an answer to Lord Eyre which could not be refuted. It was Godwin's argument, and its publication in the *Morning Chronicle*, that provided the ground on which Erskine constructed the defence.

The trial of Hardy, the first to come before the court, was made an opportunity for the Crown prosecution to reveal, detail by detail, the activities of the accused, mostly as they were reported by "witnesses who almost all confessed themselves spies and informers". All democrats must have watched in the utmost alarm, for it was believed that there were warrants ready for the arrest of hundreds of others once the verdict of guilty was announced. On November 5th, after a trial of ten days, when many, Beddoes probably among them, waited apprehensively, Hardy was acquitted. Hardy himself and Erskine, the Counsel for the defence, were drawn through the streets of Londdon in triumph. Quickly, the news reached the provincial papers; Crabbe Robinson [7] long remembered how the news came to Bury St Edmunds:

During the first trial I was in a state of agitation that rendered me unfit for business. I used to beset the Post Office early, and one morning at six I obtained the London paper with "NOT GUILTY" printed in letters an inch in height, recording the issue of Hardy's trial. I ran about the town knocking at people's doors, and screaming out the joyful words.

From Shifnal, Rosamond Beddoes [8] wrote to her aunt Mrs Whitehall to tell of the trial and acquittal of Horne Tooke, who had expertly defended himself, and to describe the crowd drawing the carriages of Erskine and Hardy. Where her sympathies lay was clear, for she was preparing to send to Hardy for shoes. She advised Mrs Whitehall, even though, as she wrote, "my aunt is an aristocrat", to take and read the newspapers. Of the other defendants, only Thelwall had to stand trial. The threat of the death penalty was swept away but otherwise, repression continued.

In view of the turmoil throughout the country, but particularly in London, it was perhaps fortunate for Beddoes that he did not settle in the capital. Even in Bristol, an introduction which would smooth his way was very useful. Such support came as a result of his friendship with James Keir. Maria Edgeworth,[9] in her account of her father's life, describes how he helped Beddoes:

While we resided at Clifton, my father became acquainted with the celebrated Dr Beddoes; and it is remarkable that this acquaintance was in consequence of the doctor's great admiration for Mr Day. This had induced Dr Beddoes to seek the acquaintance of Mrs Day, and of her friend Mr Keir. When Dr Beddoes came to Clifton with a view to setting up as a physician, Mr Keir gave him a letter of introduction to my father, who was, I believe, his first acquaintance there. My father admired his abilities; was eager to cultivate his society; and this intimacy continuing some months he had opportunities of assisting in establishing the doctor at Clifton.

Edgeworth's own unconventional attitude did not escape criticism but as a sociable man and an aristocrat his oddities could be forgiven. His friendship would certainly help a newcomer establish himself in society. He and Beddoes must soon have found they had more in common than admiration for Thomas Day. Edgeworth was well known to the Lunar Society group and, in particular, was a friend of Erasmus Darwin. He was a lively man with a talent for invention which must have appealed to Beddoes. Edgeworth recognised his abilities and foresaw a good future for this "fat little democrat" if he did not spoil it by his indiscretions.

In the eyes of his friends, whether the Edgeworth family in Bristol, Davies Giddy in Cornwall or his friends in the Midlands, Beddoes was now entirely concerned with his medical work.[10] His practice as a doctor grew and he made the principles and aims of pneumatic medicine clear in a series of pamphlets. He was busy collecting reports from other medical men interested in the use of gases and in raising funds for his projected institute. Beddoes himself had offered even before he left Shropshire to contribute all his professional fees to the scheme and proceeds from the sale of his writings were earmarked in the same way. Very understandably, in view of his distress at the violence in France and in the threatening circumstances at home, Beddoes entirely gave up political activity in 1793 and 1794. It appears that he concentrated on the experiment for which he was becoming known and for which he has always been remembered. It is something of a surprise to discover that, busy as he must have been with his medical plans, Beddoes was for the eight years between 1793 and 1801 a frequent contributor to the Whig periodical, *The Monthly Review*.[11] This writing must have been quite unknown to most of Beddoes' friends, suspected perhaps by a few. It forms an interesting link between his career as Chemical Reader at Oxford and his medical work. It must have been a demanding task but it gave Beddoes an opportunity to keep up his chemical interests and to write on new medical work which related directly to his own. It is of considerable interest in adding to our understanding of Beddoes' own work, both of its range and of the scientific ground on which his medical ideas were based. For this reason and because it reveals the breadth of interest which made him such a stimulating companion for the group of younger men who came to make his home a meeting place, it is useful at this point to give some account of his reviewing as a whole.

Beddoes may first have met Ralph Griffiths, the Editor of *The Monthly Review*, when he was in London in 1793, probably through the kindness of Josiah Wedgwood who was a long-standing friend. On an earlier occasion he had sent his partner, Thomas Bentley, to enjoy the "instructive and elegant entertainment" at Griffiths' evening parties and he may well have given Beddoes a similar introduction. Griffiths' home at Turnham Green had become a meeting place for his Dissenting and Whig friends; Mrs Griffiths too, had literary tastes and it may have been her influence that led to the *Review*'s kindliness to women writers. Beddoes, if he did visit there, would have found the company sympathetic to his views and he was fortunate to be invited to write for *the Monthly*. Though he sent in contributions for eight years, there is no record of what Beddoes was paid. He once acknowledged a draft for £12 but his letter is a hasty one and does not indicate what work he had done. At least in the early days, Griffiths had a reputation for meanness where paying his reviewers was concerned, though this may have resulted from one-sided accounts of his treatment of Oliver Goldsmith. For beginners, £3 a time seems a likely fee; but his writers, who were carefully chosen by Griffiths, were eminent scholars and men of letters. *The Monthly Review* had been enlarged in 1790 and its contributors were even more distinguished than in the early days; among them were some of the leading Dissenters such as Dr Aikin and Mrs Barbauld and they were joined by promising young Cambridge graduates. Each contributor was given for review only works particularly suited to his own interests but their articles always appeared anonymously, a point upon which Griffiths insisted. So close was the secrecy that Beddoes would not have known who else was contributing or even that two other members of Wedgwood's circle were also writing for the *Review* — Alexander Chisholme, Wedgwood's secretary, and the mathematician, John Leslie, who for a while was a tutor in the Wedgwood family. Ralph Griffiths annotated the Editor's copy of the *Review* to show who had written each article, using initials or abbreviations. The practice continued until Ralph Griffiths' son borrowed from one of *The Monthly*'s most senior reviewers, William Taylor of Norwich, a volume of the rival publication, *The Critical Review*. His copy had been annotated in a similar manner and George Edward Griffiths, realising how this might reveal authorship, discontinued making his notes. Happily for us, it was not until 1815 that Griffiths made his discovery.

Monthly reviewers worked under strict conditions. Griffiths would accept no unsolicited work; would not allow any book or article to be reviewed by a friend of the author, and permitted no "puffing". Griffiths' aim was to give his readers a clear idea of the books under consideration but he did allow his reviewers freedom to express their own opinions, even at some risk to himself, for the rule of anonymity meant that the Editor carried full responsibility for what was published. Beddoes was no greater risk than others and in the year when he first contributed to the *Review*, Griffiths was already in danger of prosecution. The extreme opinions expressed in an article on Nare's '*Principles of Government*' had been noticed and the writer, Thomas Pearne, offered, if Griffiths should come to be charged, to acknowledge his authorship and stand trial with him. Pearne was a classical scholar of Peterhouse, a Unitarian and an admirer of the French Revolution until, like Beddoes, he was alarmed by the excesses of the Jacobins. Yet in spite of this threat Griffiths still took the risk of employing a leading opponent of the government, Richard Brinsley Sheridan. It was probably because he saw that it gave him an additional means of attacking Pitt, that Sheridan wrote for the *Review*. In 1793, Griffiths escaped prosecution for Pearne's article but the threat was no new situation for him; he had probably been as near to prosecution during the war with the American colonists. Then there had been an even more dramatic incident. The American intelligence in Paris had used *The Monthly Review* to make contact with a colonist known to sympathise with the cause of Independence, George Bancroft, who was one of Griffiths' reviewers. While Bancroft travelled back and forth between London and Paris Griffiths was under suspicion for aiding a traitor. Nevertheless, the Review employed Bancroft until 1778 and re-engaged him in 1794. He can scarcely have needed the reviewing fees as he was well paid by the Home Office when he became a double agent.

All such activities must have been kept well from view and did not detract from the *Monthly*'s high reputation, though it was naturally disliked in Tory circles. Each issue [12] appeared with two sections: first came substantial reviews of important books, then a 'Monthly Catalogue', which consisted of notices of less important publications. Though these were often very brief indeed, Griffiths considered the 'Catalogue' an important part of the issue. Three times a year there appeared, in addition, a substantial 'Foreign Appendix', where continental and American works were discussed, some at considerable length. Beddoes contributed to all three sections, writing mainly, though not exclusively, on medical and scientific works.

From Volume XII, September to December of 1793, to Volume XXV in 1798, with the exception of Volume XXII, January to April of 1797, Beddoes' short reviews appear in every 'Monthly Catalogue'. It would not at first seem that in these brief notices Beddoes had much opportunity to reveal his own opinions but on occasion brevity even added point to what he had to say. His short notices are mostly of medical and scientific works but there were a few exceptions and these in particular gave him opportunities to express his own views. In 1797 it fell to him to write on a work which dealt with, in effect, wages policy — and tackled a problem still not solved. In considering, "An outline of an attempt to establish a plan for a just and regular Equivalent for the Labour and Support of the Poor ...", Beddoes conceded that Adam Smith and several other able political economists were in favour of leaving this problem "to the free operation of existing causes" but added his own view that some regulations might "be made respecting the wages of the poor which may be advantageous to them without being injurious to the wealth and prosperity of the country". Again, the publication in Oxford of an edition of the Vulgate Bible, copies of which were to be given to the exiled French clergy, gave him opportunity both to make amends for his former outburst against the refugees and to express indirectly his anti-war convictions. Describing the emigré clergy as "inoffensive and respectable ecclesiastics", he welcomed the gift to them and the "pleasure of recording a peaceful act in the midst of barbarous war". The best example of Beddoes' skill in using the short review as a vehicle for his own beliefs is in his very neat account of a speech opposing *post mortem* examinations. His entire review of 'Thoughts on the practice of carrying off bodies from churchyards etc. for dissections, dedicated to Sir J. Frederick, Bart.' reads, "We ought sooner to have made honorable mention of this judicious and spirited remonstrance. It will convince every impartial reader that the honorable Baronet's tender concern for the repose of the dead, had his motion in the House of Commons been carried into effect, would have proved highly injurious to the living". The medical and scientific works chosen for the 'Monthly Catalogue' deal with much the same range of subjects as the full length reviews and in addition occasionally throw light on his subsidiary interests, as when Beddoes noticed pamphlets on watering places or truss making. But it was in the main articles and the reviews in the 'Foreign Appendix' that Beddoes was best able to summarise and describe the work that was relevant to what he was doing and from time to time to add his own comments; their titles alone suggest this.

As was customary at the time, Beddoes was in the habit of giving a careful summary of the work under review. He did so even when he could not agree with the author, if he felt that the work was worthy of respect. Though he disapproved of Lamarck's *'Mémoires de Physique et d'Histoire Naturelle'*, as an essay of "transcendental philosophy" where the writer "did not find himself under the necessity of making a single experiment", he gave a careful account of it. Perhaps his care was occasioned partly by "the discourtesy and prejudice" of Lamarck's original audience who would not give him a fair hearing. If he felt completely out of sympathy with the viewpoint of the writer, Beddoes confined his account to lengthy quotations. He had, nevertheless, a knack of making clear the style and worth of a man's work by singling out particularly useful – or objectionable – features with a neat turn of phrase: "The experiments are made rather in the gross method of the old chemists than in the present scrupulous pneumatic way" – of W. Gaitskill's 'Essay on calculi', contributed to *"Medical facts and observations"*; or, "the utility of tables or registers, like those which occupy so considerable a part of this volume cannot be questioned [even though] deaths should not be huddled together in one item but the disease of which each individual dies should be specified" – of Thomas Reide's 'View of Diseases of the Army'. Such comments show how Beddoes' reviewing was concerned with work on the frontier of new growth in science and medicine, as in these two instances where pneumatic chemistry and the collection of statistics were involved.

Apart from works by two established medical authorities, Dr Abernethy and Dr Hunter, Griffiths gave Beddoes very modern publications for review, often work on topics related to Beddoes' own interests. He frequently reviewed studies of fevers and, in particular, accounts of epidemics. A number of such reports came from the new medical schools of Philadelphia and

New York. Reviews of American work appeared in the 'Foreign Appendix'. Among them were careful medical histories of epidemics of yellow fever and a publication by the Association of Tallow Chandlers and Soap Makers of New York. This described the comparative immunity of the members of the Association during an epidemic of yellow fever. Such occupational immunities were of particular interest to Beddoes in his search for the causes and consequent means of prevention of consumption. Beddoes' reviews also bring out the special value of accounts written by naval and military surgeons, who had unusual opportunities for observing and for recording many cases. He reviewed military reports and an essay by a naval surgeon, Colin Chisholm, which gave an horrific description of an outbreak of fever among the ships' crews stationed in the West Indies. One military report from the continent came to him; 'Intelligence concerning French military hospitals' by G. Wedekind, physician to the Army of the Rhine. But for the most part, the continental medical publications analysed by Beddoes were substantial theoretical works. He wrote on A. F. Fourcroy's '*La Médecine éclairée par les sciences physiques*'; on two works concerned with the origin of diseases, one by C. W. Hufeland, and one by A. Roschlaus. The last volume of 1798 has an unusually long review by Beddoes of the German publication of 'On the knowledge and treatment of fevers'. The author, J. C. Reil (1759–1813), Professor and Director of the Clinical Institute at Halle, aimed at showing that treatment should be founded on experimental knowledge. Reil's aim was very congenial to Beddoes, especially at that moment when he had just succeeded in opening the Pneumatic Institute for that same purpose. He added his comment, "This probably is the only way by which the truth can be attained, either in the present important department of the healing art or any other".

Reil hoped that work in chemistry, together with careful observation and experiment in medicine, would show how the various physical functions of the human body accorded with underlying chemical laws. He was also concerned with the relation between mind and body and had come to the conclusion that "natural events could be traced back to laws which coincided with those of the thinking mind". Such views had important consequences. The realisation that mental illness was caused by definite physical malfunctioning made it possible to bring about a more humane treatment of the mentally sick. More generally, these ideas suggested a change of emphasis in the education of doctors, making principles more important than empirical techniques. Beddoes' own later medical writings show him thinking in much the same way on these points. Though, unlike Reil, he was not a follower

of Kant, Beddoes pointed out that Reil's researches were important for
an understanding of human life and that they promised "to lead to a more
intimate acquaintance with the actions and composition of living animal
substance". He justified his lengthy summary of Reil's work as being made
so that doctors could judge it for themselves. He explained Reil's doctrine of
susceptibility: "In all the organs of the animal body, and particularly in those
in which the vital power exerts itself most, we distinguish a two-fold exertion;
namely susceptibility and power". As he feared all his readers would not
be convinced, Beddoes ended the review with a tribute to the German professor whose work so impressed him: Reil "is undoubtedly, as a practitioner,
a man of nice observation; as a literary man, well-read; as a theorist, one of
enlarged views; and a thinker for himself. Probably, on the Continent, he has
no equal as a speculator on organized nature; certainly no superior". In
1796 Beddoes noted with pleasure a Danish translation of John Brown's
work; two years later he suggested that the most important part of Roschlaus'
'Inquiry concerning the origin of diseases . . .' was his account of Brunonian
medicine which showed that Brown's new theories were at last revealing to
continental medicine the weakness of "humoral pathology". In the previous
year Beddoes had published his own edition of Brown's *Elements of Medicine*.
The most substantial and wide-ranging work to come to him from an American
source was Benjamin Rush's *Medical Inquiries and Observations*. Beddoes'
detailed account of this appeared, unusually, among the main articles. Rush,
Professor of Medicine at the University of Philadelphia, wrote on tuberculosis
and on the régime to be followed for healthy living, both subjects which
Beddoes himself later treated in his *Consumption, Nature, Treatment and
Prevention*, and in *Hygeia*.

   The reviews of scientific work show the same pattern. Beddoes welcomed
first-hand observation and experiment. This he praised in John Dalton's
'Meteorological observations' and in Mrs Fulhame's 'Essay on combustion
with a view to a new art of dyeing and painting'. He reviewed the two books
generously, even though he appears to have found an almost amateurish
quality in both. In 1795 when he wrote his 'Meteorological observations',
Dalton cannot have been teaching at the New College, Manchester, for very
long and the weaknesses in the presentation of his work are now easily
understood as due to the limitations of his early life in Cumberland. Beddoes
sensed that 'Observations' was the work of an author who lacked a systematic
background and though he needed to indicate this, he did justice to Dalton's
careful records and the originality of his work on Trade Winds and the
Aurora Borealis. He cannot be expected to have foreseen that in less than

ten years Dalton would already have put forward his Atomic Theory. By comparison, Mrs Fulhame should have appeared in 1796 to be the better grounded of the two. She obviously knew the debate about the nature of combustion and put forward a theory which she claimed to have originated. Her claim was disputed by William Higgins, author of the 'Comparative View of the Phlogistic and Anti-Phlogistic theories': Beddoes was not especially concerned with her theoretical grounding; he was more interested in her skill in devising very simple experiments and her contribution to the techniques of dyeing and printing. Mrs Fulhame had succeeded in producing "cloth of gold" and a silver paint which she used with good effect to draw in the rivers on maps. She gave a careful account of her "reduction" of metals by various means and of her "oxygenation of combustible bodies" which had led to her theory but she was well aware that her activities might not meet with approval: "censure is perhaps inevitable," she admitted, "for some are so ignorant that they grow sullen and silent and are chilled with horror at the sight of anything that bears the semblance of learning, in whatsoever shape it appears: and should the spectre appear in the shape of a *woman*, the pangs which they suffer are truly dismal." But Beddoes, after a descrip-tion of the experiments, continued, "We applaud the lady's persevering ingenuity, we admire her dexterity in carrying on her researches almost without apparatus, and we sincerely sympathise with her on account of that disabling and discouraging narrowness of circumstances of which she so feelingly complains. May she soon meet with 'a being such as she has heard of on the record of fame but never seen one', viz. a liberal patron". He considered Mrs Fulhame's work would have been accepted as serious scientific investigation even in the days of Priestley and Lavoisier and was the more welcome in 1796 when science seemed "no longer the order of the day". By 1796 the days of hope, for science as well as for reform, seemed to have passed away; and Beddoes like Mrs Fulhame, was struggling to find support for his work. Mrs Fulhame's careful work was support for Beddoes' conten-tion that women were capable of a higher standard of intellectual achievement than was usually allowed them and it is a happy end to her story that she was elected a member of the Philadelphia Chemical Society.

This appreciation of original work and of well constructed experiments, appearing in his accounts of scientists whose reputation was not yet estab-lished, was typical of Beddoes. But when he came to review a new German publication, *Contributions to the Chemical Knowledge of Mineral Bodies*, he had an opportunity to draw attention to the achievement of an established chemist, Martin Henry Klaproth, Professor of Chemistry at Berlin. Klaproth had had a hard struggle to obtain recognition; having begun his career as an apothecary he had had to face considerable prejudice and was treated as an "outsider". When he was finally recognised and appointed professor his work, through its scrupulous attention to exact weights, established more rigorous standards of accuracy in chemistry. *Contributions to the Chemical Knowledge of Mineral Bodies* makes clear why he became so respected as a chemist, and Beddoes, himself pre-eminently an experimenter, must have enjoyed describing this elegant account of work done by a practical rather than a theoretical chemist. His review paid tribute to original experimental work of the highest order, showing how this had advanced chemical analysis. The main body of Klaproth's book is a laboratory record of a series of analyses; the introduction explains the general method of working, listing first the observations that had to be made on any substance to be analysed and following with directions for the conduct of the analysis. Klaproth's work marked an advance in laboratory technique. His directions emphasised the care that was necessary to avoid contamination from the crucible used (which should be chosen with this in mind) and the need for a standard procedure. The fire used should always be the same — common fire without addition of oxygen — and the same degree of heat should be reached in every one of a series of experiments. Beddoes would have appreciated the fact that it was the opportunity to use the Royal Porcelain furnaces at Berlin which had made it possible for Klaproth to achieve such a consistent temperature. Most important of all, it was through Klaproth's recommendation that a substance's loss of weight during analysis came to be recorded as routine and not, as hitherto, ignored. Klaproth might well have become accustomed to such exactness during his early years working as a pharmacist. He was one of the first chemists outside France to adopt Lavoisier's theories but he chose to concentrate on experimental work and had a long struggle before he achieved recognition by the academic world of Berlin. Beddoes appreciated that Klaproth had "advanced the difficult and important art of chemical analysis". The work that Beddoes was reviewing did not appear in an English translation until 1801 but in the same year that it was published, Klaproth was made a Fellow of the Royal Society of London. Even today, and for a non-scientist, Klaproth's account of his experiments impresses by its clarity and integrity and Beddoes' review reminds us of the developments in technique and in apparatus that were important to chemists at the time.

Beddoes wrote on topics that were new or still undecided: on 'Animal electricity' and on the phlogiston debate. Whether discussing the merits of a method of filtering water or praising a careful new edition of Linnaeus'

work, Beddoes raised his reviews above the level of mere summary by drawing
on his wide scientific knowledge and, though always fair in his judgement,
writing from a distinct point of view. There is a parallel in the biological
field in J. F. Blumenbach's 'Of the native varieties of the human species'. His
study of the various human types, his collection of skulls, and the explanation of how changes in form are produced, which his collected observations
had led him to construct, can be placed in the context of contemporary
interest in classification. Beddoes reviewed the third edition of Blumenbach's
book in 1796 and admired it as an important collection of information.
He could not agree with the author's reasoning. Blumenbach's work appeared during the same years as Erasmus Darwin's *Zoonomia* (1794–1796),
which gives an evolutionary explanation of variety. Beddoes certainly knew
Zoonomia but his review does not refer to it, even if it influenced his judgement of Blumenbach's work.

The work of reviewing the reports and transactions of various learned
societies was often part of Beddoes' contribution. At home, the *Philosophical
Transactions of the Royal Society* in 1793, the *Transactions of the Royal
Irish Academy* in 1794, and the *Memoirs of the Manchester Literary and
Philosophical Society* in 1794 and 1796 were the major transactions reported.
In all of them there were papers referring to work of particular interest to
Beddoes: in the Royal Society's *Transactions*, Volta's two letters describing
Galvani's experiments on animal electricity; in the '*Memoirs*', a paper on
smelting and a description of a strange case of hydrophobia. In giving accounts
of the *Transactions of the Irish Academy* and of the Society for Philosophical
Experiments in London, Beddoes noticed in particular the works of Kirwan
and of the mathematician Dr Higgins. In his continental reviews, he welcomed
the *Journal of Natural Philosophy*, edited by Professor Gren of Halle, and
the new *Chemische Annalen* of Lorenz von Crell. From the resumption of
their publication in 1797 to 1801 when he ceased to review, Beddoes gave
careful and detailed accounts of the contents of the *Annales de chimie*.
He saw the importance of the *Annales* as a record of the vigorous scientific
activity in France, and rejoiced when it reappeared in 1797 after the break
caused by the most violent days of the Revolution. Among the work described
for the first time is theoretical work by Berthollet and Guyton de Morveau;
Fourcroy's work on Mayow and on acids and Girtanner's paper on Azote —
which Beddoes felt obliged to condemn as a "loose memoir". A striking
feature of these issues of the *Annales* is the number of papers on the application of chemistry to practical matters. Beddoes realised that these were in
direct response to national needs occasioned by war but he was clearly

interested in such developments for their own value. He gave careful attention to tanning; to steel making processes; to the making of soap; the recycling of paper and the reclaiming of copper from bell metal. The production of sugar from beet and even a detailed account of the extraction of peat in Holland were carefully noted – Beddoes would have liked to see the paper on peat "translated and circulated in those districts of Great Britain and Ireland which depend on peat for their fuel". Such a comment draws attention to the often overlooked practical side to Beddoes' nature, and in this, his last review of an issue of the *Annales*, he chose to give particular attention to a memoir by M. Proust on the practical application of chemistry, commending it as a careful account. Such a choice of items, alongside Beddoes' encouragement of Sadler; his own work on steel; Davy's work on tanning and even Coleridge's careful notes on the production of sugar from beet, made while he was in Germany, are reminders of the excitement generated by technical progress in the 1790s.

To give a balanced account of these collections of papers was an exacting task which Beddoes carried out most conscientiously. Though his general reviewing for the *Monthly* finished at the end of 1799, in the three volumes of 1800 and in the January–April volume of 1801, Beddoes nevertheless still contributed full summaries of the current issues of *Annales de chimie*. In 1801, after describing how he himself had repeated the experiments Desormes had made using Volta's pile, the last words Beddoes wrote for the *Review* expressed his hopes for chemistry:

When such experiments shall have been so repeated, that no suspicion shall remain of the communication of the acid and alkali ready formed from the containing or contiguous bodies, they will serve as the origin of a new and more subtle chemistry; in which ethereal fluids, as they were styled by Bergman, will make the principal figure by their agency, if not by their union with more palpable matter. We may expect at least to see the dispute relating to the existence of these fluids brought to an end; or, rather, philosophers agreeing in their language on these subjects. The changes which the new mode of operating will induce in the bodies, whose qualities have been most studied, will be interesting to the philosopher and in some instances (no doubt) useful to the artist. We cannot, therefore, but look forwards, with the most pleasing satisfaction, to 'those labours of the ingenious in different parts of the world', of which we shall from time to time be called to render an account.

Though Beddoes was especially well suited to review scientific and medical publications, he was from time to time asked to write on books or papers of more general interest. Sometimes these, too, relate closely to his own activities. The second essay in the series which Beddoes himself published under the title 'Hygeia' in 1802, treats the way influential men, even though not medically qualified, can contribute to health education, and part of the ninth essay describes a long-lasting case of epilepsy: both topics are foreshadowed in reviews. In 1795 Beddoes had commended the Rev. Joseph Townsend's responsible and careful 'Guide to health', especially noticing that it showed some knowledge of pneumatic medicine. This review touches on the role of the parish clergy in giving medical advice, but Mr Townsend, it is clear, was very unlike the negligent clergy [13] rebuked in 'Hygeia'. A later review, in 1798, indicates the source of Beddoes' own description of epilepsy [14] to be an account entitled: 'Diaetophilus, physical and psychological history of his seven years epilepsy'. In reviewing this account Beddoes found an opportunity to express his view that hospitals for epileptics, "ought not to remain long among the number of pious wishes". Beddoes' underlying hope was to gain more light than had "yet fallen from madhouses on the bodily and mental functions of man". While these topics still seem closely related to his professional interests, other books that came to Beddoes, such as travellers' accounts of foreign journeys, were of an even more general appeal. These ranged from a description of plants in Mauritius, by G. A. Jacobi, to Dr Townson's *Travels in Hungary* (which Beddoes criticised for the coarseness of some of its incidents). Beddoes seems to have valued most an account of a 'Journey into the southern part of Russia'. The author, P. S. Pallas (1744—1811), was a doctor and a zoologist, whose work had interested Beddoes much earlier. From the time of his appointment to the St Petersburg Academy of Science in 1767, Pallas had made a series of scientific explorations of the little-known regions of Russia, including a zoological study. From this he gradually built up his own view of the evolution of animals and plants, even proposing a family tree of organisms. It is not surprising that in 1787 Beddoes had included Pallas' works among the books he wished to purchase for the Bodleian library. Again when he wrote his essay on 'The Antiquity of the Hindus' to publish with *Alexander's Expedition* in 1792, he had referred to Pallas' travels as a valuable authority. Perhaps the most curious of such books reviewed by Beddoes is B. S. Barton's "Memoir concerning the fascinating faculty which has been ascribed to the rattlesnake". Professor Barton wondered that so many believed all the wondrous tales. Barton had collected material from many observers, including William Bartram,[15] author of "*Travels through North and South Carolina ...*". Beddoes, obviously having in mind the myth-like nature of many of the stories about the rattlesnake, described Bartram as a "man of strict veracity", an interesting comment on a man whose writings had already fired Coleridge's imagination.

A group of reviews on literary/philosophic themes may seem an even more odd assignment for a scientific and medical reviewer. The first of these is Dupuis' *'Origine de tous les cultes'* which Beddoes reviewed sympathetically in 1796, approving of the author's understanding of non-Christian religions. Beddoes' own interest in such faiths and his conviction that a man's belief was his own concern had been clear in various of his writings, particularly in the essays on Hinduism which accompanied *Alexander's Expedition*, and these were sufficient to qualify him to review Dupuis' work. Similarly, it was appropriate to give a Swiss publication, *Manuel de Philosophie Pratique* which included extracts from the English educational writings 'Poor Richard' and 'Evenings at Home', to a reviewer who had shown a serious concern with the quality of educational writings for the young. It is still a little surprising to find that in the autumn of 1798 it was Beddoes who wrote on Goethe's *William Meister's Apprenticeship* and on J. H. von Voss's pastoral poem, *Louisa*, rather than William Taylor of Norwich,[16] Griffiths' principal reviewer of German literature. Whatever the reason for Griffiths' choice, it is clear that Beddoes enjoyed both works. He contrasted *William Meister* to the *Sorrows of Werther* and praised it for being "light, airy comic but not ludicrous", having "none of that attachment to the extravagant and excessively monstrous, which has so excessively infected the modern literature of his country". *Louisa*, Beddoes found "full of motion and vivacity as well in its portraiture as in its fable". He considered that it gave a faithful picture of rural life in Germany in a style so lively that it would lose by translation. These were just the qualities Coleridge admired in Beddoes' own 'rural' tale, *Isaac Jenkins and Sarah his wife*. The understanding of village life which Beddoes showed in his story of the cottagers and their sick child, perhaps suggested him as the right reviewer for *Louisa*. Beddoes' admiration for Shakespeare would have been a pointer to his reviewing *Wilhelm Meister*. He thought of the plays as unsurpassed for the exploration of human character and his appreciation is expressed both in *"Hygeia"* and in letters with a feeling more akin to the younger 'romantic' writers of the late eighteenth century than to the 'Augustans'. His account of Wilhelm Meister's discovery of Shakespeare's works, and his comment that Goethe must have been describing his own sensations, has an air of understanding which must have come from Beddoes' own enthusiasm.

Even more interesting is the review in the summer of 1796 of yet another German work, of a very different kind. This was the German edition of

Emmanuel Kant's *Zum ewigen Frieden*, published in Koenigsberg in 1795. Beddoes translated the title as *To Perpetual Peace* and the review is almost entirely a translation of the German text. It appeared in May 1796 and was the first account of a work by Kant himself in *The Monthly*. On three earlier occasions, Sowden had reviewed writings describing Kant's philosophy; two reviews were of dissertations presented for the prize offered by the Philosophical Society of Haarlem, the third was of a translation of an explanation of Kant's ideas by J. L. Edwald, entitled '*Letters to Emma*'. Later, in 1797 and 1799, there were reviews of discussions of Kant's metaphysic written by F. A. Nitsch and by Dr A. F. N. Willich, contributed by two of Griffiths' senior reviewers, Dr W. Enfield [17] and William Taylor. Taylor also reviewed translations of Kant's writings during the course of 1798. Beddoes' contribution is outstanding as giving English readers the earliest first-hand account of Kant's work in a general periodical. It was not, as we know, quite the first account of all, for the description of Kant's doctrine of Knowledge *a priori* which Beddoes gave in his essay "Of the Nature of Demonstrative Evidence" had preceded it by three years.

The volume of Kant's work sent by Dr Girtanner in 1793 would have already made the German philosopher interesting to Beddoes but in 1796 there was another powerful reason why *Zum ewigen Frieden* should be important to him. After the outbursts in 1792—3 Beddoes did not entirely give up political activity and 1796 was the year when his efforts at protest against reaction and against the war with France reached their peak. He felt that the most objectionable parts of the "Two Acts" — the "Gagging Acts" of 1795 which limited rights of assembly and freedom of speech — had been withdrawn as a result of protest, and he pressed forward with writings against Pitt's war policy. In December 1795 he published a pamphlet, 'Where would be the harm of a speedy peace?' [18] and followed this in 1796 by two appeals to Pitt, the more substantial being his 'Essay on the public merits of Mr Pitt'.[19] Both works show how abhorrent war was to Beddoes and his conviction that ordinary people were cheated of their right even to basic necessities, by war policies. The essay on Pitt is more than a polemic; in it Beddoes attempted to analyse the purposes of society and by his criticism of Pitt to show how government should serve these. Beddoes' ideas are obscured by digressions and lengthy illustrations, in contrast to Kant's clear rules for avoiding war and his "law of nations" to regulate international relations, but he must have found *To Perpetual Peace* very closely related to his own concerns and reviewing it a congenial task. He restricted himself to a clearly organised translation of Kant's work; his only comment was to ask who would guarantee the terms of perpetual amity. At the end, Beddoes had to admit to a sense of disappointment, an admission that reveals he had not yet lost his hope for improvement in society. He wrote of Kant's plan for peace.

In this matter we were disappointed, because we allowed the title to raise expectations of a splendid project: but the author has been prudent in trusting to the operations of nature, rather than to positive institutions, for the accomplishment of the great consummation of which he treats. We regret, however, that it should require an indefinite time, and the continuance of so dreadful a process as war. We add that, according to our foresight, a *sense of justice* and not the *spirit of commerce*, is to tranquillize the dissentions of mankind. In the middle classes of society, there already prevails a suspicion that, even in affairs of state, honesty is the best policy and that powerful nations finally become the victims of the wrongs which they perpetrate. The suspicion, we imagine, will ripen into conviction, and spread by degrees to both extremes; and, this being once settled as a practical maxim, perpetuity of peace – an undisturbed succession of serene days, is secured to the harassed race of man.

In the 1790s, there may have been more readers of *The Monthly Review* who were interested in Beddoes' writings on scientific and medical themes, than in the translation of Kant's work. These reviews form a valuable record of scientific activity, but *To Perpetual Peace* has a unique interest. When it was written, Coleridge and Beddoes had already met and for a short while were partners in the protest against Pitt's government. By the spring of 1796 Beddoes, in addition to his own political writings, was helping Coleridge with his "fortnightly" newspaper *The Watchman*.[20] Coleridge was at a turning point, about to put aside his revolutionary ambitions and ideas; but for a while longer he and Beddoes shared the same hopes for peace, for freedom, for a good outcome from the changes in France and for reform at home. They were working together, exchanging ideas, and the review of *To Perpetual Peace* came just at a moment when it had most chance of impressing Coleridge. His 1794–1804 notebook has an entry on "some determinate laws, which if universally obeyed would produce universal happiness"[21] and by the end of 1797 Coleridge was asking his bookseller friend, Cottle, for a German grammar. The reviews of *Wilhelm Meister, Louisa* and *To Perpetual Peace* were only the most public indication of Beddoes' great interest in German literature. Copies of these books were to be found in Beddoes' library at the end of his life, along with many other works in German. The sale catalogue made after Beddoes' death cannot show with certainty what he already owned in 1796, but it can be assumed that copies of the works he reviewed and of other books noticed in *The Monthly* were at hand. For Coleridge, who was very hard up, Beddoes' collection would have been a valuable addition to the resources of the Bristol Library, made even more stimulating with Beddoes at hand to discuss their reading. There was also access to the *Review* itself, providing an introduction to new theological and philosophical works as well as to the scientific material which was Beddoes' special field. Coleridge appreciated the importance of the new scientific ideas and without revealing his own professional connection with the *Monthly*, Beddoes must have been a good partner for such an omnivorous reader as Coleridge. The impression of Beddoes given by his reviews suggests a man well able to communicate his enthusiasms. Forthright; widely read; intensely interested in the world around him and in practical everyday concerns; brusque with the pretentious and generous where he could recognise work of integrity on whatever scale: it would not be easy to know him and not be carried along to share his interests. When the annuity given him by the Wedgwood brothers made it unnecessary for Coleridge to accept a position as a Unitarian minister, and he was free to draw up his own programme of study, it would seem very natural that the foretaste of German writing and ideas that Beddoes gave him led to his decision to go to Germany.

The Beddoes of *The Monthly Review* is not quite the same as the man who appears in his own writings. Ralph Griffiths' editorship gave Beddoes opportunity to write on work that interested him and from this we can see a little more clearly the logic of his plans and of his preparations for the Pneumatic Institution. We also owe to the freedom the *Monthly*'s reviewers were given to express their own political and religious views, some lively glimpses of Beddoes' opinions, even prejudices. But this freedom was constrained by the discipline of giving a fair and honest review. Contributors to the *Monthly* were expected to read their books and not, in reviewing them, to take opportunity to expound [22] their own ideas. Beddoes appeared to advantage in this framework; his assessments were based on wide and up to date knowledge and his reviews show that he had been a distinguished university lecturer. His exuberance tamed, he was still able, when his duty to report carefully had been done, to find opportunity to let his unorthodoxy appear, as when he reviewed Dr Townson's *Travels in Hungary*. He scrupulously gave a descriptive summary, noted plates and map but ended:

Dr Townson can endure with true Christian patience all the wrongs that despotism can inflict on others; in his view "every upright and honest man should quietly wear the chains of the most despotic government in Europe, sooner than risk the introduction of French licentiousness". We are as sensible to the horrors of French licentiousness as this author can be; nor will we inquire how far they arose from within or without: but we think that when he has been for years a labourer with a large family or 'damned to the mines' under one of these most despotic governments then and not till then, he may allow himself to talk with complacency about wearing chains.

This reviewing, done in the years when Beddoes was intensely busy with his "make or break" scheme to test the medical application of gases, impresses both for the steady effort it demanded and for its range and vitality. It shows us Beddoes, doctor, scientist and democrat, and shows us too, how the scientific work of the day appeared to one of its liveliest — though much ridiculed — figures.

CHAPTER 8

THE ARRIVAL OF COLERIDGE/POLITICAL
AND LITERARY ACTIVITIES

The Hotwells opened up a new experience and led to unexpected developments in Beddoes' life. Here, by contrast with the Shropshire country people,
the sick were well-to-do but no less pathetic in their struggle with illness.
The spa was a favourite resort of those who sought a cure for consumption
and many from the nature of their illness were in early adulthood. Many
had come there with little hope of cure, sent by their doctors who knew they
had reached the last stages of their illness. Beddoes was moved by compassion
as well as by a scientific spirit. In its early stages, pulmonary tuberculosis
often produced an ethereal appearance and a bright glance and led to the
victims being described in poetic language, a romantic deception which led
Beddoes to indignant protest:

Writers of romance (whether from ignorance or because it suits the tone of
their narrative) exhibit the slow decline of consumption, as a state on which
the fancy may agreeably repose, and in which not much more misery is felt
than is experienced by a blossom nipped by untimely frosts. Those who
only see the sufferers in passing are misled by the representation.[1]

Beddoes, as a doctor, knew the details of the agony that was to come for the
patient and the distress that the nurse had to endure and he pointed out the
surprise and horror the romantic would feel when brought face to face with
reality.
    Beddoes later described the 'philanthropic doctor'[2] as one who "is humane
in his conduct not so much from sudden impulses of the passion of pity,
as from a settled conviction of the misery prevailing among mankind."
But in 1793 his friendship with Richard Lovell Edgeworth gave Beddoes a
personal, as well as a professional, reason to enter into this experience.
Edgeworth had with him in Bristol quite a large and complicated family
group, ranging from the baby born in 1791 to the two daughters of his first
marriage, Maria and Anna. He had brought his family to the Hotwells hoping
to find a cure for his son, Lovell, who was suffering from tuberculosis and
their situation must have been especially poignant to anyone who knew how
much the family had already suffered. The baby, child of Edgeworth's third
wife, Elizabeth, who had come to Bristol for her confinement, had been

given the same name, Honora, as a sixteen year old daughter who had died only a short while before. The first Honora had also suffered from consumption and her father had taken her to Shifnal in the hope of finding a cure. He consulted both Dr Yonge and Dr Darwin, but in vain. When Beddoes met them in Bristol, eighteen year old Anna was nursing her brother and Maria, the eldest daughter who had arrived with her step-mother, was in charge of the younger children. All accounts describe Anna's charm and liveliness and Beddoes watched with sympathy which soon led to admiration.

Life at this time must have taken on quite a different tone for Beddoes. He was free now from the conservative and ecclesiastical society of Oxford, where his university colleagues were, as a condition of their fellowships, unmarried. So far as we can see from the records, during his years at Oxford his friends outside the university were in the main his seniors. Other younger friends he may have had, but we only know of three near-contemporaries dating from his life before Bristol — William Reynolds of Ketley, just two years older; his instrument maker, James Sadler who was seven years his senior and one younger man, his closest and most enduring friend, Davies Giddy, undergraduate at Pembroke College while Beddoes was Chemical Reader. Bristol brought a great change in society and from the time of his arrival there a new and much younger group of men entered his life. This was an active time. He travelled fairly widely to find patients and quite soon after his arrival occupied himself with early versions of the "air apparatus". A new life and one much more propitious for thoughts of marriage and falling in love. Beddoes himself seems to have been aware that he might have surprised his friends, confessing "I know not whether I am taken for one insensible to women; I should think that such persons must have viewed me at a distance and have been prevented, by the exterior roughness from perceiving the internal susceptibility of impression".[3] But susceptible he did indeed prove.

Davies Giddy was the friend in whom Beddoes confided, writing about his engagement in a touchingly tender and serious style. The first letter, in May 1793, describes how Anna made light of his admiration for a girl who had devotedly nursed a brother, and how she had too, dismissed his anxiety about the difference in their ages. This serious romantic mood appears again in another letter a month later when they agree to a trial separation. Anna was to go to the family home,[4] Edgeworthstown in Ireland, to learn housekeeping and to meet the Irish gentlemen who might have been her suitors. In the same letter, Beddoes described how he rejected the idea of an exchange of promises for he had complete confidence in Anna.

The engagement stood the test. Thomas Beddoes and Anna were married at the Edgeworth's home in April 1794.[5] The fate that seemed to haunt Beddoes and turn all his affairs awry did not fail on this occasion; the vicar "was a very careless and negligent man in matter of business and . . . omitted to register the Births and marriages of many persons . . .". The matter is described by the negligent vicar's successor, the Rev. G. Keating who, with Lovell Edgeworth and Maria Edgeworth, later testified that the marriage had taken place "by licence at the Dwelling House". At the time there were no misgivings and Beddoes wrote of being most kindly received in Edgeworths-town. Letters home at the time of his marriage[6] show Beddoes' anxiety that Anna should be made to feel welcome in Shifnal. His fears are rather that his own family might be insensitive than that Anna might be too aloof and he wrote to his father from Dublin, where he was staying with the Irish chemist, Kirwan:

really for a young person to leave an house where she has always lived happily and a country where her father has been a kind of king is a very serious step. Nor could any thing hurt her so much as any kind of dislike or unkind-ness shown her by my relations. I have had opportunity of knowing her so well that I am certain of living happily with her and this is what you and my mother would principally wish for me. I think it more than probable that she will have a considerable addition to her fortune some time. But I marry her for her prudence, her good sense and the sweetness of her disposition.

He took pains to write similarly to his mother and sister, as well as to his aunt at Hopesay Farm, asking for a kind reception for his bride. He assured his aunt that she, "need not imagine that her niece will have any of the airs of a fine lady"; and his mother that Anna will not be greedy over a settle-ment. He also made clear to his mother that he never had any serious attachment to Miss Yonge. From the letter to his father it appears that lack of harmony in his own family might have been responsible for this more-than-usual anxiety on Beddoes' part. He had not, he wrote, told Anna anything of his difficulties with his own father but he urged Richard Beddoes to be kind to his own wife. Obviously, as the story of Isaac Jenkins makes clear, Beddoes had earlier formed his own view of marriage. He had lectured the reformed Isaac Jenkins on the wrongness of his broken vows to Sarah and he put it succinctly to his father:

What in the name of Heaven and earth can a man mean by rendering unhappy his own family. If hereafter has any punishment for sins, the murderer and robber ought to fare better than the bad husband or parent, and by this law I desire to be judged myself.

All the signs are that Beddoes' own marriage, so far as he could ensure it, measured up to this ideal. At the outset, Darwin's congratulations to Beddoes on "being on the road to happiness" seem well justified and Edgeworth wrote with a brevity that speaks confidence, "Anna is happy with Beddoes".

They set up house cheerfully. Anna became more relaxed with her mother and sister-in-law, inviting them to stay in Clifton.[7] She seems to have taken in her stride the presence of a bailiff in the house — he was there to distrain on the goods of their landlord — for in her letter to her husband about this she appears to be more angry with an apothecary who had spoken against him than with the bailiff. Edgeworth, in Beddoes' absence, placated the neighbours in Hope Square when they had been disturbed by the noise of hammering, presumably as the house was fitted up for the laboratory. He also found them a servant boy — Humphry, to be called Jack. Beddoes, typically, felt he must contrive to teach this boy to read and write in order to keep him employed, deciding that otherwise he would "become miserable and vicious". Edgeworth again helped the young couple by "sharing" Jack to keep him busy. Now Beddoes had a home of his own he could gather round him a number of men, mostly rather younger than himself, who responded to his enthusiasm and were excited by his scientific ideas. The talk must have ranged from politics to philosophy; from educational to medical schemes and from science to poetry. They were a fairly uninhibited band at times and this in the end laid Beddoes' serious work open to more ridicule than it deserved. For some six years, from 1795 to 1800/1, these friendships flourished. The group seems never to have had the steadiness of the Lunar Society though some of its members were comparable to the midland scientists in their achievement. Beddoes at the centre, though admired and loved, may not have had the same power as Dr Small or Dr Withering in Birmingham to bind a group together or quite simply it may have been the youth of its members that made early disintegration inevitable. Davy, Coleridge, even the staid Southey, the best known members of the group, each had a reason to leave Bristol; the brilliant Tom Wedgwood died; Peter Mark Roget and perhaps his mother too, took fright at the wildness of Beddoes' circle. All had achievements and lasting reputation which surpassed Beddoes', yet all, even Roget, eminent as a doctor in London and today famous for his *Thesaurus*, would pay him tribute. They arrived in Bristol as very young men, unknown but full of ambition. Oddly enough, the two most vivid and original among Beddoes' Bristol circle, Samuel Taylor Coleridge and Humphry Davy, were each introduced by a friend whose later reputation presents him as, by comparison, staid and conventional.

Beddoes must have seen in the various young men who, between 1795 and 1800, appeared in this group the same enthusiasm and ambition that he himself had felt in the late 1780s at the outset of his own career. By temperament he was ready to respond and encourage. On their side, Coleridge and Southey would have found in Beddoes that rare kind of senior, one who could understand their hopes for a new society and, when occasion demanded it, join in active protest against reaction. They can hardly have failed to know how he had been, and, despite his growing reputation as a doctor, still was handicapped by his reputation as a 'revolutionary'. He had learnt discretion but he was still an active campaigner for peace and reform.

Coleridge soon came to be the most active politically but, in 1794, Southey,[8] who invited him to Bristol, was at least as restless and dissatisfied with society as his visitor. He was a precocious rebel, expelled from Westminster School at seventeen for writing verses in protest against corporal punishment and critical of the Headmaster. This did not prevent him from being accepted as a student at Balliol. Like many others, he was disgusted with the official curriculum and found his own occupation — writing verse. From 1793 he concentrated on his verse drama, *Joan of Arc* — another of the many protests at the time against tyranny and war. Southey at 20 was among those who identified with the efforts of the revolutionaries in France to establish a just and free society and who condemned England's part in the war against the new Republic. His choice as heroine for his play, of the French Maid who with her dreams of a regenerated France and her indomitable ardour defeated the treacherous and cynical English, was a daring challenge to all that English conservatism was doing. Much the same mood of rebellion and despair threatened to overwhelm Coleridge, Southey's contemporary, during his time at Jesus College, Cambridge. Coleridge had even disappeared for a while from the University and enlisted in the 15th Dragoons as S. T. Comberbake — a self-expulsion parallel to Southey's experience at school. The unconvincing pseudonym, coupled with a glaring lack of practical ability, quickly revealed Samuel Taylor Coleridge as a most unusual recruit. His brother George soon found him, bought him out of the regiment and arranged his return to Cambridge. But the trial and expulsion of William Frend only strengthened Coleridge's sense of oppression. In June, 1794 Coleridge set out with a friend from his school days at Christ's Hospital, Joseph Hucks, on a walking tour in Wales. It was the Christ's Hospital network that led them to visit Oxford en route. The schoolfellow Hucks and Coleridge were planning to visit there, Robert Allen, introduced Southey, and in this accidental way was made the introduction which led

Coleridge to Bristol. The visit of Hucks and Coleridge to Oxford lengthened into a stay of some weeks; the four young men, joined by Southey's fellow west countryman George Burnett, must have spent the time "setting the world to rights"; more than that, planning how they might give their ideas reality. Coleridge arranged to visit Southey at his home in Bristol, on his return from the tour in Wales.

Their ideas began to take a definite form and as described by Southey in 1810, the sequence of events was that

The American plan not having been formed until after I left Oxford . . . it was communicated to Robert Allen — poor fellow — by letter from Bristol.[9]

This "American plan" was that they should form a community where they could realise their ideas of freedom and simplicity. According to the plan, possessions were to be held in common and work so shared among members of the community that there would be leisure for literary pursuits and for putting into practice their ideas on education.

When Coleridge came to Bristol for the first time in 1794 this scheme was taken a step further by his engaging to marry Sarah, sister of Southey's wife-to-be Edith Fricker. Southey was devoted to Edith, a childhood friend, but Coleridge entered into the engagement with Sarah because marriage with Edith's sister would further their plan. A third Miss Fricker was to marry another member of the party. Coleridge was then twenty-two and Southey twenty. In 1795 Coleridge returned to join Southey and George Burnett and make further plans for the society which they planned to set up on the banks of the Susquehannah. There is something about the phrase, "on the banks of the Susquehannah", and the name by which the venture became known — Pantisocracy — that has come to suggest the fantastic and improbable. Nevertheless, this great river valley which crosses Pennsylvania and leads to Chesapeake Bay was not all wild virgin territory, but an attractive place of settlement; it was here that Joseph Priestley had made his home when he left England. Southey took care to inform himself about the country and discovered that it was fertile and free from dangerous animals. Though the scheme to start with equal numbers of young men and young women might seem outrageous, the young people concerned carefully organised themselves as married couples as part of their plan for leaving England. Nor was the idea of a settlement in itself so preposterous and to Robert Lovell, the Quaker member of the party, the land of Pennsylvania cannot have been entirely unknown. Had it ever been realised, Pantisocracy might have become like many another settlement, with the responsible

Southey carrying most of the burdens — but this was to come into being in Keswick, not by the Susquehannah; the scheme was doomed by the naivety of the young men themselves. Cottle, the Bristol bookseller who helped Coleridge and later published *The Lyrical Ballads*, was clearly horrified by their impracticable ideas,[10] he, though the same age as the 'Pantisocrats', was of a more cautious nature. One thing he was right about, and he was relieved to discover it: when he was asked for a loan towards current expenses he realised he could dismiss worries about what would happen on the other side of the Atlantic. There was not enough money for life in Bristol, let alone for equipping gentlemen settlers and Cottle could continue to help them with an easy conscience. We have Cottle's account in his *'Reminiscences of Coleridge and Southey'*, where he makes some reference to Beddoes[11] and his 'Gaseous Institution'. How much more we could wish him to have told: whether Beddoes discussed with them their schemes for educating first their servants and then, presumably, their children; whether he listened to their literary discussions — after all his own 'Domiciliary verses', written in 1795, show a knowledge of new tendencies in poetry; or whether he gave medical advice. Stock is no help, he determinedly ignored even the existence of Coleridge and he very largely passed over the political unrest of the years immediately following Coleridge's arrival.

After the acquittals in the State Trials the government renewed its harassment of the democrats. There was general dislike of the war with France, which was going badly, both on the continent and in the West Indies. The building of barracks and stationing of troops in the towns continued and caused alarm, even distress. On this issue Beddoes felt strongly and had most incautiously suggested that, given "the burden and danger attending a large standing army the citizens ... should provide themselves with arms, if it were but to avoid the absurd injustices of a law which condemns them to pay for depredations they have no adequate means of preventing".[12] Bad harvests and the war combined to bring about serious shortages of food. Political protests continued; after the trials the holding of mass meetings took the place of the hope of summoning a National Convention. In London the Corresponding Society organised a carefully planned meeting, to take place on October 26th, 1795, shortly before the opening of Parliament.[13] An enormous crowd gathered in Copenhagen Fields, Islington, where a number of platforms had been put up for the speakers who included John Thelwall, recently acquitted in the treason trials. The meeting had been called to protest against Parliamentary corruption which, in effect, disenfranchised the nation and frustrated the expression of deeply felt grievances. The meeting was well organised and orderly and it ended with the reading of an 'Address to the Nation' and a 'Remonstrance to the King', both approved without dissent and both quite lawful. It was, nevertheless, seen by the government as a dangerous threat to order. Then, on 29th October, came an attack on the royal coach when the King went in procession to open parliament. This was considered to be a direct result of the meeting and gave the government even better occasion to take repressive measures. Without delay, Lord Grenville in the Lords and Pitt in the House of Commons introduced the Treasonable Practices Bill and the Seditious Meetings Bill. Together these were so worded that they would bring about a complete denial of free speech and severely limit the holding of meetings – even of serious lectures.

Early in 1795 both Southey and Coleridge had set about providing for themselves by embarking on a series of public lectures. Southey gave a well ordered though somewhat conventional course on historical topics; Coleridge lectured on politics and religion. Cottle's [14] comment was that he needed the money to get married. Inevitably, Coleridge's lectures [15] came increasingly to deal with current topics and there are a number of threads linking him and Beddoes at this time. They shared, metaphorically and in the end literally, the same platform on public affairs and there are verbal echoes and details of style which suggest not so much formal collaboration as the enjoyment of exchanging ideas. Coleridge gave his Political Lectures in the spring of 1795, publishing the first of them in March, but he delayed the publication of the later ones until December when, after revision, they appeared as 'Conciones ad Populem'. The lectures on religion followed in May and June and on 16th June there was a single lecture on the slave trade. In Bristol, as Clarkson had written in 1789, this was a topic "in everyone's mouth. Everybody seemed to execrate it, though no one thought of its abolition". There cannot have been much change in the seven years since Clarkson's fact finding visit to the city, though there were circles in Bristol where honey and East India sugar were used instead of imports from the West Indian plantations. Coleridge drew on Clarkson's work but he did not confine himself to a narrow attack on slavery. In addition he analysed the evils inherent in colonisation, a line of argument Beddoes also had developed, though rather more sketchily, in his earlier writings. Coleridge's lecture came just at the time when the Abolitionists were making their last efforts to gather support.

After 1796, when another attempt was made to introduce anti-slave trade laws, the public meetings and the activities of the committees fell away as it became clear that to press their cause would embarrass Pitt's government. As the threat from France grew, the Abolitionists came to feel that the government needed unqualified support. It seems that Beddoes, too, may have been active in the anti-slavery movement in 1795. Stock describes how, in this year, medical practice and work for the Pneumatic Institution occupied most of Beddoes' time and he then breaks into the narrative to give some account of Beddoes' political views, adding,[16] "When the question of the abolition of the slave trade was in agitation he went from Bristol to Shrewsbury to attend a meeting there and almost for the only time in his life, addressed a public assembly on behalf of the wronged children of Africa." This appears to overlook Beddoes' part in getting names in Shifnal for the Shropshire petition in 1792, but as Beddoes was very unlikely to have gone from Bristol to Shrewsbury in that year, it does seem that Stock is describing a different occasion and, despite the lack of detail in his account, we can think that in 1795 Beddoes, as well as Coleridge, was active in support of the abolitionist movement.

Cottle noted that none attended Coleridge's lectures but those who were already in agreement with his views, a comment made nearly fifty years later in his '*Reminiscences*'. Coleridge himself, writing to George Dyer [17] during the course of the lectures, described the opposition he encountered. He had "endeavoured to disseminate truth" but found that "the opposition of the Aristocrats is so furious and determined, that I begin to fear, that the Good I do is not proportional to the Evil I occasion — Mobs and Mayors, Blockheads and Brickbats, Placards and Press Gangs have leagued in horrible conspiracy against me. The Democrats are as sturdy in support of me, but their number is comparatively small". Beddoes assuredly was one of the Democrats and if he was in the audience at the lectures he must have heard much that he approved. Coleridge gave a reasoned account of the French Revolution and vigorously opposed the war. In the later lectures he took up the topical subject of the political prisoners in this country and the threat to liberty inherent in two Bills before Parliament. It was very understandable that Coleridge warned his hearers to be vigilant lest they lose their freedom — 'The Sleep of Nations is followed by their Slavery'. Such a warning Beddoes had already given in substance in his 'Letter to a lady' and on 17th November, 1795 he renewed his plea, this time in words that clearly echo Coleridge:

A nation which slumbers over its rights will be fortunate if it awake not in fetters.

The words are from Beddoes' protest against the two Bills,[18] 'A Defence of the Bill of Rights'. The pamphlet appeared on the same day as a great public meeting at the Bristol Guildhall occasioned by the attempt on George III's life on 29th October. Such meetings were called throughout the country to express the general feelings of horror. In Bristol, the Mayor presided and the meeting passed "a very loyal and dutiful, and affectionate Address" congratulating the King on his escape. The conservatives were determined to prevent this meeting being used to protest against the war and against the two Bills. A proposal to send a petition to Parliament was defeated but not without difficulty: there were "vociferations from all quarters". Coleridge began "the most elegant, the most pathetic, the most sublime Address". He was shouted down, but at the end of the meeting he did succeed in making one more appeal to end the war. Another Bristol doctor, Edward Long Fox, and Robert Lovell "were purposely and in a manner personally" prevented from speaking. In the end authority quashed even the mildest expression of hopes of peace and closed the meeting peremptorily. Consequently, at mid-day on November 20th the Bristol citizens again met at the Guildhall "to consider the Propriety of Petitioning Parliament against certain Bills now pending in the House of Commons, by which Bills it is conceived that the Bill of Rights will be invaded". Dr Fox was elected to the Chair and thus it was he and not Beddoes who bore the brunt of subsequent conservative hostility. A Bristol banker, Mr Savery, moved two resolutions, the first abhorring the attack on the King and the second presenting the petition. This petition was seconded by Dr Beddoes who spoke effectively about the dangers of the two Bills. Later on, Beddoes persuaded the meeting to keep to the orthodox course of asking the town's own M.P.s to present the petition to Parliament and Coleridge withdrew his suggestion that Charles James Fox and Richard Brinsley Sheridan be asked to speak for them. Beddoes' intention was that no opportunity should be given for criticism of the conduct of the meeting. Certainly, the report in the Bristol paper, the *Star* described it as most orderly and serious, in spite of provocation from a military parade outside and the distribution of leaflets urging people not to attend.

Beddoes re-issued his 'Defence' on 19th November and then again on 21st with a postscript about the second meeting. The strength of Beddoes' feeling is clear and equally impressive is the orderly, concise way in which he marshals his arguments. His style was forceful and authoritative, in contrast to Coleridge's oratory. He was proud of the orderly conduct of the meeting, of the honest way in which signatures were collected for the Petition. For himself he said:

I deprecate all violence. I have no talents for pillage . . . . Peace and liberty are my dearest wish. I shudder at the idea of confusion. In this spirit, decrying anarchy at the end of an avenue of oppression, I protest against that revolution of law which threatens our liberties.

It is very understandable that Beddoes should endeavour to ensure the orderly conduct of the meeting and that he should emphasise his hatred of anarchy, for without doubt he and his friends were putting themselves at risk in taking part in such a demonstration. Though in London Hardy and his companions found a brilliant lawyer to defend them and a jury prepared to give a Not Guilty verdict, in the provinces opinion was less liberal and conviction, if it came to a trial for sedition, a serious possibility. The protest by the Bristol 'democrats' is in comparison with the 'Treason trials' of 1794 and the later and even more famous Peterloo massacre,[19] forgotten, but the two Bills are still known by Beddoes' scornful name for them: the Gagging Acts.

A short pamphleteering skirmish followed this battle. Widespread agitation had brought about the withdrawal of the most oppressive clause in the Bills and Beddoes, encouraged by this success, decided to press his opposition to the war. On December 9th he issued his vigorous, "Where would be the Harm of a Speedy Peace?" Not content with a general protest against the government's war policy, he asserted roundly that it was not the duty of England "to attempt to apportion out the continent of Europe, nor of Grenville and Pitt to appoint themselves 'Inquisitors general' of this planet the earth". Little wonder that an indignant Bristol citizen, known only as A. W., felt impelled to reply, two days later, in a 'Letter to Edward Long Fox M.D.'. Though he only named Fox, whom he attacked in offensive personal terms, it is clear that A. W. had Thomas Beddoes equally in mind.

A reply was not long in coming: the 'Answer to the Letter to Edward Long Fox M.D., by C. T. S.'. The author's line of argument, his style, his hatred of war both in principle and for the opportunities it gave for financial corruption, his sorrow over Burke's abandonment of the defence of freedom, all recall Beddoes' arguments. Echoing phrases such as "The Almighty has not commissioned the people of one country to try and punish those of another . . ." emphasise the connection to the point of suggesting collaboration between C. T. S. S[amuel] T[aylor] C[oleridge] and Beddoes. Edward Long Fox [20] was as original and independent a character as Beddoes and the two doctors seem well suited to join in leading these political protests. A Quaker and a Cornishman (for which reason A. W. dubbed him a foreigner and C. T. S. rejoiced that he was not a native of "this city of blood"), Fox had early schooling in unconventional behaviour. At the end of the war with

America he had been sent to Paris to search for the owners of a French ship which had been captured by a packet part-owned by his father. Joseph Fox had put aside his share of the prize money and his son was sent to find means of repaying it. In 1785 the young Edward traced and paid one of the French owners and then deposited the remaining £600 with the "treasury of the invalid seamen of Paris". So he accomplished his father's aim, "not to partake of any profits which may arise from war". Beddoes must have known Edward Long Fox, at least by repute, as early as September 1793 when protest against the continued charging of tolls on the Bristol bridge had turned into a serious riot. Dr Fox had intervened and, although not allowed to use the Guildhall for a meeting, he succeeded in bringing about a settlement. The writers of the pamphlets obviously remembered this incident and Dr Fox's part in it. Small wonder that the two doctors were on the same platform in 1795 and that their alliance so offended A. W.

These meetings did not end the collaboration between Coleridge and Beddoes. Once his two courses of lectures were ended Coleridge planned a journal, to be called *The Watchman* and, in order to avoid Stamp Duty, published fortnightly.[21] Coleridge needed both a source of income and a new means of expressing his humanitarian and political convictions and it is clear that Beddoes found he could wholeheartedly collaborate in this venture. Here was, in fact, an opportunity to fulfil an old ambition. On an earlier occasion inaccurate and alarmist rumours about events in France had brought home to Beddoes the need for a free press, in particular a free provincial press, where the truth could be published. In November 1791 he had written to Davies Giddy that he was "extremely desirous to establish such a paper", continuing, "It will be a losing project at first and when I solicited my democratical acquaintance to join in bearing the expense they said, 'you will inflame'. It was in vain for me to reply that I should destroy the tendency to inflamation and that what was always beneficial would every day become a more urgent necessity".[22] Though there was no overt censorship of the press, government agents and government publications frustrated freedom to criticise and inform and the need for independent reports was no less in 1796. The first number of *The Watchman* appeared in March 1796 to serve the same purpose as Beddoes had had in mind:

That all may know the truth and the truth may make us free.

Coleridge developed this theme of Pitt's tax on newspapers. He poured scorn on Pitt's justification of the tax on the ground that newspapers were mere luxuries, and his criticism is in detail close to Beddoes' writing on the same subject. Beddoes not only sympathised but gave invaluable practical help with each of the ten issues — all that Coleridge was able to publish in spite of great efforts to muster subscribers.

The Watchman was not unlike other "democratic" publications though its tone was, on the whole, moderate. It was concerned with the same issues that Beddoes had written on earlier and, not surprisingly so soon after the passing of the two Acts and the intensification of the war with France, conducted a persistent attack on Pitt. Beddoes, as we have seen, was writing independently to criticise Pitt's policy and in 1796 published both a short pamphlet, "A letter to the Rt. Hon. William Pitt on . . . the present scarcity" and a long and elaborate essay on the 'Public merits of Mr Pitt'. It seems likely that an attack on Pitt in the April 19th number of The Watchman was also written by Beddoes. The two publications, the letter and essay, provided useful material for The Watchman — two substantial reviews and an analysis of the Essay. Beddoes' ideas about the food shortage and about taxes may also have been helpful in suggesting topics for separate articles. The main concerns of The Watchman were quite naturally slavery and the efforts to abolish the Slave Trade and the importance of peace with France. These are vigorously treated but without any particular details that suggest collaboration with Beddoes. It is rather in the general news items that we see the possibility of his influence. Reports of the bias of magistrates, in particular in Birmingham; indignation against the alliance between church and government in opposition to France; concern that the cost of newspapers, by forcing men to read the ale-house copies, increased drunkenness: Beddoes in letters and published writings refers constantly to these topics. There are even little news items which could well have been contributed by him, the account of a puzzling case of hydrophobia and the notice of the appointment of Dr Carmichael to a Birmingham hospital. The proposed tax on hair powder was another subject of common interest. The Beddoes circle had been having fun at the expense of this tax as early as 1795 when Anna's sister Emmeline had thought of the idea of wearing a bandeau of brown silk ribbon with gold lettering, "Pitt for ever . . ." and it would be good to think Coleridge joined in the joke as well as publishing the serious essay on the tax that appeared on 2nd April, 1796. A less tenuous link exists in both the first and the last number of the paper. To open The Watchman, Coleridge wrote a review of prospects for peace between France and England which included an account of discussions in the French Assembly. Since he did not know French, the report from Paris must have been translated for him, in all likelihood by Beddoes. In the dismal task of putting together the last Issue, No. 10, Beddoes was again at hand to help by contributing an extract from his *In Defense of a Bill of Rights*.

Details which suggest help from Beddoes or topics which might well have resulted from discussions with him, appear in every one of the ten numbers of *The Watchman* but the over-all tone of Coleridge's paper, rather than such details, shows how close he and Beddoes were. They were both in sympathy with the general liberal protest against the trend of events and the ten issues of his paper show how consistently Coleridge planned the contents to support this protest. It is possible to follow through from issue to issue items which keep the horror of war and the obduracy of the government in opposing peace moves before the readers. Debates in the English Parliament are paralleled by accounts of the deliberations of the French Assembly and by repeated reminders that France might be prepared to negotiate. In the strength of his feelings Coleridge is very close to Beddoes; the most emphatic report was on 19th April, 1796 when *The Watchman* followed an analysis of the insincerity of the English government in its negotiations with France, with a vivid account of the horrors of the campaign in the Low Countries during the winter of 1794–5. One particular detail had so seized Coleridge's imagination that it is found again in *The Destiny of Nations*,[23] a poem written during 1796, in part while he was at work on *The Watchman*. The verse is indifferent stuff, some of it written for inclusion in Southey's verse drama *Joan of Arc* but, by turning it into an incident in Joan's life, it faithfully reproduces a scene from the battlefield of 1795:

An unattended team! The foremost horse
Lay with stretched limbs; the others, yet alive
But stiff and cold, stood motionless, their manes
Hoar with the frozen night-dews.   [Ll. 197–200]

. . . feebly, with slow effort pushed,
A miserable man crept forth: his limbs
The silent frost had eat, scathing like fire.
Faint on the shafts he rested. She, meantime,
Saw crowded close beneath the coverture
A mother and her children – lifeless all,
Yet lovely! not a lineament was marred –
Death had put on so slumber-like a form!
It was a piteous sight; and one, a babe,
The crisp milk frozen on its innocent lips,
Lay on the woman's arm, its little hand
Stretched on her bosom.   [Ll. 205–217]

No part of *The Destiny of Nations* appeared in *The Watchman* though it does show how closely related were the poetry and the politics. *Lines on the Present State of Society* were on the other hand included in the issue of 25th March and are just as strongly felt an attack on oppression and poverty, though their rhetorical tone has become less effective with the passing of time than the story of the peasant family. Oppression and want is another subject to which Coleridge persistently returned with feelings of anger very similar to Beddoes'. In the main, Coleridge, like many editors of his time, got much of his material from already published sources, but in addition to Beddoes himself he had help from the Bristol circle, Lovell, Southey and Thomas Poole who contributed an anti-slavery article; also from Dr Edwards of Birmingham. Dr Edwards, who wrote as 'Phocion', sent a report of, possibly, the first men to be charged under the two new Acts. Phocion, in a second letter, gave an indignant account of the misconduct of the trial and the ill-treatment of the prisoners, one of whom was kept in fetters. He wrote in the hope that "Britons will rouse from their lethargy and with united and irresistible voice demand ... the restitution of their rights". Beddoes, like Coleridge, would have approved this report as a vindication of their protests in the winter and of his own, *In Defense of a Bill of Rights*. Coleridge throughout had support from the Dissenting community. On his fundraising tour in January 1796 he had preached at the High Pavement Chapel, Nottingham, and had quoted from Beddoes' 'Letter to ... Pitt ... on the Present Scarcity'; and one of the contributors to the last *Watchman* was Roscoe, of Liverpool, where the demise of Coleridge's paper was much regretted. In all, it seems that Beddoes encouraged Coleridge in his venture and helped him while it lasted, not just in detail but as a sharer in a common endeavour.

In one issue of *The Watchman* Colerige gave the first hint of a change in his feelings. The *Remonstrances to the French Legislators* which appeared on 27th April is still part of the advocacy of peace, but in it Coleridge no longer expressed the view that France was the peace-maker; instead he felt it necessary to plead with the Republic not to continue the war. He opposed the annexation of the Netherlands and urged France to realise that if she continued to wage war it would be for the sake of ambition, not of liberty. Two years later, this change and a more sombre and moral view of politics became clear in the poems *France: An Ode* and *Fears in Solitude*. Beddoes had already faced up to the violence that, rather than dying away, had increased in France under the Jacobins; but the change in his attitude took a different form.

Each issue of *The Watchman* carried some verse, Coleridge himself writing mostly on political themes. The lines on 'The Present State of Society', published later as part of 'Religious Musings', describe powerfully the horrors of war, the distress of ordinary people and the debasement of religion which "Worked whoredom with the Daemon Power" and a vitriolic sonnet (later regretted by Coleridge) takes its place in the attack on Pitt. There are lighter verses too, and it was in *The Watchman* that Beddoes found the lines, also by Coleridge, romanticising the consumptive patient, lines which he considered so heartlessly inappropriate.

Beddoes'[24] *Domiciliary Verses*, if Stock is to be believed, was written in 1795 which places it very close in time to two poems by Coleridge, the lines *To a Young Ass*, written in October and published in December, 1794, and 'Reflections on having left a place of Retirement', composed in 1795. These three poems taken together certainly give the impression that Beddoes had as friendly a knowledge of Coleridge's poetry as he had of his struggles with *The Watchman* and was not above teasing his young friend. The poem is the neatest of Beddoes' verses:

Invitingly yon single-storied cot
Peeps o'er the frosted heath. The broad, brown door,
Scaled of its white-wash, is so low that he
Who steps in upright, steps in jeopardy
To smite his forehead. Two projecting walls
Fence in the roomy fire-place. Close by each
Is set an oaken bench, on whose hard sides,
His sore impatience many a lubber loon,
Keen for his meal, has notched. Here, when silently
Coating the green and lozenged panes, thick snow
Bedims the scanty daylight, nestles the snug
Family, delighted up the chimney's shaft,
Illumining the chasm, to trace the spark's
Ascent; or touch with timid finger-tip
The faggot's hissing ooze, and snift the fumes.
I knew an Irishman; to England he
Came every spring a hay-making; and much
Would praise his cabin. By a bog it stood,
And he had store of peats. Without a chimney
Stood the little cabin. Full of warmth and smoke,
It cherished its owner. The smoke he loved,
Loved for the warmth's sake, though it bleared his eyes.

Now when the North-East pinches, I bethink me
Of this poor Irishman; and think "how sweet

"It were to house with him, and pat his cur,
"And peel potatoes mid his cabin's smoke."

In *To a Young Ass*,[25] Coleridge, watching the tethered donkey and her foal,
thought of the miserable future these two ill-treated creatures would have
to endure and drew a political moral which Beddoes would have approved.
He remembered the donkeys' owner, in all likelihood equally deprived, and,
his thoughts still on the Pantisocratic settlement, wished for a better world
for them all. But, free from moralising, the lines

Oft with gentle hand I give thee bread
And clap thy ragged coat and pat thy head.

have a directness which seems to have caught Beddoes' fancy. His Irishman
come to England "a-haymaking" breaks into the musings on the single-storied
cot with its low door, just as in "Reflections on having left a place of Retirement" "Bristowa's citizen" interrupts Coleridge's happiness in the "pretty
Cot" so low that the

tallest Rose
Peep'd at the Chamber window.

Coleridge's mood of light-hearted happiness turns to deep feeling and serious
reflection; Beddoes remains with observation of the English labourer's cottage
and the Irish cabin that he may well have remembered from his visit to
Edgeworthstown. Resemblance between the two poems is halted. Though
he wrote in fun, Beddoes shows that he had understood the new style just as
well as he had caught the tone of Erasmus Darwin's couplets. The *Domiciliary
Verses* were not published until 1799 when Cottle included them with a
passage from *Alexander's Expedition* in the first issue of the Annual Anthology. Southey,[26] who edited the Anthology, was very angry. The *Lyrical
Ballads* had appeared only months before and Southey did not want to
"offend Wordsworth by publishing anything designed as ridicule"; indeed he
considered Beddoes had acted with "indelicacy and a kind of arrogance".
Cottle's explanation was that Beddoes had refused to contribute anything
at all if the *Domiciliary Verses* were rejected "and that *he* did not wish to
offend *Beddoes*".

Some two years later Coleridge, watching by his own fireside,[27] trans-
formed the homely to magic:

the thin blue flame,
Lies on my low burnt fire, and quivers not;
Only that film which fluttered on the grate,
Still flutters there, the sole unquiet thing.

Beddoes' description of the snug labourer's family on a winter's night, the thick snow, "Coating the green and lozenged panes" and their simple pleasure in watching the sparks and smelling the hot oozing sap, though it shows real sympathy never aims beyond homely realism mixed with humour. Their paths were diverging; each was leaving the overtly political world, Coleridge, as he hoped, to concentrate on poetry and Beddoes to devote himself to pneumatic medicine. Yet they clearly still had common interests: this must be the explanation of Beddoes' *Receipt for a Legendary Tale*,[28]

In woods immersed beside a lake
A very ancient castle take
And plant there-in a damsel fair,
With eyes sky blue and auburn hair;
But chief be sure she's ever prone
To heave the sigh, when left alone,
At tales of tender woe to swoon,
And, pensive, watch the pale-faced moon.
Next, pray provide a hoary sire,
All snow without, within all fire;
Once served by dames, admired at courts
And always first in knightly sports.
But now, past life's meridian hour,
Not wanting will so much as power
To poise the lance and the sword to wield
And bear aloft the blazoned shield,
A calm sequestered life he leads,
Confessing still or counting beads.
Not that, as yet, the good man's soul
Yon other world possesses whole;
Two earthly things his cares divide;
His daughter much, but most his pride.

In 1797 Coleridge was writing the first part of his magical poem *Christabel.*[29] He was never able to finish the poem and what he had written was not published until 1816 with the second part, written after his return from Germany. All Beddoes' "ingredients" appear in the poem with its mediaeval setting and its mysterious wood. But everything is transmuted, just as the conventional "pale-faced moon" is replaced by,

The thin grey cloud is spread on high
It covers but not hides the sky.
The moon is behind and at the full
And yet she looks both small and dull.[30]

Coleridge himself, writing of the genesis of *Christabel*, claimed, "there are such things as fountains in the world".[31]

There are as well, many entries in Coleridge's Notebooks which reveal familiarity with what Beddoes was doing. Some are just domestic details. In 1796 Coleridge and Sarah, with their first child, Hartley, set up home in a cottage in the village of Nether Stowey. This, though further from Bristol, brought Coleridge within walking distance of William and Dorothy Wordsworth at Alfoxden and so made easy, friendship and collaboration. Domestically, by contrast, it was an uncomfortable time. Coleridge discovered that it was more difficult than he had imagined to make his plot of land keep them in vegetables, and Sarah found the care of a small baby and housekeeping in a tiny cottage with a smoky chimney almost more than she could manage. Coleridge had advice from kindly Thomas Poole who had found him the cottage but Sarah was critically observed by Dorothy Wordsworth. Visits from Coleridge's friends must have added to Sarah's trials. Prescriptions and a comment on diet were jotted down at this time and seem very likely to have come from Beddoes who might well have tried to give practical advice. Such personal notes as are scattered among these entries[32] show that Coleridge was reading Beddoes' 'Considerations on the uses of Factitious Airs' (the fifth part was published in 1796) and that he was familiar with the work of Dr John Brown whose life Beddoes was editing. There is an admiring description of the Edinburgh medical school, "the finest medical school in Europe", which also has obvious connections with Beddoes. In 1796 Coleridge's reading of Andrew Baxter's 'Inquiry into the Nature of the Human Soul' led him to think about the nature of dreams and to explore Erasmus Darwin's theories. This was a topic in which both Darwin and Beddoes had a professional interest; much later in *Hygeia* Beddoes treated it in detail. It certainly was a subject discussed in Beddoes' circle when Davy and Tom Wedgwood were included but the most evocative treatment of the dream world is in Coleridge's poetry. *Christabel* with its dreamlike quality and the story of Bard Bracy's dream, was in Coleridge's mind while he was in Somerset. The Day Dream poems and the poignant *Pains of Sleep* come in 1803. He may have joined in abstract discussion and have read Beddoes' *Hygeia*, but to Coleridge the dream world was often tragic reality.

One of the most interesting of the entries in the Notebooks is about F. A. Nitsch, Professor at Königsberg and author of 'A general and introductory view of Professor Kant's principles'. Nitsch lectured on Kant in London in 1794—6 and it is possible either that Coleridge heard him lecture or that he had reports from Thelwall who certainly met Nitsch in London. Whatever exactly took place, Coleridge had in Bristol an excellent opportunity to follow up his curiosity about Kant's doctrines or indeed to discover his writings. Nitsch's book was likely to have been to hand at Beddoes' home and Beddoes ready to discuss it, for it had been interestingly reviewed [33] in the summer of 1797. Beddoes' writing on works by German scientists and doctors must have brought into his library new books where Coleridge would have found an introduction to current philosophical ideas. At one point his interest was particularly caught by the work of Blumenbach whose 'Of the Nature of Human Species' Beddoes had reviewed so carefully. In 1800 [34] Coleridge seriously thought of making an English translation of Blumenbach to the dismay of Davy who urged that his overriding duty was to write poetry. All these discoveries at Bristol must have been stored in Coleridge's mind for the later years when he gave himself more to philosophical speculation than to poetry, when he had faced the loss of his "Shaping Spirit of Imagination".

At this time, in late 1797, Thomas and Josiah Wedgwood came to a decision about the practical furthering of their educational plans. [35] A variety of circumstances and people, and their own careful observation, led them to choose Coleridge as most likely to be able, with preparation, to put their ideas into action. The annuity which they agreed to settle on him enabled Coleridge, in Hazlitt's phrase [36] "to devote himself entirely to poetry and philosophy", and "in the act of tying one of his shoes", to dismiss his plan of becoming a Unitarian minister. Education must have been very much a topic at Bristol; Beddoes, it will be remembered, had written on it and in 1796 had corresponded with Thomas Wedgwood about the educational toys which they considered valuable; he had a scheme for having such toys made. Education can hardly have escaped taking a place alongside politics in the many discussions between Beddoes and Coleridge. Characteristically, when it came to the point, Beddoes was more practical than Wedgwood. Though his philosophy was essentially the same, more objective considerations originally led him to the subject: the wish to counter teaching based on religious views which failed both to give ordinary people the knowledge needed to keep themselves healthy and to develop a harmonious society. When he came closer to Wedgwood in concentrating on the education for "the affluent", Beddoes was still concerned with the physical well-being and health of children and, whatever the theory, did not in practice see it developed in the same rigid way as did Wedgwood. Probably neither his character nor his experience as a doctor would allow him to accept what amounted to turning education into unimaginative training. This is to anticipate a little, for it was not until the publication of *Hygeia* in 1802 that Beddoes had time once more to write on education. Coleridge certainly read *Hygeia* but seems not to have commented on the chapter on education. In 1797–8 the immediate result of the Wedgwoods' generosity was that Coleridge was free to travel and study and it may have been at Beddoes' suggestion that the first use he made of the annuity was to go to Germany in order to learn the language, read the German philosophers and attend their lectures.

Coleridge, with William and Dorothy Wordsworth, left for Hamburg in September 1798. It has been pointed out that the planned journey to Goslar and the Harz mountains followed an earlier journey made by Tom Wedgwood. But Coleridge chose to remain at Göttingen and to attend lectures there. Beddoes, we know, had links with the University there and knew its work; Coleridge would probably have seen the scientific journals and papers from Göttingen in Beddoes' library. It seems very likely that Beddoes influenced Coleridge's decision to go to Germany and in particular led to his choice of Göttingen.

Beddoes' energy must have been prodigious in these years. We see him bustling through the town, hurrying on foot to visit his patients; driving out into the country on professional visits or, towards the end of the period, treating patients in his own home. He was busy at his desk writing the essays that would justify his plans and bring in contributions for the Institute; corresponding with doctors and scientists; preparing and writing his reviews, this alone an exacting task. With all this, he had energy to spare to discuss education with the Wedgwoods and to think of teaching his servant to read; to make new friends with his home as their meeting place and to keep up a lively correspondence with Davies Giddy. It is not surprising that the crisis of 1795–6 brought him once again into political action. As foreign affairs had roused him to unrestrained protest in 1792, so his feelings were whipped up by developments in England. The 'gagging acts' of 1795 were the occasion for Beddoes to make one more demonstration of his opposition to the government. It would not have been surprising if he had found time for no more than participating in short-lived protest meetings and supporting Coleridge's efforts to found an independent journal. But not content with the pamphlets immediately connected with the protest against the two Bills, Beddoes set himself to make clear in detail the grounds for his criticism of Pitt. His 'Essay on the public merits of Mr Pitt' has already been referred to in the accounts of Beddoes' own reviewing and of Coleridge's *Watchman* but, as Beddoes' major political statement, it is worth examining in more detail, together with the 'Letter to Pitt on the present scarcity'.

After the publication of this Essay, Beddoes' political writings as such came to an end and his hopes for reform were united with his medical work. This time, in contrast to October 1792, he chose his ground judiciously, for opposition to the two Acts had considerable support. To some extent also, he restrained the manner of his protest. The monarchy and the monarchical system of government was nowhere attacked in the Essay and nothing appeared that could justify the charge that despite his protestations of loyalty, Beddoes was a republican. Beddoes' main protest was against the autocratic powers of the ministers, corruption and poverty. The attack, as the title claims, is entirely directed against the *Public* activities of Pitt – as *The Watchman* reviewer writes, "The Author wages war with the Minister and nowhere degrades his cause by stepping from the Senate or the Cabinet to Holwood House."[37]

Beddoes did certainly set himself to survey in some detail Pitt's career from 1781, when he succeeded his father, the Earl of Chatham, in office, to 1796 and the consequences of the war against France. His theme was the disappointment, so bitterly felt by many people, that the statesman who had once promised reform, should have become the leading advocate of reaction. This same disappointment Coleridge expressed in his sonnet, 'To Pitt' in 1796 in terms so violent that in 1797 he withdrew the poem from publication. To Pitt, no mercy should be shown, the "dark Scowler" who had "kissed his country with Iscariot mouth". Beddoes did not hide his scorn and anger but he examined Pitt's career systematically. Separate chapters are devoted to Pitt's policy towards the American colonies; to India and to Ireland; he described Pitt's early efforts at parliamentary reform, so disappointingly abandoned, and the practical consequences of his taxation system on the daily life of the people. A memorable example is the threat to the health of poor people which Beddoes saw in the window tax which obliged them to close up windows in their already ill-ventilated houses. There is a mass of detail; sometimes Beddoes' own quirky ideas and observations stand out, as when he advocated a "furniture of science" in every town so that people might be educated to understand the physical world; or when he drew attention to the "tiller of the ground ... doomed to the greatest hardship and the most scanty allowance [so that] they practise methods of preventing the increase of offspring". Sometimes we notice a vivid phrase which has not yet lost its power. Writing of the unhealthy state of London he quoted Dr Hunter's "new poisons are arising every day"; and of Ireland he observed, "Without being admitted to participation in prosperity, Ireland is coupled with England as a yokefellow in adversity".[38] It is impossible not to admire Beddoes' energy in collecting and organising all this information but the reader comes to feel he has to fight his way to find a path through an overgrown wood.

Beddoes himself may have realised this for sometimes he chose to put his argument in surprising forms. The war against the American colonists is condemned by dramatising in heroic couplets Burke's attack on Lord North in the Commons:[39]

'Let slip the dogs of war', the Premier said;
Each skip-jack peer inclined his stately head.
For North his commons passed the ready vote,
To North the preacher lent his liquid throat,
Masked in unspotted lawn, with solemn breath
The fawning bishop blessed the work of death.

The verse ends with a description of the catastrophic collapse of the English power:

Back on his head the ills he broods recoil!
Invasion threats his Country's peaceful soil.
Dismay with gorgon aspect strides the land,
And points where hostile prows o'erlook her strand;
Her haughty navy veils its pendants low
Cries to the gales, and scuds before the foe.
Such glories Fate's all-just decrees impart
To nations light of faith and proud of heart.

The ostensible subject was the American war but Beddoes must have had in mind the unexpected successes of the French in 1796. Another 'diversion' is an exposure of the harshness of the Poor Law cast in the form of a dramatic scene; and for good measure there is a mock parliamentary speech. Beddoes possessed powers of invention, but not a sense of the appropriate. Dr Aikin,[40] who reviewed the essay for the *Monthly*, began with a tribute to Beddoes' acuteness of research and introduced him as a "Bold and original thinker", but ended ambiguously: "he is anything but dull; and the reader who may not be convinced by his arguments, will scarcely fail of being entertained by the copiousness of his research and the vivacity of his remarks".

The chapter which remains most vivid is the ninth, which concerned the degraded conditions of labourers in both town and country. Beddoes made his mood clear by his 'text'.

How blest the land whose sons are fed
By honied words in place of bread.

But he put aside irony to give a moving account of the miseries of the poor. He had begun the Essay by calling attention to the special opportunities a doctor had to observe and judge the conditions in which people live, and this chapter is written in that spirit. His authorities are medical writings and the reports and the novels of Henry Fielding. Beddoes described the pestilences [41] that so frequently visited cotton works; the palsy that was found among the makers of certain metallic toys; the small-pox and the "Fevers which like the fires of Vesta, eternally rage among the poor of some manufacturing towns". His account is as harsh as any to be found among the more familiar nineteenth century reports and, being so well documented, shows how early such conditions arose. Here the most attractive trait of Beddoes' character comes to the fore, his humanitarian and unsentimental sympathy with the poor. He reminded the reader, in the words of Henry Fielding, that "The sufferings of the poor are indeed less known than their misdeeds. They freeze and rot among themselves — they beg and steal and rob among their betters".[42] And his own summing up would have won the assent of many of the democrats:

A little addition of comforts to mediocrity, a little diminution of pain from penury would have established the public security for an indefinite term. No alarms would have disturbed the banquet of opulence.[43]

The pamphlet is not entirely descriptive. Like Paine in the *Rights of Man*, and like Kant in *To Perpetual Peace*, Beddoes gave an outline of his aims. It is the only occasion on which he is so clear. He listed the "Common concerns of Mankind" in a series of questions:[44]

(1) How far am I secure against false alarms, fraud and violence?
(2) Do circumstances which I cannot control threaten deprivation of the accommodations and necessaries of life?
(3) Do unjust laws, encroachments on freedom by persons in power, or other public impediments prevent my hands from executing what my head has innocently devised?
(4) Is the distinguishing bounty of nature to man frustrated by infringements on the privilege of speech?
(5) Are the contributions of the people faithfully applied to the public service?
(6) Do the fruits of my industry or possessions go to delude the weak, bribe the corrupt, and slaughter the innocent?

(7) What share of the blessings of nature do my countrymen at large enjoy?

(8) Are their sufferings on the increase or decline?

(9) Are intoxication and the grosser vices on the wane? In charity of thought and action are they superior to their predecessors?

(10) Do the less instructed begin to distinguish their preservers from their destroyers? Or are they still ready, at the beck of a minister to bawl for a war, and light bonfires for victories which will bring the rot of famine upon all the human creatures that have the misfortune to belong to them?

(11) What are the causes of the growing happiness or misery, the improvement or the degeneracy of the community?

The 'Essay on the public merits of Mr Pitt' takes us into the England of the 1780s and 1790s, probably all the more vividly for being "tinctured with pretty strong prepossessions". It reveals Beddoes: his vigour; his industry and wide interests; his hatred of war and oppression; his humour and his anger. It is the complement to the reviewing: there we see the chemist and the medical scientist; in the essay we find the physician and the humanitarian.

For Beddoes as well as for Coleridge, 1797–9 was a time of change. At least the financial difficulties holding back the realisation of his plan were removed. While Coleridge was working with Wordsworth on the *Lyrical Ballads*, Beddoes was looking for an assistant for his laboratory. Humphry Davy arrived in Bristol in October 1798 and no-one could have foreseen, when Coleridge returned from Germany and met the young Cornish poet and chemist at Beddoes' home, that the next months were to be a time of unparalleled inspiration for chemistry and for poetry.

CHAPTER 9

THE PNEUMATIC INSTITUTE/HUMPHRY DAVY

The engagement of Humphry Davy was the culmination of the five years' work which went towards the setting up of the Pneumatic Institute. It had begun when Beddoes stayed with William Reynolds at Ketley and Erasmus Darwin, Dr Yonge and William Reynolds himself made the first subscriptions to the plan for a Pneumatic Establishment. From the outset Beddoes saw that he had both to raise money and to gain assent for the scientific respectability of his project. As a student and then as a teacher of chemistry he had already realised the significance for medicine of the work of Priestley and Lavoisier and, as he himself had shown, of Mayow. Once he arrived in Bristol, committed to the plan for a Pneumatic Establishment, he was under the necessity of making clear, to scientists and to lay people, how the experimental work of these pneumatic chemists had shown the nature of respiration; that in breathing, oxygen was taken from the air and absorbed into the blood and that this chemical knowledge was of far-reaching importance for medicine. In Beddoes' words, it would give "Command of the elements which compose the animal substance". To do this he would need to draw on experimental work, in some of which he had himself taken part, as a student under Black or in his own re-working of Lavoisier's experiments in Oxford — made possible by Sadler's skill as an instrument maker. This he could strengthen by reference to his own wide and critical reading of the literature which extended from his early study of Mayow to new books and reports. He realised that the public needed to be assured that care was being taken to establish which gases were harmful and which were beneficial and attention should be given to finding convenient means of using the gases in the sick room. The accusation of being impetuous and ill-organised is often made against Beddoes but he obviously had in mind a well-worked-out programme; he wrote at first to prove the validity of his ideas and then to explain how he intended to put them into practice.

In the eighteen months from July 1792 to December 1793 he set out in three publications the theory which underlay his scheme for the treatment of illness by means of gases and supported this with case studies, mostly contributed by medical colleagues. He was clearly determined to lose no time in establishing himself in the new field and worked with tremendous energy,

145

for in the same period he wrote four papers on medical problems for the Philosophical Transactions of the Royal Society and began his work for *The Monthly Review*. Beddoes realised how inter-related were the financing of his project and an open explanation of its scientific basis. He was, too, on principle[1] convinced of the absolute necessity of doing away with secrecy in the treatment of illness and he took every opportunity to attack the secrecy with which the traditional doctor surrounded himself.

The first of these publications was 'Observations on the Nature and Cure of Calculus, Sea Scurvy, Consumption, Catarrh and Fever'. Published in 1793, it must have been begun in 1792, for the preface is dated from Oxford in July of that year. In parts it has the excited, over-optimistic tone heard in his other writings of that difficult time and reveals the same aggressive mood:

An infinitely small portion of genius has hitherto been exerted in attempts to diminish the sum of our painful sensations; and the force of society has been exclusively at the disposal of Despots and Juntos, the great artificers of human evil. Should an entire change in these two respects, anywhere take place every member of society might soon expect to experience in his own person, the consequences of so happy an innovation: and should the example be generally followed, there is no improvement in the condition of the World, for which we might not hope from the bloodless rivalship of nations.[2]

This hardly seems the opening of a scientific work. Yet the link between the three, at first sight separate, sections of 'Observations' is precisely the demonstration in each one of the importance for medicine of recent developments in chemistry. The opening essay on Calculus describes a new remedy for the stone which was clearly an application of the recent discovery of carbon dioxide. Beddoes reported the use by an unnamed doctor of a "solution of fixed vegetable alkali supersaturated with carbonic acid or fixed air".[3] This soda water needed a fairly fragile apparatus for its preparation and when it had been made, room to store the bottled liquor. In one of Beddoes' patients it caused dizziness (he was struck by the likeness to some sparkling wines) and he devised an alternative formula. In this the white powder which results from exposing crystals of natron or sal sodae to the air was to be bound by soap, or any other cement. It might if so desired be flavoured, to form a pill. Beddoes was pleased with the cheapness and simplicity of this form of the medicine, which could be "adapted to the poor, who are by no means exempted from calculous disorders". It seems that the well-to-do might drink soda water to ease their pain, while the poor had to swallow pills made from washing soda and soap. Though this treatment of stone was — and had been ever since Joseph Black's experiments — connected with the chemistry of gases, it is not quite what we now think of when we use the expression Pneumatic medicine. The next section of 'Observations' deals with diseases arising from deficiency or excess of oxygen and is a systematic treatment of Pneumatic medicine in its better known form.

Of the four conditions grouped together in the second part of 'Observations' two, sea scurvy and obesity, were in Beddoes' view caused by a deficiency, and the two others, consumption and catarrh, by an excess of oxygen. The two deficiency diseases led Beddoes to consider whether the remedy might be found in a diet which would supply oxygen: raw [4] or frozen meat or fresh vegetables. He felt unable to follow up this possibility: "We are certain," he wrote "that the blood, in the first instance, and afterwards the solids, are oxygenated by means of the lungs. They may acquire this principle by means of the stomach; but we have no direct experience of their doing so." [5] All he could conclude was that a right diet might have contributed to the healthiness of the sailors on Captain Cook's ships and it might be of some little help in the early stages of consumption. But the more hopeful line of investigation was to notice the correspondence between living in the fresh air and absence of scurvy. Lack of oxygen could often be remedied quite simply by living more in the open air and he considered scurvy could be avoided by paying attention to this. Beddoes had referred to the freedom from scurvy among the sailors on Cook's voyages and, to support his contention that this was due as much to fresh air as to diet, he cited the findings made by Dr Trotter in a study of scurvy on slave ships. Dr Trotter [6] observed that while mortality from scurvy was high among the men who were confined below decks for 15 out of the 24 hours, there was an insignificant occurrence of the disease among the women and boys who were allowed on deck. It is hard not to feel that the detail of the slaves "stowed spoonwise and so closely locked into one another's arms that it is difficult to move without treading on them", is reported as much by Beddoes the opponent of the slave trade as by Beddoes the scientist. Consumption and catarrh, of course, needed more drastic treatment. For them Beddoes advocated altering the atmosphere by the substitution of one of a number of harmless gases for part of the oxygen. Here we have pneumatic medicine in its most popularly recognised form. As explanation of this Beddoes gives an account of Priestley's and Lavoisier's experiments with air and concludes,

nothing would so much contribute to the rescue of the art of medicine from its present helpless condition as the discovery of the means of regulating the atmosphere. [7]

The last paper in 'Observations' is by Dr Girtanner on 'The Laws of Irritability'; it follows Beddoes' own accounts and reinforces the view that oxygen is of primary importance: "Oxygene is the principle of irritability; it unites with the blood during respiration; it is distributed to every part of the system by the circulation, and it combines with stimulating substances, with which the different parts of the system come in contact." Even more relevant to his plan is Girtanner's belief that oxygen is absorbed in the blood and that the venous blood is oxygenated in the lungs during respiration especially as this was a view which was not universally held at the time. At the same time Beddoes made clear his reservations [8] about the Brunonian system, in particular his criticism of the idea of "negative stimuli"; it seemed to him "vain to say (what Dr Girtanner repeats) that the depressing passions are only the abstraction of the stimuli of the exciting passions." The human sympathy, as strong in Beddoes as his pursuit of scientific explanation, rejected the suggestion "that sorrow should be to indifference what darkness is to twilight".

Once his plan had been made known, 'Observations' was published as the initial statement of Beddoes' conclusions about the nature and possible treatment of this group of diseases. He quickly followed it with two more publications which together form a logical development of his thesis. The good reception of the first work, the confidence shown by his friends and the move to Bristol obviously had a steadying effect, and he felt able to dedicate his next publications openly to two of the most eminent men of his day, Dr Erasmus Darwin and Professor Joseph Black. They are markedly better organised and calmer in tone.

The 'Letter to Dr Erasmus Darwin on a new method of treating pulmonary consumption' is in the main a medical work. It opens with a vivid description of consumption which shows how imperfectly the condition was then understood. Beddoes had no precise means of identifying tuberculosis and he considered 'catarrhal' and florid conditions equally to be consumption, though he proposed to deal principally with the 'florid' type. This he believed to be a condition of hyper-oxygenation and he gave a description of the symptoms, the bright eyes and florid complexion; the violent cough and the raised acuteness of the senses. In this 'Letter to Darwin' we find the clearest account of how Beddoes came to make the connection between his medical observations and his chemistry. He referred to his own work on Mayow,[9] to his intimacy with Dr. Edmund Goodwyn, and to his knowledge of experiments with animals at Edinburgh, leading up to the way his appointment at Oxford led him "to acquire as minute a knowledge of the properties of elastic fluids as possible." Finally, his observation of the remission of tuberculosis during pregnancy [10] gave him the idea of the application of this work to medicine. He expressed his "firm persuasion . . . that the system might be as powerfully and as variously affected by means of the lungs as of the stomach." Though he knew of others working on the same idea, Beddoes claimed that he alone worked to a guiding principle. One of Beddoes' patients, a boy, the son of Dr Crump, had been treated by means of diet and the breathing of airs. Beddoes insisted that "the father, convinced of the utter insignificance of all the ordinary means of cure, cheerfully consented." With obvious disappointment Beddoes described how he had to face the fact that difficulties had arisen from the boy's attendants being unaccustomed to the breathing apparatus. The boy died, but Dr Crump felt that the treatment might have succeeded and in any case, gave relief, for he said "the poor boy used frequently to ask me for some of Dr Beddoes' breath." This failure made clear the need to study the various gases that might be needed. His programme included finding the means of producing a range of gases, ensuring that he had sufficient air for any need, and had the ability to mix the airs with one another and with the atmosphere in measured proportions. The obviously sad and disappointing case of Dr Crump's [11] son, emphasised the practical problem of finding suitable apparatus to administer the gas. For producing the gases, Beddoes had constructed "gazometers" such as Lavoisier and Van Merum used and he told Darwin that he had made both a model apparatus [12] and a portable one for use by patients confined to bed.

The 'Letter' to Darwin ends with a careful account of another trial of gases, an experiment that Beddoes made on himself in breathing oxygen. His detailed observations and his realisation that there was a danger that in such an experiment he might deceive himself — he gives Mr William Reynolds, Mr Joshua Reynolds and Dr Yonge as objective witnesses of what happened — show how ill-founded were the charges of charlatanism made against him. Beddoes "oxygenated" [13] himself by breathing equal parts of oxygen and nitrogen daily for seven weeks, 20, 30, 60 minutes a day, though only four to five minutes a time. He considered that the experience produced an insensibility to cold that would be permanent, as might be the change in his complexion "from a uniform brown" to "fairer and somewhat florid". This gave him the idea that oxygen, once a suitable breathing apparatus had been devised, might be used as a cosmetic. Beddoes' detractors considered that he had permanently damaged his health and adopted a judgemental attitude at his early death.

Beddoes could not refrain from an over-enthusiastic claim that his treatment would be efficacious in a wide range of ills. Dr Darwin made no comment on this but where consumption was concerned, urged Beddoes, "Go on, dear Sir, save the young and fair of the rising generation from premature death; and rescue the science of medicine from its greatest opproprium." [14]

The collection of letters which appeared in December 1793 was dedicated to Black with an expression of admiration that "neither years nor celebrity the bane of vulgar minds", made Beddoes' former teacher unreceptive to new ideas. Beddoes made clear his reasons for early publication of his theories, which were that he wished to gain support for a proper trial of the new medicine and that he was very anxious to prevent harm being done if 'airs' were used by those who did not understand them or were unfamiliar with the apparatus for administering them. For the first time he put forward the desirability of making trials not just as occasion arose and by the uncoordinated work of a number of doctors but "in an appropriate hospital under the management of an able and impartial physician".[15]

Before presenting the letters on various cases of consumption, Beddoes gave an account of work done on the composition of air by Lavoisier and reported in a paper to the Paris Society of Medicine in 1785, "Observation on the alteration produced in the Air of Places where a great number or persons are assembled". Beddoes realised that this account was not easily accessible and he reported Lavoisier's experiments in detail, "rendering the chemical terms conformable with those in the French nomenclature adopted since 1785". Beddoes felt he had to comment on this, for the new terms were still not fully accepted – Priestley in 1790 referred to them as poetry and wished for reports in "plain prose". Beddoes' account, apart from its value in making available in England an important contribution to the study of the composition of the atmosphere, was basic to his argument. He saw it as "intimately connected with the study of Diseases that may be cured or relieved by breathing different airs". In particular, it made clear the very narrow limits within which it was necessary to remain in any experiment that involved varying the composition of air to be breathed. Beddoes faithfully reported Lavoisier's belief that the atmosphere consists of 27–28% oxygen to 72–73% nitrogen, an analysis with which Priestley's experiments agreed.

This work is appropriately addressed to Black for in it Beddoes wrote as a chemist. The letters that follow his summary of Lavoisier's work are all from doctors who were treating patients with gases and they are collected here and presented to Black to demonstrate the application of pneumatic chemistry in the actual practice of medicine. One of the most interesting is from Dr Withering, encouraging Beddoes with the promise that "philosophers will urge you to proceed, from a conviction that should you fail in your higher aims, you must extend the boundaries of science". He had himself used carbonic acid air obtained from a bubbling vessel placed in such a way that the draught of air up the chimney drew the vapour across the patient. Dr Withering considered that the air from kilns might be beneficial; he was prepared to pay attention to such folk cures and "disposed to believe that opinions generally prevalent have some sort of foundation". Most of the letters are less modest; they make horrifying reading and show how wild and desperate were the 'cures' which Beddoes' contemporaries were driven to try. This, and the feelings of pity they arouse, do much to make us understand Beddoes' sense of urgency in his search for an effective treatment for a disease which was tragically common in his time. The letters he quotes are a collection of un-coordinated reports from individual doctors but there were others, such as Dr Percival in Manchester, who aimed at more systematic trials and even had in mind some form of infirmary. Beddoes' distinction lay in combining scientific investigation with medical trials and in stressing that his institution was to have a limited life.

It was unfortunate that Beddoes accepted without criticism the reports of doctors whose interest in pneumatic medicine was enthusiastic rather than scientific, for it was their work, not his own, that deserved to be considered irresponsible. Much as he wished to show that there were already grounds for believing in the value of airs, he harmed his case by bringing forward these reports.

The three publications of 1793 are in effect the parts of a single work. They form a progression explaining and supporting Beddoes' aims and make clear how carefully the Pneumatic Institute "experiment" was thought out. In spite of passages of excited enthusiasm in 'Observations' they do much to counter the idea that Beddoes was a wild eccentric. Among the many doctors interested in pneumatic medicine he was the one to put himself at risk:

Although it is evident that none but beneficial consequences can result to the public from the prosecution of my design, yet it requires very little knowledge of the world to perceive the danger to which I am exposing my reputation. It is impossible to engage in a new and arduous undertaking without incurring ridicule and obloquy. Of course I must expect to be decried by some as a silly projector, and by others as a rapacious empiric.[16]

There is a further interest in these three related papers in that they make clear that the medical experiments in which gases were used to treat various conditions were part of a wider interest in the part played by oxygen in the organisation of the body. This is most apparent in the first and least well organised of the three – Observations on the Nature of Scurvy etc – which may have been in Beddoes' mind before the plans for a Pneumatic Establishment had been formed.

After a short interval, Beddoes published in the autumn of 1794 his 'Considerations on the medicinal uses of factitious airs'. The time for advocacy had come to an end and he now set out to describe how his ideas would be put into practice. A new edition of 'Considerations' appeared quickly in two parts, the second by James Watt. Beddoes began with a clear statement of his 'proposal': "to ascertain the effects of these powerful agents in various diseases," and "to discover the best means of procuring and applying them". After recalling the work of the pioneers in the study of the atmosphere – Priestley, Scheele, Cavendish and Lavoisier – and their demonstration that atmospheric air contained both nitrogen and oxygen, Beddoes cautiously made a further observation. He had noticed the presence of "a little carbonic acid air".[17] He realised the possibility that this was accidental and added the comment that it was present "though no fires burn or animals breathe near". This must be a very early observation of the presence of carbon dioxide and emphasises the importance of the new apparatus, the improved air pumps and the accurate balances which Beddoes had been so keen to have made for him. Good apparatus and his own experimental skill must have been brought into use in his experiments on the breathing of various gases. Beddoes had earlier warned that experiments with human subjects should be undertaken with great caution and he used kittens, 'whelps' and rabbits, Various mixtures of gases were breathed and there were further experiments to see how this affected an animal's reactions when it was immersed in water, noting the time taken to drown or to be revived. They are not pretty reading and those who are opposed to experiments on animals will be relieved that Beddoes too felt they were 'cruel experiments'[18] and that he discontinued them as soon as he had evidence to prove "the power of factitious airs to affect the human frame". He observed, as well as the effect on the lungs, "the necessity of oxygen air to muscular exertion" and advised his readers to keep still if in trouble – it is "best not to struggle in the Black Hole of Calcutta". From this experience he had come to conclusions about the proportions of the various mixtures of gases he proposed to use in his treatments. This he worked out in detail and set down in the form of tables, at the same time warning that the preparation of such 'atmospheres' was not to be undertaken by amateurs.

The second part, by James Watt, marks the end of Sadler's collaboration with Beddoes. The problems involved in producing gases to exacting standards of purity seem to have been beyond Sadler who devised ever more complicated apparatus. It may be that his interest was not in work with medical cases. He probably left Bristol in 1795, since in 1796, when he was appointed as chemist at the Board of Naval Works, he was already Barrack Master at Portsmouth, drawing a salary of £400 a year. There cannot have been any breach in friendship, for Beddoes later employed Sadler's son (though in a different sphere). The reason why Watt joined Beddoes is by contrast quite clear. After the death of his daughter from consumption in June 1794, Watt felt, as he wrote to Erasmus Darwin, that "when an evil is irreparable the best consolation is to turn the mind to any other subject that can occupy it".[19] He had chosen to make "an apparatus for extracting, washing and collecting of poisonous and medicinal airs" which he wished Darwin to try. He sent another apparatus to Dr Beddoes as well as "a short list of . . . hints".

James Watt was anxious about Beddoes publishing an account of the apparatus while it was still untried but it is clear that his collaboration was greatly to Beddoes' advantage. The apparatus that Watt had designed was capable of being used in practical, clinical situations. He gave a very clear account illustrated, as might be expected, by technical drawings, of its construction and of the way it should be used. From this stage onwards, James Watt provided the solution to a number of technical problems. He tried out various substances that might be used to produce gases and it may have been at his suggestion that Humphry Davy, when he set out to investigate the properties of nitrous oxide, used ammonium nitrate to prepare the gas. In developing the apparatus, Watt devised means of cooling the gas and of making it possible to regulate the quantity that was delivered. Watt also saw that it would be necessary to have a simple method of collecting the gas, for in practice the apparatus would often be used by relatively untrained helpers. He constructed a "bellows", a gas holder formed of two cylinders, one sliding within the double walls of the other and, as it filled with gas, controlled by a counterpoise.[20] In this manner, by the removal of the weight, the gas could be automatically driven through the outlet, either into storage containers or into "breathing bags". He continued to adapt this apparatus for use in different circumstances. His most elaborate invention was the sealed breathing chamber – like a Sedan chair – which he made when Davy wished to observe the effects of more prolonged breathing of nitrous oxide. Furthermore, Watt could of course undertake to manufacture all that was needed. With some tact he put forward the firm of Boulton and Watt as possible makers of the apparatus – "The price shall be as moderate as we can make it". In spite of the technical nature of his account, Watt gave sympathetic attention to the needs of the patient which he must have known all too well. He described a "beehive" to go over the head of anyone too weak to breathe from the bag containing the gas. This bag was made of oiled silk and Watt also gave detailed instructions for removing the unpleasant smell, which he explained could be absorbed by rolling the silk in powdered charcoal.

Beddoes and Watt collaborated in the next part of 'Considerations', Part III which appeared in 1795. Beddoes brought forward cases to show the success of the treatment and Watt was mainly concerned to give warning about the need to dilute the gases. In October 1796 the last two parts, IV and V, deal partly with the same division of material: Beddoes giving accounts of many varied cases in which gases had been tried and Watt describing the apparatus. In this volume can be found a description of Guyton de Morveau's success in using hydrochloric acid as a disinfectant which Beddoes had seen on his visit to the Dijon hospital, eighteen years earlier. Watt had by this time simplified his apparatus, making it portable. He had even been able to cost it and give prices for apparatus of various sizes, for ancillary equipment and for Exeter manganese and iron filings, to be used in making the gases as well as for the Cornish china clay for the lute. By this time Beddoes was full of confidence:

Mr Watt goes on improving the air apparatus and I endeavour to improve the powers of the airs. He has just printed an account of a simplified apparatus which . . . will suit a kitchen or parlour common fire. My opinion of the benefit which will be derived from the scheme is daily confirmed by the facts.[21]

And he made this last edition the occasion to explain not merely his practical aims but what might well be described as his medical philosophy. As well as his determination to test gases he "desired to be instrumental in diffusing a taste for the most useful species of human knowledge and in converting nations into *Humane Societies*. . . . This is the art of living; for whose reception men's minds can only be prepared, by being familiarised with just ideas concerning animal nature; and whose precepts can only issue from the Shrine of Hygeia." [22] In reply to those who would trust entirely to common experience, he justified his whole procedure in two almost aphoristic sentences:

It belongs to science to sort facts.

Experiment gives permanence to what is discovered empirically.

In spite of the seriousness of his intentions and the open, detailed exposition of his plans in 'Considerations', Beddoes met with opposition and misrepresentation. He could not resist replying and "contributed" an ironic letter and a witty poem to Part III. Now that the argument is over, these are enjoyable jokes and also useful evidence of the climate in which Beddoes was working; but they are so out of place in a serious work that they must surely have damaged his hopes of gaining wider support. Soon after his arrival in Bristol it had seemed that Beddoes might find help from the "establishment". The Duchess of Devonshire [23] had visited him in Hope Square in 1793 and had been enthusiastic about his plans. She it was who put forward the idea of a hospital. The Duchess did not stop at encouragement; she wrote to ask the support of Sir Joseph Banks, President of the Royal Society. Sir Joseph totally rejected the possibility of supporting a known revolutionary.

Since Banks himself presented Beddoes' two papers on 'The affinity between Basaltes and Granite' to the Royal Society in 1791, and since in March 1791 and May 1792 Beddoes' Royal Society papers on "the conversion of cast into malleable iron" had been written in the form of letters to Sir Joseph Banks, this indignation was clearly the result of Beddoes' activities in the late Summer of 1792. The Royal Society, as well as Oxford, closed its ranks against sympathisers with the French Revolution. Joseph Priestley too met the same rejection. In October 1793 he wrote to Dr Withering [24] from London, "As to the *Royal Society*, I see myself regarded in so unfavorable a light by the most considerable members of it that I never go near them", and not for the first time he complained that a candidate he supported for membership of the Royal Society had been rejected because he was a "democrat in politics". He deplored this "party spirit" as " highly unworthy of the society and injurious to the interests of philosophy", and his letters in 1792 and 1793 show how strongly the tide of feeling was running against the "friends of liberty", even in the scientific world. When Priestley had to struggle to justify compensation for the loss of his laboratory and irreplaceable books – he had to ask Wedgwood for details of the price of the laboratory equipment he had supplied – it is little wonder that Sir Joseph Banks would not support Beddoes' scheme. Sympathy for the ideas of the French Revolution may indeed have had a more direct effect on Beddoes' endeavours to raise money for the Pneumatic Institute than on his leaving Oxford.

The steady unfolding of his theories and plans in these publications was the outcome of much hard work at the details of the scheme. Throughout the time of preparation Beddoes had the active support of a very faithful group of friends. Dr Withering, under whom Beddoes had worked while he was still a medical student encouraged him over "Observations on . . . Calculus" and Beddoes was heartened to see that this, the first exposition of his theories, was quickly sold out. Dr Parr came forward with a suggestion designed to bring Beddoes increased prestige, that he should edit the works of Dr John Brown. This was not a very congenial task but Beddoes was probably the more willing to undertake it as it would provide some financial help for Brown's widow. William Reynolds continued to give help in a practical form, supplying 'manganese' (manganese dioxide for making oxygen) when Watt was ready with his apparatus for producing gases. Even more sympathetic support came from Davies Giddy. Since the early days of his arrival in Bristol, Beddoes had confided all his fluctuating moods of anxiety and hope to Giddy; his success at finding patients; his worry at the high price [25] of the house he wished to buy; his fear that the war with France would devalue money and mean that fewer people would visit Clifton. Giddy was banker, financial adviser, and perhaps even more valuable, confidant. As Beddoes' former student, he was uniquely able to show that the idea of pneumatic medicine was not hastily conceived and when the 1795 edition of 'Considerations' was in the press, Beddoes appealed to Giddy to strengthen it by writing an account of having heard Beddoes explain his theories in 1791.[26]

The help [27] that James Watt gave was by no means limited to solving the technical problems of the apparatus for making and administering the airs. From the time of Jessie Watt's death until the summer of 1795 Beddoes and Watt were in constant correspondence, exchanging letters frequently, at least every week during the rest of 1794. James Watt had written to Darwin and Beddoes almost immediately after his daughter died and on 9th June Beddoes replied in a letter showing how very moved he was at hearing so soon of Watt's readiness to help and how much he valued the opportunity of corresponding with Watt. By 17th August Beddoes had received an apparatus. From then on, Watt was actively helping Beddoes, not only contributing to 'Considerations' but correcting proofs, and being consulted over the plates for the diagrams. Beddoes was eager to do justice to these, and to pay for shaded rather than outline diagrams if Watt thought this preferable. Beddoes consulted Watt about finding someone to make the apparatus and after this came the offer that Boulton and Watt would undertake the production. Before the Pneumatic Institute itself came into being, apparatus was being made or repaired for some of Beddoes' patients. Beddoes himself had very probably made some apparatus for the first cases, for James Watt junior wrote that his father wished to express admiration of Beddoes' ingenuity. An apparatus was presented by Watt to Glasgow Royal Infirmary and another sent by James Watt Junior to Philadelphia. The firm had printed 250 copies of leaflets illustrating the apparatus and giving instructions for its use. James Watt Junior sent Beddoes six copies to pass on to his patients who were already using an apparatus. The help which at first had been directed towards solving a technical problem and which from the beginning Beddoes recognised as of enormous value coming as it did from the foremost inventor of the day, grew into a true sharing of interest. Watt's son contributed an enthusiasm not unlike Beddoes' own, but combined with good business sense and James Watt himself gave Beddoes wise guidance as well as technical advice.

Both James Watt and his son were active in collecting subscriptions for the proposed Institute and in encouraging Beddoes to make sure that the names of scientific supporters were known. James Watt junior, indeed, had ambitions for the scheme beyond anything Beddoes planned and thought of London as a better place for the proposed Institute. He had hopes of going to London to gather "aristocratic" support – he brushed aside Sir Joseph Banks' rejection of Beddoes and refusal to receive further representations on his behalf. Beddoes did not fall in with this plan. He wrote that he was too responsible for critically ill patients to leave Bristol but perhaps he did not wish to encourage too much activity by James Watt junior whose revolutionary fervour was well known. Beddoes admitted that his own political activities at the time of the petitions against the Two Acts harmed the cause of "the airs". But the younger James Watt proposed advertising in *The Monthly Review* that Boulton and Watt' had stock of the apparatus and he took the opportunity of the firm's correspondence with Germany, to send copies of "Considerations" to Crell and Gren and others he knew.

In August 1794 Beddoes had been able to tell James Watt that he had the active support of Thomas Wedgwood.[28] Beddoes wrote to Wedgwood on 12th August 1794 about the important development for his schemes. He described how difficult he had found it to help James Watt's daughter for lack of a proper apparatus to administer airs and how "Mr Watt set himself to invent a domestic apparatus for procuring and administering airs". From then on, Beddoes had frequent correspondence with Wedgwood too, telling him, as he did James Watt, of scientists who supported the scheme and passing on reports of cases where airs had been beneficial. Besides his disappointment over the Duchess of Devonshire's failure with Sir Joseph Banks, Beddoes had had no response from Sir Charles Blagden (secretary of the Royal Society); he supposed this would be because "the trials" (i.e., the treason trials) "have engaged almost every person's whole attention in London". As the scheme progressed, drafts of his appeal for the Institute were submitted to Wedgwood and he was consulted about Trustees to manage the receipts from sale of publications and the Institute itself. Beddoes offered, if the Trustees desired it, to have a more neutral doctor than himself as Director. Having kept him so well informed, Beddoes hoped that Wedgwood would lend his name to the scheme and help publicise it in London.

We can see how closely this group were working together. Darwin, who had written so enthusiastically in support of Beddoes, now had a share in improving Watt's apparatus, suggesting how the gas might be delivered from the bellows without disconnecting these from the furnace. He was full of ideas for further uses for the apparatus, to "supply the world with new Materia Medica,[29] to be drank in by the lungs" and proposed to show the one he had received to the "Derby Philosophers". On his side, Beddoes was using a "rotatative couch" to swing a patient with tubercular lungs. This was a method Darwin had advocated, thinking it a better way of inducing sickness than any medicine. (The sickness would produce absorption of fluid.) Beddoes asked Watt's help in making the couch move quietly; Watt did in fact make drawings for such a piece of medical apparatus. It is clear that Beddoes was prepared to try a variety of treatments in his search for anything that might alleviate the condition of tubercular patients and was prescribing a careful diet, tonic or gently stimulating medicine as well as using oxygen, hydrogen and carbon dioxide. He stressed to Watt that in using airs he was testing a hypothesis and by no means putting forward a cure. It is unfortunate that as time passed, Beddoes no longer made this distinction so clear to the public.

It is not surprising then, that the names on the published list of supporters were those of Beddoes' scientific and medical circles — the Darwins, the Reynolds, the Wedgwoods. The London doctors were absent even though one, Dr Saunders of Guy's Hospital, had offered to allow Beddoes to work in his practice in the hospital. Medical support came from Edinburgh, from the Royal Medical Society and from Dr Alexander Monro; and among the names of more recent medical friends, were Dr Currie of Liverpool and Dr Ewart of Bath. Beddoes respected Dr Ewart's trials of oxygen as a treatment for tumours and had had discussions with him on the subject of pneumatic medicine. Dr Ewart's brother was ambassador at Berlin and this may have been another route by which Beddoes' work became quickly known in Germany.[30] In the end however, the decisive large contributions came unexpectedly. Lord Lambton, in gratitude for Beddoes' medical care, left him a large bequest. Thomas Wedgwood, both patient and friend, gave £1000. This gift was made in the best spirit of generous patronage:

I think I shall contribute as the attempt must be successful in part if it only goes to show that airs are not efficacious in medicine.

Beddoes now had financial security and a well-appointed laboratory, and by the end of August 1798 was awaiting the arrival of his young scientific assistant. Here once more Davies Giddy proved his true ally, sending him from Cornwall the unknown, nineteen year old Humphry Davy.

Beddoes had already heard of the young Cornishman.[31] Davy, working on his own as well as he could and using home-made apparatus, had made experiments to test the findings he had read in Lavoisier's '*Traité Élémentaire*'. He was apprenticed to the Penzance surgeon, Mr Borlase, but though his youthful and ambitious schemes of study not only included the sciences needed for medicine but ranged from theology and languages to mathematics, nothing had focussed his attention until the discovery of Lavoisier's work released his genius for devising and carrying out experiments. It was fortunate indeed that at this moment Mrs Davy took as a lodger James Watt's son, Gregory. Apart from the influence of his father's scientific and technical interests and friends, Gregory Watt had himself studied chemistry at Glasgow; he was too an enthusiastic democrat. Now Davy found a friend with whom he could discuss his ideas and experiments and by great good fortune it was someone who could at the right moment be a link with Thomas Beddoes. At about the same time Giddy invited Davy to read in his library, where he found a copy of 'Considerations'. As an appendix, Beddoes had included a paper by L. S. Mitchell which put forward the view that nitrous oxide was a poisonous gas, causing fevers and even the plague. It was this part of 'Considerations' that above all caught Davy's attention. Straight away he made a series of experiments which clearly disproved it. The friendship of Gregory Watt and Davies Giddy led to the accounts of these experiments and of work on heat and light being sent to Beddoes. So it was that, when his search for an assistant began, Davy was already in Beddoes' mind.

It was vital to Beddoes' plans to find the right superintendent for the laboratory. Before making an appointment he consulted his friends and Giddy suggested Davy. This was acceptable as the account of the experiments had already impressed Beddoes. Davy was not yet twenty and it is scarcely surprising that even though he saw the advantages of Bristol he was a little nervous about the decision he had to make. He had never left Cornwall before; in becoming apprentice to Borlase he had already taken the first step towards a medical career and would hesitate to jeopardise this. Protégés of well-intentioned families at that time often found themselves in an anomalous position. Davies Giddy's other young friend Thomasin Dennis,[32] for example, was made very miserable when for a short while she lived in the Wedgwood family. Giddy had suggested her as a collaborator in Josiah Wedgwood's educational schemes but the ladies of the family treated her as an inferior. Beddoes found Davy's anxiety surprising. He was confident that, though the resources of the Pneumatic Institution could not provide a large salary, he was offering a valuable opportunity for Davy to further his studies and experience in medicine. He sensed that the difficulty could not be removed by writing and asked Giddy to see Davy; Giddy must have given the necessary reassurance and all that remained was for Davy to be released from his indentures. Mr Borlase was generous in cancelling the agreement; having found his apprentice "a youth of great promise" he "would not obstruct his pursuits which are likely to promote his fortune and fame". For a while after going to Bristol, Davy seems to have continued to plan a medical career. Oddly enough, it was in the end Dr Beddoes who deflected his interest from medicine and gave him the opportunity to concentrate on chemistry, for in his turn, a little over two years later, Beddoes too released Davy from an agreement entered into. This enabled him to leave the Pneumatic Institute to take up his appointment as a chemist at the Royal Institution.

Davy's fears were immediately dispelled by his welcome in Bristol. He stayed at first in 3, Rodney Place, where Beddoes had made his home after his marriage, moving to this fashionable new crescent high up on Clifton Downs.[33] Davy found the kind nature hidden by Dr Beddoes' stern manner and Mrs Beddoes put him at ease. There is a note of relief in the happy letter he wrote to his mother reporting how he was received:

Our house is capacious and handsome; my rooms are very large, nice and convenient; and, above all, I have an excellent laboratory. Now for the inhabitants, and, first, Dr Beddoes, who, between you and me, is one of the most original men I ever saw — uncommonly short and fat, with little elegance of manners, and nothing characteristic *externally* of genius or science; extremely silent, and in a few words, a very bad companion. His behaviour to me, however, has been particularly handsome. He has paid me the highest compliments on my discoveries, and has, in fact, become a convert to my theory, which I little expected. He has given up to me the whole of the business of the Pneumatic Hospital, and has sent to the editor of the *Monthly Magazine* a letter, to be published in November, in which I have the honour to be mentioned in the highest terms. Mrs Beddoes is the reverse of Dr Beddoes — extremely cheerful, gay and witty; she is one of the most pleasing women I have ever met with. With a cultivated understanding and an excellent heart, she combines an uncommon simplicity of manners. We are already great friends.

Anna Beddoes, only five years older than Davy and less busy than her husband, found time to show him round. Already in the family were the two young Lambton boys whose education Beddoes was organising in his own original way so the household cannot have been too intimidating. There were walks on Clifton Downs and a romantic admiration for Anna Beddoes who must have sympathised with Davy's excitement at his new life. From Davy's accounts we see a cheerful, happy home; a place where people could easily form friendships and exchange ideas. Davy in his turn made a vivid and lively impression. Cottle described the sense of concentrated intelligence that all who talked with him felt. Among Dr Beddoes' circle of friends Davy seems quickly to have gained poise and confidence. Anna's aunt Charlotte who visited him not long after he was settled in London seems to have been equally impressed by the wonders of the Royal Institution and Davy's improved "style" — nevertheless he could still get "into the depths of metaphysics in the middle of Bond Street."[34]

When Davy arrived in October 1798, he found negotiations still in progress for the purchase of a house suitable for a small hospital and his notebook shows his excitement, "By the ardent and incessant exertions of Dr Beddoes the Pneumatic Institution is at length on the point of establishment. The design of this great object has been repeatedly pointed out by him — to investigate an important branch of medicine which has heretofore been little considered."[35] Soon he was helping move the apparatus from the laboratory at Hope Square where Beddoes' experiments had been done. Once all was in order, Davy began his own work of producing and testing gases.

The house[36] where Beddoes planned to receive his patients, to set up a laboratory and provide accommodation for a superintendent is in the northwest corner of Dowry Square, No. 6. Its door is hard up against the corner, at right angles to the narrow frontage with a single window on each floor which looks along the north side of the Square. No. 6 is at first sight a puzzle. It appears impossibly small for Beddoes' plans and is much less imposing than the double-fronted houses which form the north side of the Square. These face out over the central garden and, since the Square is open on its fourth side, they must, at the time of the Pneumatic Institute, have had a fine view down to the Hotwells Parade. No. 6 in fact stretches out to the rear and continues the line of the west side behind and at right angles to the north

side. It is as if it were intended to make a side of a second square behind
Dowry Square. When the house is seen from the garden it is clear that it was
not only large enough but well adapted for Beddoes' purposes. It seems very
likely that in the space between Nos. 5 and 6 which is now a builder's store
yard, there was stabling and that in these outbuildings Watt's apparatus for
making the gases was housed. The house was even larger than the present
No. 6, for the cottage at the end, now quite independent, though probably
built a little later than the main house, must have been a part of the whole.
It was a pleasant central place that Beddoes had found, at the bottom of the
steepest part of the hill. Young Humphry Davy, bringing the apparatus from
Hope Square almost directly above, could have come down a steep foot-
path and across the rear garden. This would be the garden where, as he told
Coleridge, "I have removed my furniture into the garden amidst the straw-
berries and am now writing under the shade of an apple tree. Thus I begin to
claim relationship with nature." [37]

We have from Beddoes, in his 'Notice of some Observations made at the
Medical Pneumatic Institution' which appeared as early as 1799, some indication of the medical side of the Institution. He could not, he said, receive more
than 8–10 patients in the house. He referred to an apparently large, though
unspecified, number of 'out patients' – "invalid paupers – who afforded an
opportunity of trying the effects of digitalis, and other substances which
we supposed might possess similar virtue, and on a very extensive scale in
consumption, and of verifying, perhaps of essentially improving the new
treatment of syphilis, it constantly offered us the choice of patients, who
could have no hope from common remedies and by consequence might be fit
subjects for factitious airs." [38] There is, unfortunately, no further record or
description of these out-patients and what does remain a little puzzling about
the house is how Beddoes and his assistants contrived to deal with this large
number.

The writings leading up to the establishment of the Pneumatic Institute
largely dealt with cases treated by the use of various gases but Beddoes had
considered how the oxygen level might be altered by other means, such as
diet, injection or even poultices. His concentration on respiration and his
observation of the suffering of the many victims of consumption led him
to put first that form of tuberculosis and to leave in second place his own
theory, following on Mayow's, that oxygen was for some reason essential for
muscular activity. The early results of trials of gases altered the whole course
of his plan. Beddoes had at first intended to test a variety of gases; he had
originally good hopes of hydrogen, but very quickly he identified nitrous

oxide as the most interesting of the gases. Twenty four volunteers breathed the gas and what most impressed Beddoes was the involuntary movements of the subjects. He had to reconsider the medical use to which the gas might be put. He also came to see that he might not have gone so far as he had hoped towards the practical use of gas in medical treatment. He realised, possibly with some regret, that they were still at the stage of fundamental research, as his firm statement at the opening of his account of their work shows; they were, he wrote,

applying chemistry to the elucidation of animal nature, principally by pursuing the connections between the properties of elastic fluids and the condition of life.[39]

The man body of the 'Notice' describes the effects of breathing this gas on the twenty four volunteers, Davy and Beddoes himself included:

after the first moments of surprise it was impossible not to recognise the expressions of the most extatic pleasure. I find it entirely out of my power to paint the appearances such as they exhibited themselves to me. I saw and heard, shouting, leaping, running and other gestures, which may be supposed to be exhibited by a person who gives full loose to feelings, excited by a piece of joyful and unlooked for news.[40]

Now he noticed that no weariness or languour followed the breathing of this gas and that only one subject — a hysterical young lady who for weeks afterwards suffered from a series of fits — felt any ill effects. Beddoes accordingly decided to try it for a new purpose:

the consideration of the whole of the phenomena emboldened us to attempt the cure of the palsy and diseases proceeding from a defect of nervous energy.[41]

After such a long period of preparation costing much time and effort, and after the hopes raised by the early trials, the dream of curing consumption was to all intents and purposes abandoned. Beddoes' description of the work at the Institute ended defensively:

I easily gave up my first hypothesis respecting consumption; but what is of principle importance, I have reason to think more favourably than ever of certain gases or vapours in this disease. By combining their application to that of the foxglove and analogous remedies, I am mistaken if multitudes will not be preserved from premature death.

Between these statements comes the series of descriptions of the effect of breathing nitrous oxide as experienced by Beddoes' friends. The accounts are often lively; Beddoes described himself as "bathed all over with a bucketful of good humour", and wrote that "Mrs Beddoes had frequently seemed to be ascending like a balloon up the hill to Clifton." Beddoes did recognise that it is necessary to take precaution against "the delusions of imagination" but the whole account is remarkable for its unscientific nature. The reports of the three paralysed patients who had shown improvement are brief. There are no objective physical tests of improvement and no repeat examinations after a passage of time, though he does suggest the desirability of experiments in which "animals shall be subjected to long processes". It is difficult to understand why Beddoes rushed such an account into publication; possibly he was provoked by prejudiced criticism from those he describes as, "reptiles that plant themselves on the high road of improvement try to hiss back all who would advance". Yet as Director of the Institute he needed to produce at as early a stage as possible a report of the work being done. He many have hoped that it was only an interim report. Once more, he spoiled a worthwhile case by lack of restraint.

Beddoes was able to report that, with the approval of the Committee, he had secured for the Institute "a superintendant equal to my wishes and superior to my hopes". Davy certainly entered into the spirit of Beddoes' "scheme of pure scientific medical investigation". His first task was to find the best means of producing nitrous oxide and to test the effect of breathing the gas, both mixed with air and in a pure state. Once he had found that pure nitrous oxide was safe, the next step was to test the quantity of gas that could be tolerated. The decisive experiments were made in April 1799 and on the 17th came their climax when, with Dr Beddoes watching, Davy breathed four quarts of the gas using a silk bag and closing his nose. He continued to experiment with other gases, some dangerous, throughout the summer and autumn. The details of these experiments [42] belong more to Davy's story than to Beddoes', but the last is particularly memorable. For this, Davy needed James Watt's "Breathing Chamber . . . like a sedan chair" into which he could be shut and where he could take his temperature and pulse. Watt was not without misgivings. He wrote to Beddoes about the plan and warned Davy to "take care your mixtures be not too strong". Davy carried on with the plan and after his experiences on 26th December wrote a classic description of the sensations:

I lost all connection with external things, traces of vivid images rapidly passed through my mind and were connected with words in such a way as to produce perceptions perfectly novel. I existed in a world of new connected and newly modified ideas. I theorized, I imagined that I made discoveries. When I was awakened from this semi-delirious trance ... I exclaimed 'Nothing exists but thoughts'. The universe is composed of impressions, ideas, pleasure and plain.[43]

He realised the significance of the work and summarised it: "We are discovering a new science, the science of corpuscular motion".

But Davy did not confine himself to observing the experiences of members of Beddoes' circle and recording their reactions when they breathed nitrous oxide. The move to Bristol gave him a laboratory worthy of his talents. He had apparatus better than any he had known previously. He must have benefited from working in an experimental institution organised for research with an experienced chemist whom he could consult, though from the beginning Davy was left responsible for the scientific work. He set about a systematic study of the various compounds of nitrogen and oxygen and of the properties of nitrous oxide. He devised his own experiments and repeated carefully those of earlier chemists, using quantitative methods in both synthesis and analysis. This work he published independently in January 1800 as, 'Researches chemical and philosophical chiefly concerning nitrous oxide, or dephlogisticated nitrous air, and its respiration'. This was the publication which brought him immediate reputation as a chemist, cancelling the bad impression which had been made by the over-hasty inclusion of the early work in 'Contributions to Physical and Medical knowledge principally from the West of England' edited by Beddoes.[44] Davy stressed that this new work was "written throughout on the strictly inductive plan, with a total rejection of ... abstract speculation or hasty generalisation". It was probably to ensure that this character of the work should be recognised, that Davy allowed it to appear without any reference to Dr Beddoes or to the Pneumatic Institute. It is easier to sympathise when we remember Davy's consciousness of his own powers. A sense of the shadow cast on the Pneumatic Institute by political prejudice may also have been in his mind: he knew, when he told her of his new appointment in London, that his mother would be glad to see him "getting amongst the Royalists".[45] In a dedication to Dr Beddoes which was never published he asked that his 'Researches' would be received "as pledges of more important labours".

There were four parts to 'Researches': the first two entirely an account of the chemical study of the gases, the third on respiration by animals and the fourth the accounts of respiration by humans. This last part, of course, is a parallel account to the one given by Beddoes in his 'Notice'. Even here

Davy's restraint; his methodical approach and his concentration on the work
in hand unmixed with polemic, contrast with Beddoes' descriptions. They
had as subjects for the breathing of the gas a quite unusual group of men and
women. As is well known, the poets Coleridge and Southey were among
them. We can compare two accounts which Southey gave of his experience.
To Thomas Southey he wrote on July 12th: "O Tom such a gas has Davy
discovered, the gaseous oxyde! O Tom! I have had some of it, it made me
laugh and tingle in every finger tip . . . I am sure the air in heaven must be
this wonder working gas of delight". For Davy's record of the experiments he
described his experience:

In breathing the nitrous oxide, I could not distinguish between the first
feelings it occasioned and an apprehension of which I was unable to divest
myself. My first definite sensation was a dizziness. a fulness in the head, such
as to induce a fear of falling.This was momentary. When I took the bag from
my mouth, I immediately laughed. The laugh was involuntary but highly
pleasurable, accompanied by a thrill all through me; and a tingling in my toes
and fingers, a sensation perfectly new and delightful. I felt a fulness in my
chest afterwards; and during the remainder of the day, imagined that my
taste and hearing were more than commonly quick. Certain I am that I felt
myself more than usually strong and chearful.
    In a second trial. by continuing the inhalation longer, I felt a thrill in my
teeth; and breathing still longer the third time, became so full of strength
as to be compelled to exercise my arms and feet.
    Now after an interval of some months, during which my health has been
materially impaired, the nitrous oxide produces an effect upon me totally
different. Half the quantity affects me . . . .

The difference in style suggests strongly that Beddoes' friends, the subjects
who breathed the gas, clearly distinguished between serious experiment and
light-hearted amusement, while Beddoes' critics chose to ignore the scientific
trials. Coleridge's report was detailed and well-considered:

The first time I inspired the nitrous oxide, I felt an highly pleasurable sensation of warmth over my whole frame, resembling that which I remember once
to have experienced after returning from a walk in the snow into a warm
room. The only motion which I felt inclined to make, was that of laughing
at those who were looking at me. My eyes felt distended, and towards the
last, my heart beat as if it were leaping up and down. On removing the
mouthpiece the whole sensation went off almost instantly.
    The second time, I felt the same pleasurable sensation of warmth, but not
I think, in quite so great a degree. I wished to know what effect it would have
on my impressions; I fixed my eye on some trees in the distance, but I did
not find any other effect except that they became dimmer and dimmer, and

looked at last as if I had seen them through tears. My heart beat more violently than the first time. This was after a hearty dinner.

The third time I was more violently acted on than in the two former. Towards the last, I could not avoid, nor indeed felt any wish to avoid, beating the ground with my feet; and after the mouth-piece was removed, I remained for a few seconds motionless, in great extacy.

The fourth time was immediately after breakfast. The few first inspirations affected me so little that I thought Mr Davy had given me atmospheric air: but soon felt the warmth beginning about my chest, and spreading upward and downward, so that I could feel its progress over my whole frame. My heart did not beat so violently; my sensations were highly pleasurable, not so intense or apparently local, but of more unmingled pleasure than I had ever before experienced.

Beddoes' medical colleagues gave very professional reports and others as objective as Josiah and Thomas Wedgwood and John Rickman took part. Since their letters are very characteristic of the writers it is a surprise to find the practical Thomas Poole reporting "all the faculties absorbed by the fine pleasing feelings of existence without consciousness". Not that Davy fails to convey the excitement of the whole episode, especially when he describes his own near-fatal experience – which ended with his often-quoted gasp, "I do not think I shall die". Both Beddoes, in his 'Notice of some observations . . .', and Davy in 'Researches' agree in stressing the intensely agreeable sensation which breathing nitrous oxide gave. Beddoes suggested that here they might have found a new source of pleasure for humankind and Davy admitted that he breathed the gas every day for a considerable period just for the sensation of being filled "with inward transports" and "clad with a new born mightiness around". In our own day in the history of LSD, we have seen the same association of serious study and indulgence. That the gas could induce short periods of loss of consciousness was also noticed by both. Beddoes wrote of the children's author, Mrs Barbauld, experiencing faintness and even of Dr Kinglake, who should have been a calmer subject, falling into a "perfect trance for about a minute". Davy alone recorded the analgesic property of the gas and described how he ceased to feel the pain of his tooth-ache. As part of the published report of his work on nitrous oxide, Davy unequivocally pointed to the possible use of the gas in surgery. This was not put forward as part of his series of experiments but appears almost as a footnote. After posing the question, "Does not sensibility more immediately depend on respiration?", he put forward the suggestion that "As nitrous oxide in its extensive operation appears capable of destroying physical pain, it may probably be used with advantage during surgical operations in which no great effusion of blood takes place". The reservation must have arisen from Beddoes' theory on which the work at the Pneumatic Institute rested: that nitrous oxide was a means of "lowering" the oxygen in the blood. Davy, it is obvious, had both the experimental skill and the determination to work to a rigorous standard which would win him respect in the scientific world. He established without doubt the independence of his work. Though his success showed up vividly the shortcomings of Beddoes' own work once the Institute had been set up, it is also clear that there was an intimate connection between Davy's work and Beddoes' thinking and that the Institute's original scientific aims were not abandoned.

Though contemporary opinion concentrated on the ludicrous aspects of "Pneumatic Revelry", these observations of unconsciousness and more especially of anaesthesia remain today as the most teasing and fascinating part of the story of the Pneumatic Institute. The question that will not go away is why neither Beddoes nor Davy followed up this discovery. There is too the wish to tidy up the history of anaesthesia by establishing a link between Davy's work and Hickman's experiments in 1824 in anaesthetising animals by using gas. This second question was being asked as early as 1856 when Dr John Davy, Humphry's brother and biographer, was consulted about the possibility that "Dr Hickman's researches originated and were acknowledged to have originated in the suggestion" from Humphry Davy. It still remains open.[46] Beddoes, faced with similar speculation about the discovery of oxygen, was quite sure that it was possible for more than one man to make the same discovery independently and it may be necessary to be content with this as an explanation.

When Davy left Bristol early in 1801 to become an assistant lecturer at the Royal Institution in London, he was expected to give public lectures and carry out research; he had, as well as to make a good first impression, to work to increase the prestige of the Royal Institution as a body dedicated to applying science to practical ends. The considerable programme of research he undertook in addition to duties in the laboratory and in editing the *Journal*, was quite sufficient to absorb his attention. Beddoes as a doctor, with a vivid awareness of the horrors [47] surgical patients had to face, might more than Davy have been expected to follow up this discovery of the anaesthetic properties of the gas. He had after all, with Watt's help, overcome the preliminary difficulties and found a means of administering gas. But Beddoes too was busy; in addition to a considerable medical practice he had the problem of how to carry on the Pneumatic Institute. He was generous in releasing Davy to accept Count Rumford's [48] invitation to go to London and he remained interested in Davy's work, but his "scientific medical investigation" had not produced the results he had hoped for and Davy's departure must have left him despondent. Just at the time when, in general, enthusiasm for pneumatic medicine was waning, Davy's careful work showed what would be involved in continuing the study. Without Davy it could scarcely be within Beddoes' powers to continue the scientific activities of the Institute.

Beddoes had other reasons for anxiety. Miss M_____'s experience had been alarming. Breathing the gas had induced a fit and for several weeks afterwards the young "martyr to this course of experiments", as Beddoes himself called her, had had recurrent attacks. She had needed constant care and treatment with bark [49] and opium. By contrast, many found breathing nitrous oxide such an ecstatic experience that they came forward, as Davy records, merely for pleasure. For a short while breathing the gas became a craze in certain circles in Bristol. There were ludicrous incidents: Cottle remembered − and even seems to relish the memory − a young lady who dashed out of the house and ran along the street leaping over a dog as she went. The *Anti-Jacobin* published verses ridiculing the adventures at Clifton and when Davy later repeated the experience of breathing nitrous oxide in his early lectures at the Royal Institution, the laboratory, the fashionable amateur scientists and the ludicrous aspects of such experiments were the subject of Gillray's cartoon, 'Scientific Researches! New Discoveries in Pneumaticks − or − an Experimental Lecture on the Powers of Air'. On the other hand, Beddoes was accused of being a heartless experimenter. When he planned to leave Oxford, Beddoes' reputation was high. As one who, by his 'Memorial', had helped to bring it about, he must have gained in prestige and standing if a chair of chemistry had come into being in Oxford, even though he probably would not have wished to hold it himself. All this was lost when he became marked as a seditious person. The idea of the chair was abandoned and he went into the wilderness. Beddoes had courage, as his political activities in Bristol show, but, in spite of his scorn for the medical establishment, he may well have been reluctant to jeopardise his second career. He cannot have failed to recognise the political thrust behind the satires produced by the propagandists of reaction which gave widespread publicity to the comic and frivolous side of his venture. For humane reasons too, he must have hesitated, fearing the possibility of some tragic accident. The risks of continuing were too great.

By the summer of 1802 there was also a hard, practical reason for giving up; as Southey described: "The Pneumatic Institution continues. The name should be changed as they do little with gases, on account chiefly of the expense of experiments. Beddoes now chiefly supports it".[50] Southey was writing to his friend John Rickman the statistician who was at work on the first Census. The letter was primarily to introduce Thomas Poole, forewarning that Davy would probably bring him to Rickman who had certainly met Davy at the Pneumatic Institute, and recorded for him his experiences in breathing nitrous oxide. Poole was to become a most valuable assistant to Rickman in recording the census material; so here, even in the last days of the Pneumatic Institute we have another instance of the value of the informal network around Beddoes.

This part of the story is always tinged with regret that it is not possible to attribute the effective discovery of a useful anaesthetic to the work done at the Pneumatic Institute and, as a sort of consolation prize, the discovery of Davy is produced as Beddoes' most important achievement. Clearly, Davies Giddy was the first to have noticed Davy's promise; but Beddoes' well-equipped laboratory with the means at hand of producing the gases he wished to study, made possible Davy's work on nitrous oxide. There was even more for him in Bristol than an improvement in material circumstances. Friendship with a man who was among the leading chemists in England and who was conversant with continental work must have encouraged Davy and broadened his view. The scheme of classification of chemical substances which Beddoes published in 'Contributions to Medical and Physical Knowledge from the West of England' (1799), rounds off our knowledge of his work as a chemist and makes clear how much Davy had to learn from him. He had prepared the scheme to illustrate his public lectures and divided bodies according to their reactions towards oxygen. Davy's hasty speculations and Beddoes' over-enthusiastic adoption of his ideas have become the best remembered parts of 'Contributions ...' and have contributed much to Beddoes' reputation for over-hasty publication but the division of the bodies into groups able and not able to combine with oxygen is particularly interesting. Among those unable to combine, Beddoes put the alkalies and alkaline earths and silica. An early appreciation[51] of Beddoes' work as a chemist suggests that:

He appears to have been the first to recognise the similarity between the alkalies and alkaline earths, for he classifies them together under a general heading of alkalies, an advance upon Black's theory. Another theory that he puts clearly, although only as a speculation, is that the alkalies may be unable to combine with oxygen because of 'some peculiarity of the union of their elements or because they *are already combined with oxygen*'. This was certainly a speculation in advance of contemporary thought and one that foreshadowed his pupil's (Davy) future decomposition of the alkalies. Another theory yet more advanced is (p. 223), 'The existence of iron in such variety of plants and animals; and of manganese in some plants, suggests an opinion that these metals are compounded by the organic powers'.

This reminder of Beddoes' achievements as a chemist draws attention to the range of work he envisaged for the Pneumatic Institute. There were other young men working there in the early days, while Davy was investigating gases. Dr John King, and for a short while Dr Peter Mark Roget, conducted physiological investigations and Dr William Clayfield produced an improved breathing apparatus. There was, during Davy's time in Bristol, a new and exciting topic of scientific interest: Volta's discovery of the galvanic pile. Beddoes had already in the *Monthly Review* written on Galvani's work and on 'animal electricity'; he might himself have suggested, even as far back as his chemical lectures in Oxford, that strong forces of electricity might be used to split compounds.[52] Now it was in Beddoes' laboratory that Davy began experiments with electricity. Galvanism was the subject Davy chose for his first and dramatically successful course of lectures when he arrived at the Royal Institution.

There was one other respect in which Beddoes may have helped Davy. The application of chemistry to industry had always interested Beddoes. Though since leaving Coalbrookdale he had taken no active part in such work, he was well informed about what was being done and in particular his reviews of *Annales de Chimie* had made him recognise how the pressures of war had stimulated applied chemistry in France. The Royal Institution had been set up in London for the purpose of assisting in the application of chemistry to the needs of industry and agriculture. The sort of topic that interested Rumford and the Managers of the Institution could very likely have been under discussion at Clifton. Davy's first reasearch assignment, after his opening lectures, was to investigate the chemistry of tanning. Whether or not Beddoes had ever revealed a particular interest in this, his father's occupation, Davy had certainly met at the laboratory Thomas Poole who had a tannery at near-by Nether Stowey and who was very active in the search for means of improving the tanning process. Poole, who befriended Coleridge, was well placed to help Davy with information and introductions. It was not only by providing a laboratory that Beddoes encouraged Davy's development as a scientist.

Chemistry and medicine were for Beddoes not simply intellectual discipline. He was convinced that the discoveries being made in chemistry would lead to a new social order. Science, freed from misuse when despotic governments were finally overthrown, could be used for the benefit of all men.

Above all, men would be liberated as they came to understand the laws of their physical being. Beddoes must have had something of the spark of poetry in him, not so much in the verses he wrote as in his nature. In many ways a man of the late eighteenth century, for a short while as that century ended his home and his enthusiasms encouraged the very spirit of romanticism. He gave encouragement and support to the idealistic young Coleridge during his early days in Bristol. Then his home where science and poetry met, made possible the vivid and creative friendships between the young men who gathered there. It was fortunate that Davy was at first welcomed as a fellow poet. Southey encouraged him to write, published his poems in The Annual Anthology in 1799 and urged him, "Do not lose the habit and feeling of seeing all things with a poet's eye".[53] When Southey visited the Pneumatic Institute Davy "had to tell of some new experiment or discovery and of the views which it opened for him"; and when Davy returned the visit he would find Southey had "a fresh portion of Madoc for his hearing". The meetings gave each of them a chance to tell of his progress and Southey described the mood of elation, "The bag of nitrous oxide with which he generally regaled me", Southey writes, "was not required for raising my spirits". Later, when he left for Portugal, Southey asked Davy to revise and publish his 'Thalaba'. But for Coleridge the meeting with the young poet/chemist was an awakening as he felt the excitement of the new discoveries in chemistry and the unity of science and poetry. Davy was already aware of the philosophic problem of putting experience into words: he wrote, "In theorising we make use of the terms connected with feelings our ideas are indistinct and placed together solely for the purpose of examining their relations". And later, "the use of physical science is that it gives definite ideas".

When he returned from Germany, Coleridge took part with enthusiasm in the new activities at Beddoes' home. This is the time of the Notebook entries on subjects that might concern Davy, such as the medical school at Edinburgh, and it shows Coleridge following Beddoes' work, his translation of Brown's 'Elements'. More importantly, he responded to Davy's thoughts; a letter about the publication of Davy's 'Researches' by Longman passes on to the underlying significance:

I wish in your researches that you and Beddoes would give a compressed history of Human Mind for the last century, considered only as the acquisition of ideas or new arrangement of them. Or if you won't do it there do it for me and I will print it with an Essay I am now writing on the Principles of Population and Progress.[54]

After the mind-expanding experience of breathing nitrous oxide — "a more unmingled pleasure than I had ever before experienced" — Coleridge proposed "to attack chemistry like a shark". His enthusiasm continued after Davy was established in London. He attended Davy's [55] lectures, noting accurately the colours of the flames in the experiments and the producing of "a spark with the electric machines", for he hoped to find in the experience a new source of imagery. Davy exerted himself to encourage Coleridge to work entirely at his poetry and to abandon the idea of translating from German, urging, "You were born to connect man with nature by the intermediate links of harmonious sounds and to teach him to disconnect his feelings from unmeaning words."[56] It was Coleridge who drew Wordsworth's interest in the association of ideas, feeling and pain to Beddoes' attention and who consulted Davy about the laboratory that Calvert, Wordsworth's friend and neighbour, proposed to build. Chemistry appealed to Wordsworth [57] as an intellectual pursuit not closely connected with deep passion but Coleridge wrote to Davy that he was attracted to this study, "for its own sake and in no small degree likewise my beloved friend that I may be able to sympathise with all that you do and think". Yet the most objective and accessible evidence of the spark leaping from science to poetry, comes to us not in the published work of Coleridge but in the writing of Wordsworth who, during his stay in Somerset, had remained in seclusion and had never met Davy. Coleridge had asked Davy to help him in preparing the second edition of *Lyrical Ballads* and his request was quickly followed by a formal letter from Wordsworth that Davy should undertake the necessary editorial work on the manuscripts and proof sheets. In the exchange of ideas that must have followed, Davy and Wordsworth discovered that rather than standing opposed to one another as scientist and poet, as Wordsworth felt when he wrote his preface to the first edition of *Lyrical Ballads*, they were engaged on the same search.

And thus the Poet, prompted by this feeling of pleasure, which accompanies him through the whole course of his studies, converses with general nature, with affections akin to those, which, through labour and length of time, The Man of science has raised up in himself, by conversing with those particular parts of nature which are the objects of his studies. (*See Note 57.*)

This sense of shared feeling described in the Preface to the second edition led Wordsworth to look forward to a time when the discoveries of the scientist would be "proper objects of the Poet's art".

When we try to think of what might be the continuing results of the Pneumatic Institute and of the friendships made there, Beddoes' trial of pneumatic medicine; his efforts to produce, test, and regulate in a scientific manner in the laboratory, the gases he hoped to use; together with his collaboration with James Watt in the production of practical breathing apparatus, all must excite admiration. This work at the Pneumatic Institute was an attempt to put into action particular chemical and medical theories yet, since the key facts about the causation of disease were not known, it was bound to fail and to survive only as a historical curiosity. Beddoes saw that he had failed. It is sad that he can hardly have known what his achievement was. Some of the friendships made at Bristol were enduring, with Davy as the constant figure: Davy and Coleridge; Davy and King; Davy and Southey; Davy and Thomas Poole. Most significant of all was Beddoes' own friendship with Coleridge and the opening up for him of new intellectual interests. The sustaining of this circle of friends, making opportunity for growth and exchange of ideas, was Beddoes' most important achievement. Such an ebullient figure can hardly be said to remain in the background, but in these two years his greatest achievement was — being a chemist he might appreciate it — as a catalyst.

CHAPTER 10

PREVENTIVE MEDICINE

Dr Beddoes must have enjoyed these few years, when his home was the meeting place for such a lively group of young men and when he was able to help and encourage them. It is true that these are the years for which he is most often remembered and in which he achieved something of lasting influence; and it is easy to let the breathing of nitrous oxide dominate the story. The youth and subsequent fame of some of those involved and the excitement when they found that being subjects in a serious, scientific experiment led to such an unexpected and thrilling pleasure have always been the most memorable and attractive part of the story of the Pneumatic Institute. There is romance of a different kind in Davy's intense ten months of effort when he produced the work that won him recognition as an experimental chemist; one of the rare occasions when genius and opportunity met. It is easy to let these dramas overshadow the scientific work which Beddoes did in preparation; his work as a chemist and his studies of gases would alone have been valuable to Davy as preparation for his own work. In medicine Beddoes' efforts may seem to have led him into a cul-de-sac; but even here his thinking went beyond the limited purpose of using gases in the treatment of diseases. He understood the basic principle that it was possible to give medicine by inhalation rather than by the stomach or by other means. Beddoes would not have claimed, as his description of cases contributed to "Considerations on ... Factitious Airs" makes clear, to have been the first to try this method but he became its most systematic and respected advocate. Dr Cartwright draws attention to this:

Beddoes was the leader of the school of medical men who experimented with the method of applying medication to the bodily organs by means of the lungs; and this is the essence of inhalational anaesthesia. Again and again Beddoes makes it clear that he had grasped this idea, that he was using the lungs as a more convenient and more easily controllable channel than the stomach.[1]

Here is the medical parallel to Davy's observations of the physiological effects of breathing gases, and it is a reminder that Beddoes was as skilful and independent as a doctor as he was as a chemist. His reputation was sufficient

175

for men in public life to put themselves under his care even during the time of the experiment with pneumatic medicine. John Wedgwood,[2] writing in 1800 to his brother Tom who had just left for the West Indies, told how Sir James Mackintosh had sent a paralytic friend to Clifton "to consult Dr Beddoes" and how "Mr Bollingsbey, the author of the survey of Somersetshire who has been dangerously ill of consumption is very rapidly recovering his health by use of foxglove". In the later history of the Institute, it is important to remember that by the time he had brought his 'Pneumatic Establishment' from being a dream into operation, Beddoes had been in practice as a doctor for more than five years. Clearly he now felt a strong sense of responsibility as a medical man; he could not run the Pneumatic Institute merely as a research establishment but must do his best for his patients. He had fore-seen that he would need to use existing methods of treatment, but in the event, the balance between orthodox and experimental methods was not as expected. "In the very infancy of the establishment", according to Stock, "the gases were administered in a smaller proportion of cases than might have been expected . . . they were comparatively rarely resorted to. Some remedies, however, to which the attention of the medical world was only then beginning to be particularly directed were . . . fairly tried and upon a very extensive scale".[3] Beddoes later categorically denied that any "aerial medicine" had "ever been tried in syphilitic cases at the Pneumatic Institute".

At the Pneumatic Institute, Beddoes had at last created the conditions where he could put into practice his belief that "experiment gives permanence to what is discovered empirically"; yet, once Davy had arrived, Beddoes seems to have left his young laboratory superintendent to continue the investigation of gases and to have devoted himself to the more medical aspects of the work of the Pneumatic Institute. During these years his publications show him advocating a method of attacking disease which was, for its time, as modern as was his interest in pneumatic medicine. He saw that the careful, systematic and long-term compilation of data could be the base from which to work in medicine and between 1797 and 1800 he worked in this way in relation to two major conditions: syphilis and pulmonary tuberculosis. Beddoes had such high hopes of treatment with gases that he never seems to have realised fully the significance of this second approach, though it must have occupied much of his time and attention. Numerical analysis had begun in the seventeenth century with Sir William Petty's 'Observations upon Bills of Mortality', when life expectation and the operation of chance had been subjects of study. In 1761, with the publication of J. P. Süssmilch's 'The divine ordinance', the numerical study of population began, even though

Süssmilch's object had been philosophical — to use "the constancy of the numerical relations of the vital state" to demonstrate a divine plan. By the time Beddoes came to work seriously on the collection and classification of facts which would throw light on consumption and syphilis, he had access to the outstanding collection of social data made by Sir John Sinclair[4] for the Statistical Survey of Scotland which appeared in a series of volumes between 1791 and 1799. Beddoes was obviously familiar with this work and with Sinclair's methods. He most probably also knew John Rickman's preliminary work, begun in 1800, for the first Census, especially as Thomas Poole was invited by Rickman to help with the returns. In spite of his obvious interest and his application of these methods to medical problems, there is no indication that Beddoes foresaw that this was, in the future, to be a major way of attacking disease. The drama, sometimes the tragi-comedy, of the Pneumatic Experiment has diverted attention from Beddoes as one among the pioneers of the method of the systematic survey.

The history of the investigation of a new treatment for syphilis begins in 1795 with a report in Part III of 'Considerations on the Medical use and production of Factitious Airs'. The description of the use of oxygen gas in the treatment of this disease was one of many diverse reports brought forward by Beddoes to support his proposals. It was sent by Dr Helenus Scott, one of the East India Company's doctors, from Bombay. Two years later, Beddoes published a collection of fifty case studies[5] of syphilis, put together by Mr Hammick of the Royal Naval Hospital at Plymouth. After reading Parts IV and V of 'Considerations', Mr Hammick had treated his sailor patients with nitric acid conveyed in a lemon-flavoured syrup. The distressing side-effects of treatment with mercury, the traditional medicine, and the low rate of cure led to his readiness to try the new remedy. Although he could report both apparent cures and absence of side-effects, Mr Hammick felt it necessary to warn that "Future experience must determine whether the cures wrought by nitric acid will be permanent or not." Beddoes follows these accounts with an appeal to the medical profession. His aim was to "induce a large part of the medical profession to unite in an enquiry of the highest interest." In this way evidence could be quickly collected and he hoped for reports of a thousand cases. He emphasised the need for such evidence and set out precisely the form the reports were to take in order to provide valid material. They were to give date; symptoms; note on the permanence of the cure; a description of any symptoms which might lead to knowledge of the cause of failure. Most important: these cases were to be recorded separately from those given other treatment. He ended:

It is desired that all technical terms and quantities may be written at length and the whole in a fair hand.

Any profits from the sale of the collection were to be divided among charitable institutions devoted to the care of patients suffering from venereal disease.

Beddoes asked for reports to be sent to him in January 1798 but their publication was delayed until 1799, difficulties in Ireland having held up the forwarding of material from Dublin. The conditions in Ireland so delicately referred to were of course the rebellion of 1798 and the threat that French troops would succeed in coming to aid the Irish. During this period of violence even his wife's home at Edgeworthstown had scarcely escaped attack, even though Edgeworth was a well liked and resident landlord. The accounts in this second collection,[6] 'New reports concerning nitrous acid in the venereal disease', are much fuller. They show that besides giving nitrous acid as a medicine, some doctors had used it in the same strength solution to bathe limbs; in a compress and even as a bath. The first letters are from Dr Helenus Scott; Beddoes omits the letter No. I as having already been published in 'Considerations'. Dr Scott's contributions are of special interest for they make clear the connection between this treatment and the pneumatic medicine. Scott believed that the effective principle was oxygen; this was also Davy's belief. Consequently, Dr. Scott tried other gases and other ways of "oxygenating the body", for example he gave "black calx of manganese" combined with a "drink well acidulated with acids of vitriol or nitre". This was in the hope of producing oxygen in the digestive tract. He had to report this treatment was unsuccessful. Difficulties over supplies of nitre by the East India Company had hampered Dr Scott in his efforts to obtain pure oxygen and he hoped "The patriotic Dr Beddoes has now put it in our power, to get without trouble this aeriform substance". He was quite clear that what they were working on was the improvement of "the art of healing . . . by the beautiful discoveries of chemistry".

Beddoes re-issued in the 1799 volume of 'Reports' his earlier appeal for the collection of information. Then in 1800 he published his last set of 'Communications respecting the use of nitrous acid . . .'. This was a more formal work, opening with a 'Preliminary Discourse' by Beddoes. Here we find his familiar attack on the conservatism and self-seeking of the members of the medical establishment. He longs for a "condition of society [which would] permit medicine to be zealously practised under an exemption from immediate mercantile principles".[7] The interest of this work lies not in the collection of further cases, though these descriptions do give a vivid and pitiable account of the effects of this cruel disease; it is to be found rather in the evidence it gives that Beddoes considered carefully his criteria for records. In this respect it is quite unlike his vagueness in the 'Notice of some observations made at the Medical Pneumatic Institute'. He suggested work in military hospitals where the giving of medicine would be under responsible observation and pointed out that military surgeons had the authority to continue to inspect patients and to collect a large number of cases. He stressed the form the observations should take; the need to keep the experimental treatment uninvolved with other treatment, e.g., with mercury. The spoken word he dismissed as "too fugitive" for reports and he underlined the need for accurate, persistent observation; he warned that reports of such observation should not be general and unqualified and positive until there had been many experiments, over a long time and confirmed by being repeated independently.

The reports in these publications refer both to nitric and nitrous acid and reveal that there was clearly a struggle with practical difficulties. Dr Scott devised a method of giving the syrup by means of a tube through which the patient sucked. Some patients found this tedious and tried to discontinue the treatment. Dr Scott thought of making a paste of nitric acid and earth of alum — this still made the patient's mouth sore. Beddoes treated many cases of syphilis and clearly worked with determination to find a cure for this terrible disease in the face of indifference and hostility. With his usual over-optimism he hoped that a treatment once found would be useful in other conditions; he names scurvy, dysentery, and other forms of ulcers. To us these are illnesses with quite different origins but when we remember that Beddoes connected them all with a deficiency of oxygen his thinking does not seem so foolish. Here the perseverance which he showed in preparing the way for the Pneumatic Institute, which has been overshadowed by the unprofessional excitements that were connected with it, is clear. So too is his appreciation of the careful organisation of any investigation. And this work to find a remedy for a disease which could not excite the same widespread sympathy as consumption is a reminder of the sincerity of his conviction that a doctor's compassion is better exercised in efforts to alleviate general suffering than in pity for individual cases. He saw the progress being made by the military and naval surgeons and the desirability of applying their methods to civilian problems. At the time there was no organisation to whcih he could turn and he still had to face the task of collecting data himself. Once more he was defeated, but the interest of the attempt lies in the method he advocated. The methods he used extend our understanding of Beddoes' character and work.

This same approach to the solution of a medical problem is the hallmark of Beddoes' 'Essay on the Causes, Early Signs, and Prevention of Pulmonary Consumption' which was published in the same year as 'Observations made at the Medical Pneumatic Institute'.

The survey method is here used to make a "map" of consumption. By collecting and classifying facts Beddoes intended to gain a clearer picture of the disease's causes [8] and nature and, more important still, to establish what he calls "a doctrine of exemption [which] might furnish something useful by way of a moral". In the whole of this 'Essay' Beddoes makes only one slight reference to treatment with gases, and then only to admit that he had little hope of it:

It becomes me to acknowledge that the trials that have been made of factitious airs and vapours, seem as yet, very far from having produced anything like a successful method of treating consumption. The utmost that can be said is, that certain vapours and stabling the patient with cows are not unpromising when the method is tried on a proper scale. [9]

He turned to treatment with digitalis, even though he realised that it could be a dangerous medicine, and he ended the 'Essay' by stressing that the only real hope lay in prevention. This striking difference between the 'Observations' and the 'Essay' springs in part from their being written for different readerships. The 'Essay', unlike the 'Observations', was not for medical men but was intended "for the use of parents and preceptors". At the same time it goes far beyond homely advice.

He first sets out figures [10] to show the incidence of the disease. To collect these, Beddoes employed the same method Sir John Sinclair was using in his pioneer statistical survey of Scotland; this was to ask the clergy for an account of their parishes. In 1798 Beddoes had written [11] to Sinclair expressing admiration of his work and asking if he might be allowed to see a copy of the 20th volume in advance of publication, and had quickly been sent the copy. Beddoes was so impressed by Sinclair's work that he considered spending a summer in Scotland. He seems not to have made the tour but to have concentrated on work nearer home. From one Shrewsbury parish, Beddoes was sent, from the church register, the number of deaths and their causes and in Bristol he found a "pastor" willing to inquire from house to house in his parish. This helper estimated that the population was about

10 000 and he was able to produce figures for the years 1790–1796. These two reports revealed that "consumption or decline" accounted for from a quarter to half the deaths. Beddoes drew attention to several possible sources of inaccuracy but even so these figures emphasise the need for such an inquiry.

Beddoes had clearly been to much trouble to assemble evidence for any possible connection between pulmonary tuberculosis and climate. He had reports from British possessions — the West Indies and Bengal — and from favourite resorts of invalids — Egypt, Italy, Madeira and Portugal — and came to the conclusion that a mild climate did little good for either natives or visitors; from his own experience he believed that such benefits as did appear derived more probably from the sea voyage than from the new climate. A less negative result came from his analysis by occupation. He found that butchers, catgut makers, soap boilers and fisherwomen were exempt from consumption and that a whole series of trades where workers were exposed to dust or where the work was sedentary or confined to ill-ventilated premises presented a high risk. Among these he named his particular bête noire, needle grinding, where danger from dust had arisen since the introduction of steam-powered grindstones; the clothing trades; and "fifers on board ship".[12] Needle grinders he could easily have known in Shropshire where from the early days this was a common small-scale industry and the comment on fifers is one of the few details that remind us that the Hotwells was next to the busy port of Bristol. Beddoes ranged widely for his information. Some he collected himself, from the valleys about Rhyader in North Wales and from Ireland; one report came from America and another, perhaps the most interesting, from Peter Mark Roget who contributed a vivid first-hand account of the fishwives of Musselburgh and of the sailors of the east coast of Scotland. Beddoes felt that a much larger body of information needed to be collected. He was cautious about the reason why butchers, etc. seemed to escape consumption; pneumatic theory led to the supposition that they breathed some gas that protected them. He took note of a butcher's wife who was sure she was in no danger:

Lord bless you sir, die of a cough! Why I never heard of such a thing! every one knows 'the smell of meat keeps off infection'.[13]

and who went on to describe how her husband had often "taken sheep into gentlemen's bedrooms". But Beddoes did not fail to notice that both fishermen and butchers could easily have a better-than-average diet. So all he could put forward was that

Certain classes are less liable than others to consumption, either because the exhalations to which they are exposed preserve the lungs in a healthy state, or because they acquire from their mode of life a habit less susceptible of the complaint. [By contrast,] . . . the puny by descent, by diet, by sex, by occupation [are] the most exposed to infection.[14]

These were not the first, nor were they to be the last, occasions on which Beddoes collected statistics, but these two works, on syphilis and consumption, are the most sustained examples of his interest. The praise he gave to Sir John Sinclair's surveys is proof of the value he sets on a well-conducted survey. "We possess in the Statistical reports", he wrote, "a document more precious than can, I believe, be produced concerning any other country," and in commenting on the variable quality of the individual reports collected he observes that "each account is so completely checked by the rest that the causes of the condition of the people are perfectly apparent". As early as 1792, Beddoes had shown interest in statistical analysis of the causes of death; he did not follow this up but the noticeable difference between his attempts at numerical analysis in the years 1795—8 and the work done in the following three years, shows how considerably Beddoes was influenced by Sinclair's Statistical Survey. Unfortunately Beddoes' own work does not give confidence that he was able to work to the standard he saw was needed.

The second half of the 'Essay' is devoted to practical, down to earth advice, particularly relating to the case of young people. In the 'Essay', Beddoes' system has all the more authenticity as, following on from the 'Observations', it can be seen to rest on good evidence. The importance of a nourishing diet, not abandoning vegetables but containing a good proportion of meat; of fresh air; of suitable clothing and of maintaining an equable temperature of 60°F are all clearly explained; this last is supported by an entertaining description of the healthy Dutch women who keep the same temperature indoors and outdoors, whatever the weather, by means of layers of petticoats and the use of little personal warming stoves on which they rest their feet. Again there is a "human touch" when he comes to point out the importance of physical activity; Beddoes shows a real understanding of children, asking that parents and teachers should make good use of young people's natural energy and curiosity, not imposing "exercise" but making activity purposeful. In the end, Beddoes' hopes rest on a better informed public which at last will reject belief in medicinal springs and in useless medicines, "those syrups and balsams which it cannot be certainly denied, that they are infallible in all such bad coughs, as would go off of themselves".[15]

Little if any of this advice is new in Beddoes' writings. Hints of it can be seen in 'The History of Isaac Jenkins' where temperance is urged not on moral grounds but for the practical reason that it would enable Sarah and Isaac to feed their family better and make their cottage warmer and cleaner. A little threepenny pamphlet published in 1792, the same year as 'Isaac Jenkins', makes just the same points. Its language is simple and the illustrations homely, as is promised by the title 'A guide for self preservation and parental affection or plain directions for enabling people to keep themselves or their children free from several common disorders', but the principles seen in the 1799 'Essay' are already there. Beddoes urged the poor parents to whom this early booklet was clearly addressed to get as much knowledge as they could and to encourage in their children a lively interest in natural things. They were to put first the provision of good food and clothing for their family and to understand the importance of a cool, even temperature. As in the later work, he stressed above all the importance of cleanliness and fresh air and felt so keenly about this point that even in this popular writing he produced a little table of figures to show how the death rate in the Dublin Lying-in Hospital was reduced by "a contrivance to change the air in the rooms". After his early observations in the hospital at Dijon Beddoes had, it seems, continued to take an interest in ventilation, a topic which was receiving attention at the end of the eighteenth century. It seems to have waited for the Crimean war, with the hospital at Scutari and the influence of friends near government, for its importance to be recognised.

It is clear that Beddoes attached much importance to these efforts to educate people, both rich and poor, in health matters and to suggesting ways of maintaining good health, for he persisted in them over a long period. His concern took a practical form not long after Davy left Bristol, when the Pneumatic Institute became a Dispensary and Clinic for outpatients. Gradually Dr King took over the duties of doctor there and, as well as treating the patients, kept records and statistics of cases. The value of such dispensaries had been in Beddoes' mind as early as 1791 when he argued against the establishment of an Infirmary in Cornwall. At that time he explained how much more financially practicable a Dispensary would be, in addition to its avoiding the high risk of infection which was ever present in hospital treatment. Then as ever, he was not afraid to affront public opinion, for an Infirmary was a much more traditional charitable cause than the outpatient provision he favoured. In Clifton he was independent. His Clinic had moved to the Quay by 1803 and continued there long after Beddoes' death. Some at least of his contemporaries understood that by this plan

Beddoes was developing his Pneumatic Institute into something different
from an ordinary hospital. The change in organisation was followed by
the publication of 'The Rules of the Medical Institution for the Sick and
Drooping Poor', a handbook for his humblest patients which brought from
the reviewer in *The Medical and Physical Journal* [16] the comment that "There
is something truly original in everything that comes from the hand of Dr
Beddoes". The account emphasised that "the purpose of the Institution is
singular" in that it was set up "for the prevention of some of the most
widely wasting diseases"; and it promised that such a new conception was
bound to have a "formidable" appeal to the imagination. Some hundreds
of patients were treated in Beddoes' own lifetime and the work there may,
in practical terms, have been the most lasting of Beddoes' own achievements.
There is today, no more trace of the Medical Institution in Dowry Square
than of the Pneumatic Institute, though Beddoes might have approved of
the City Day Nursery that now stands on the side of the square to which
it moved. Dr F. F. Cartwright went to much trouble to trace the later history
of the Institution and has shown that its work continued for longer than
might have been supposed:

The branch establishment in Little Tower Court, Broad Quay, was flourishing
as late as 1817, nine years after Beddoes' death; it was still called the Medical
Preventive Institution and its purpose was the same "to prevent disease; and,
when possible to defeat by anticipating its attacks". The Dowry Square
Institution was still extant in 1809; it is possible that in 1812 the buildings
were acquired by the Clifton Dispensary which was founded in that year;
the Clifton Dispensary moved at least once more, and in 1823 found a home
at 13 Dowry Square where it still [in 1952] flourishes. There can be no
absolute certainty, for Dowry Square has been renumbered on more than
one occasion, but the balance of probability lies in the supposition that
we have not only the dead bricks and mortar of Beddoes' Institution as a
memorial to him, but a living successor to his foundation in the shape of
Clifton Dispensary carrying on today as part of its work, those very ideals of
preventive medicine which he almost first of all men preached and practised
a century and a half ago. [17]

How clearly Beddoes understood the problem and the needs of the poor
people for whom the Institution was intended; how well he knew their
anxieties as well as the practical difficulties of their lives is very clear in the
'Rules' of the Medical Institution.

It is a remarkable work. There is no change in the practical measures that
are brought forward but, even more than in his earlier writings, Beddoes

shows a keen sympathy for poor people and understanding of their life. He knew how easy it was to despair when a long course of treatment was needed and to ensure that his patients continued their attendance Beddoes insisted on a deposit of 2/6, held until treatment was finished. The 'Rules' described vividly the need for confidence between patient and doctor:

The experience of former medical charities shows that the sick are constantly flying off before they have a chance of due benefit. I take this to be a want of proper understanding. All has hitherto been conducted in a style of authority. It has been too much mere dumb show between doctor and patient.[18]

This, he realised, was why poor people had confidence in the advice given by their neighbours who "speak to the sick in their own language". In renewing his description of the importance of sensible ways of living, Beddoes introduced in these 'Rules' details which would be familiar in many circumstances. The rich would recognise (supposing they ever read this work) the description of nursemaids, "swarming abroad like bees in Summer",[19] when the Spring weather was still cold and causing their charges, "whether carried or creeping by the help of their own feet", to be "all alike pierced through by the blast". The poor would sense how well Beddoes understood their predicament, at its most poignant when a baby is born, as they read, "Now however poor you may be, you might surely contrive to stop the crevices and panes for a time, or find out some snug corner for the mother and child".[20] The importance of cleanliness is more than ever stressed; Beddoes brought forward dramatic accounts of how disease was encouraged by the filthy condition of emigrant ships and of the bedding rolls which soldiers had to keep "at the ready". When he came to making suggestions for everyday life he knew how far down he had to start; "well disposed and careful poor people can do no better service to their family than in having as many beds as they can provide and the charitable can do nothing more kind than assist them in this respect".

He wrote with great feeling and with determination to make his poor patients understand the importance of his warnings and advice. Next to 'Isaac Jenkins' these 'Rules' are the most vividly written of all Beddoes' work and the one where he most sincerely directed himself to his readers without exhibitionism. First and foremost he wanted the simplest of his readers to appreciate good health. Knowing how many of them would be familiar with limbs distorted by scrofula he described how "nothing can be imagined more beautifully contrived than the joints upon which our limbs turn". He went on to warn them of the horrors of the surgery that would follow neglect:

See the scene that is to follow − the surgeons in their dresses with grave countenances, and doubtless, though inured to such work setting about it with a heavy heart ... the sand strewed to catch the falling blood ... the sponges set to suck up that which oozes from the dissevered palpitating flesh ... the patient carried to the table ... the instrument to prevent his bleeding to death fixed upon the limb ... the operator hiding the knife he holds under his apron, while he begs the patient to draw his nightcap over his eyes that he may not see the stroke.[21]

His wish to make the least educated understand what a tubercular lung was led him to ask them to look at the lights (i.e., lungs) seen hanging in butchers' shops and then to imagine in them the bodies like grains of sand which become plums, then pigeon's eggs and then bags of pus. The effects of 'hell-fire preaching' were described with a vividness which was obviously intended as a warning (though it may well have merely added to the dangerous fascination): women, he warned, "become ill by listening to those preachers who dominate over-tender consciences and weak minds by bawling out *hell* and *damnation*! The poor, especially women, have no enemies like these fanatics, in their hands many become bad wives and bad mothers and not a few fit only for the madhouse".[22] Some of the lively writing is, happily, more gentle: infants carried abroad should have veils over their faces; doctors should be allowed "to survey and project like *landscape gardeners*" though often they are "unhappily not aware that the habit of human creatures is as susceptible of improvement as the surface of this earth".

This comment shows how, as well as helping his readers to improve their own health individually, Beddoes was concerned with the well-being of the community as a whole. He tried to interest his readers in this wider concern and explained his work simply but without condescension. He made clear the way he and others had set about collecting information and tried to give an idea of the numbers of people involved:

walk through the numerous streets and mark what a multitude of passengers you meet. Next think of the infinitely greater multitude within doors at the same time. Will you not be startled ... (that) so many are destined to perish in Great Britain every year under the gripe of this Giant Malady [i.e., consumption].[23]

He encouraged them to feel that progress could be made by showing how statistics showed that there were good results from the work of the Institution. People could, and should, take responsibility themselves for improving health by being more observant of their children. In particular, they should face up to any family tendency to tuberculosis and when they knew of such, not try to hide illness but be especially careful to seek advice. There were two means of making progress which people were in general often reluctant to recognise and Beddoes tried to remove the popular fears. He dealt in detail with the new protection against smallpox promised by inoculation. Beddoes himself had approached this with caution, but by 1803 he was convinced and ·wrote to the *Medical and Physical Journal* in support of an award to Dr Jenner.[24] He tackled a more difficult prejudice when he tried to persuade his readers to allow *post mortem* examinations.

Throughout, these 'Rules' are designed to educate rather than merely exhort and to develop a positive attitude and encourage confidence that something could be achieved by simple means. It seems a good statement of Beddoes' medical philosophy that, writing of the medical profession, he regrets that "we bestow our labour and admiration upon cumbersome contrivances and only arrive after long beating about, at simple utility".

· Beddoes' liking for down to earth measures had appeared in his 'Letter to the Rt Hon. William Pitt on the means of relieving the present scarcity and preventing the diseases that arise from meagre food' which he wrote in 1796 during the time when he was actively opposing Pitt. Though it was written as a political pamphlet, the 'Letter' is largely devoted to practical suggestions. There is description of a "broth machine" (pressure cooker) with sketch and practical information; of his two horses who became "very fond of potatoes", so illustrating that one crop could be grown to feed both man and beast; of the possibility of making more use of "oats and pease" and of reducing the amount of barley used in the manufacture of both beer and distilled spirits. He wrote of the possibility of avoiding waste of resources by relying more on the direct use of vegetable food,[25] instead of 'processing' it by turning it into beef:

Every ounce of beef contains the quintessence of many tons of grass, hay and turnips, together with part, or whole, of several other vegetables. If by any cheap culinary process these productions of the earth could be made food for man, it is evident that our pressing wants would not only be relieved but we should have provision for a boundless increase of population.

Beddoes next considered additives which might make food more nourishing — garlic or opium, perhaps. In 1796, his hope for amelioration of the harsh conditions he described rested on political action. The 'Rules', by contrast, present a programme largely independent of politics and show what he hoped could be done by medicine and by encouraging self-help. They also show, with the 'Observations on ... consumption', the justice of his claim,

"A large portion of human misery passes under close medical inspection".[26]
His survey of the various occupations, even his understanding of the effects
on the health of the Highland Scots of the importation of Lancashire cottons,
which came to replace the wool more appropriate to the climate, indicates
how he came to suggest to Pitt that among the possible causes of human
misery "may not some be political".

Beddoes' Medical Institution did not receive the recognition it deserved
and only a visiting physician from Vienna gave it careful study.[27] Dr Frank
was one of the many who celebrated the Treaty of Amiens in 1802 and the
ending of war by setting out on a foreign tour. In Paris he met Richard
Lovell Edgeworth, enjoying the same freedom to travel and to meet interesting people, and was given an introduction to Beddoes. Once he arrived in
Bristol, Dr Frank was so impressed that he decided to prolong his stay and
to give up visiting other institutions in the city in order to devote more time
to Dr Beddoes. On his return to Germany he gave an account of the aims of
the Pneumatic Institute and of its development into the Medical Institution
in his 'Medical Tour' and it is from a translation of the work, made by
Beddoes' assistant Dr King, that we learn how much the German doctor
enjoyed his visit. In particular, he remarked on the size of Beddoes' library,
and its unique collection of German works and periodicals. The two doctors
are reported to have found common interest in two very different German
writers, Schiller and the physician Reill.

His position as a doctor and the scientific work he was doing for the
Pneumatic Institute made it entirely natural that Beddoes should arrange
public lectures on matters of health and on chemistry. What at first might
seem an independent activity, almost a relief from more serious concerns,
emerges as a further effort to encourage that understanding of the physical
laws of life which in Beddoes' view was the means of securing happiness both
for the individual and for society. The first of the lectures was to be given
in 1797 when Beddoes arranged for Dr Scott and Dr Bowles to give 'A course
of Popular Instruction on the Constitution of the Human Body'. Beddoes
himself wrote an 'Introductory lecture' which he dedicated, significantly,
to:

The Absent Promoters of this Attempt
The Bishop of Sodor and Man
Thomas Coutts, Esq.
Benjamin Hobhouse Esq., M.P.
Lord Lansdowne
Lord Stanhope

James Watt Jnr., Esq.
and
Particularly
Thomas Wedgwood Esq.

Beddoes may have been giving due acknowledgement for financial support and singling out Thomas Wedgwood as the most generous of his patrons and the one who most shared his ideas on education, but it can hardly be accident that Benjamin Hobhouse, Lord Lansdowne and Lord Stanhope were all leaders of the radical and Francophile Whigs whose activities had been selected by Burke for criticism; and that James Watt was a known sympathiser with the Revolution. It is worthy of note that Beddoes turned to what appeared to be purely academic lecturing just when political expression had been limited by the 'Gagging Acts'. The more practical radical leaders, notably Thelwall [28] who made a valiant effort with his lectures on Greek and Roman History, saw that discreet and carefully worded education was the only means left open to them for keeping alive their movement. 'The Watchman' ended, pamphleteering dangerous, Beddoes too felt that something perhaps might still be done in this way. Beddoes would appear to be promoting a favourite form of entertainment, public lectures. Such lectures or series of lectures could range from the popular and superficial, to serious treatment of a scientific or literary topic. Nothing in Beddoes' introductory address could be construed as in the least subversive but dedication to such a group, coupled with certain hints in the lecture, suggests that Beddoes' aims were not entirely apolitical.

As is fitting in an introductory lecture, Beddoes treated the subject in a general way, stressing at the outset the importance of knowledge and of systematic study. He surveyed the successes of medicine in the past – the conquest of infectious fevers and the banishing of diseases that are due to dirt, slipping in the comment that it is the well-off and not the poor who have benefited from the introduction of inoculation against small-pox. His list of continuing evils includes, as we might expect, consumption and venereal disease but other objects of attack reveal Beddoes' own particular views: he described the delicate constitution of women and the feebleness of so many of the well-off. Both, he considers, could easily be remedied by more sensible styles of women's dress and less self-indulgent habits in general. He made his usual vigorous attack on the "poison of fermented liquors" and on quack medicines but less expected is his discussion of opium adiction, which Beddoes described as widespread in the ports. This is as much social commentary as medical and Beddoes' remedies too are those of a reformer.

He urged that institutions for treating the "indigent sick" should be set up on a new pattern and in his peroration, made a strong plea for "Preventive medicine the destined guardian of infancy, youth, manhood and old age, adapted to the interior of families". In addition he stressed the real benefit that would come from a greater understanding of man's nature. To bring this about he advocates not piety and true religion but the study of Locke, Hartley, Sydenham and Darwin's 'Zoonomia', very much the syllabus of non-traditional education. Beddoes' tone was restrained; his information down-to-earth. It rested with each of his hearers to find, or to be unaware of, any tendency to criticise the established order.

In one respect, this series of lectures was of itself innovatory. Beddoes urged that the study of anatomy would be valuable and interesting to women as well as to men. He had as "a favourite portion of (his) original design" the wish to invite women to a part of the series. Carefully chosen, this could be quite acceptable; he insisted that [29] "Women, for example, attend without scruple lectures in which the eye is demonstrated ... mothers, the most delicately educated, brave disgust for the sake of their children". Beddoes had regretfully to give up this part of his plan, but he did arrange a second series. He wrote to James Watt junior on 2nd January, 1798, "The ladies' anatomical lectures succeed perfectly. The number exceeds 30 and I hope will reach 40. McIntosh who was at yesterday's — the 1st lecture owns that such a course may be rendered highly instructive without any alloy of disgust". The 1797 series had been planned to be useful to Bristol medical students as well as to the general public. It would be pleasing, though unfortunately perhaps a little exaggerated, to nominate Beddoes' 30–40 pioneering ladies as our first women medical students.

In 1798 Beddoes himself lectured on chemistry. None of the series has survived but letters [30] to James Watt and Davies Giddy show how eagerly he prepared for this new development in popular education and how well he was able to share his enthusiasm with his friends. He had already set up a good laboratory and now engaged an Irish "lab. assistant", a Mr Boyd. Beddoes was very careful to make clear in advance exactly what he expected from his technician: Mr Boyd should be prepared for practical work and demonstration and he would need to be able to construct apparatus. A theoretician was no use. Although Mr Boyd had been recommended by a friend, Beddoes suggested that it would be useful to have a reference from Dr Kirwan. He was clearly determined to be very sure about this appointment for he made inquiries himself from the mathematician, Dr Higgins. Beddoes clearly had high hopes of the lectures; he went to much trouble about them and his preparations give a vivid impression of the standard of scientific activity Humphry Davy would have found when he came to work at Beddoes' laboratory. Some specimens [31] were still in the care of William Reynolds and these Beddoes now arranged to have sent to Bristol and, with James Watt acting as intermediary, he had apparatus made for him by Matthew Boulton; he mentioned in particular a chemical lamp. Boulton also lent his precious specimens of Herculaneum glass; much to his surprise, Beddoes found James Watt offering to send him one of Wedgwood's copies of the Barberini vase. He could hardly believe his good fortune. "Are you joking about the Barberini vase?" he wrote, "Is it not too dangerous to transport? I mean the Wedgwood imitation". Beddoes also asked Watt for various copies of Crell's 'Annalen'; for a copy of a drawing by Black in his father's possession and for material from Dr Keir. Even so, Beddoes characteristically could not get out a full syllabus before the lecture course started; but he did feel it necessary to have printed at the outset an explanation of the changes in chemical nomenclature. Beddoes seemed glad to return to his chemistry and it appeared that his lectures were designed to show the application of chemistry in manufactures. Stock [32] particularly comments on the success of illustrations from the working of iron and Beddoes did indeed have a copy of the Barberini vase to show his audience; he may have used it to illustrate technical developments in the manufacture of decorative pottery.

The success of the lectures pleased Beddoes and he planned to give more in the following year. There was to be a geological series and thinking well ahead, he wrote to ask Davies Giddy [33] to employ captains of the Cornish mines to look out for specimens for him. Beddoes admitted that this subject would have no direct practical value for his audience; his hope was that the lectures would attract people of opposite political convictions who would "acquire in common a number of agreeable ideas – and the effect may be to spare some acts of barbarity in the times that are approaching". For a short while, Beddoes had dreams of something of greater permanence than occasional series of scientific lectures and enthusiastically set about plans for a Bristol Philosophical Institute. Humphry Davy was invited to give a lecture on chemistry; he was excited and gratified by such an invitation which was given very quickly after his arrival in Bristol. This came to nothing. Any gathering of people was suspect at that time and even academic lectures did not escape. Beddoes' reputation would certainly not advance any scheme for an Institute.

Though the scheme for a Philosophical Institute had to be abandoned, Beddoes kept in mind the need for education. "It can, I think, do no good to have people linger under easily avoidable diseases"[34] he wrote in January 1802, "and to excite a relish for the study of human nature is the most likely way to find the outlet from the dire dilemma if one exists − I have a very definite idea of the principle on which this outlet will be discovered but I do not understand the *moyen d'execution*". He may have had in mind the series of essays on health matters which were published later under the collective title *'Hygeia'*. These are eleven long essays, treating in depth the whole subject of maintaining health.

After *'Hygeia'*, Beddoes wrote no more, apart from his very odd *'Manual of Health'*, on the subject of preventive medicine, but he continued his efforts along particular lines. The 1803[35] numbers of *'The Medical and Physical Journal'* show him working vigorously to solve the problems presented by influenza. The previous year must have been marked by a particularly widespread epidemic and there were very differing views about the nature and cause of the illness. Early in the year he was writing on this subject and debating with Dr Edward Long Fox whether it was an infectious disease. He once again turned to Davies Giddy[36] for help in setting up a survey; he sent him a questionnaire to be used in collecting information from various small towns in Cornwall and asked that a clerk should be engaged on the work. He wanted to find whether "the disorder is excited by the atmosphere: dates are evidently of the first importance" and records were to show whether the wind had any effect. Beddoes, with his usual enthusiasm, felt that the opportunity to establish facts should not be missed. He wrote to the *Journal*: "A good deal of communication with medical men about the Influenza makes me feel that the next epidemic of this kind, will find our successors as uncertain in some essential respects as our predecessors have left us .... One of the first points of view in which I regard every disorder is its prevention."[37] It was important to know whether influenza was contagious for if so, "acid fumigations bid fair to stop the progress of the complaint" and it would be right to try this generally and not only in individual families. He invited reports from doctors and asked that they should attend to four points: the exceptional cases which led to the belief that the epidemic could be 'ascribed to the atmosphere'; the distinction between catarrh and influenza; a comparison with illnesses known to be infectious; and the possible correlation with age and class − 'mere enumeration' was not sufficient.

He received such a large number of replies that in the summer he felt obliged to write to ask for time to make some systematic arrangement before publishing. The letters that appeared in August came from all parts of the kingdom and even from Ireland; some of the senders appear elsewhere in Beddoes' life. Dr Rolfe sent a report from the 11th April Admission Register of the Pneumatic Institute; his nephew Henry, then a medical student, sent a report made by the surgeon at Edgeworthstown. He believed that the influenza had been caught from a parcel sent from Dublin; the only one to suggest a source of this kind. In spite of Beddoes' attempt to organise the replies, most of his respondents were unable to limit themselves in the way he had suggested. Beddoes on his part made no attempt to extract information, giving the letters in full. They provide a vivid account of an epidemic of influenza; its sudden onset, duration, and in some towns the overwhelming numbers affected; a few of the letters end with an apology that the account had been curtailed by the doctor himself, and his household, succumbing. Following his aim of determining whether influenza arose from "contagion" (which seems to have been a general term and not distinguished from infection), or whether it was the result of something in 'the atmosphere', Beddoes divided the replies into contagion, atmosphere and 'don't know'. The result was indecisive but the replies make clear that in 1803 the true answer was "cause not known".

There was confusion too about the nature of influenza, whether it was merely a catarrhal illness or whether it was distinct. Some doctors believed that when influenza was fatal it was because it had turned into typhus, a reminder of the difficulty at the time of making clear-cut identification. Each doctor was eager to describe his own method of treatment and only the two army surgeons who replied attempted a numerical analysis of their observations. Many doctors describe vividly the "langour" and "dejection of spirits" which follow the attacks; and the sudden onset which made so many doctors feel that it was impossible for it to have been passed on from person to person. Just one correspondent, Dr T. Peaal [38] of Aberdeen, qualified his view that influenza was contagious by the observation that it could only be caught by contagion once in a lifetime. By this time inoculation had become popular — Beddoes himself had written in 1802 to support a national subscription for Dr Jenner — yet it seems Dr Peaal did not attach any great significance to his observation.

The replies made clear the difficulties of conducting surveys in the early years of the nineteenth century, when no methods had been established and when such lengthy letters were sent. Combined with delays, this led to Beddoes' abandoning his project and early in 1804 the Editor of the 'Journal' published a note that "various interruptions" had prevented Dr Beddoes from "forwarding the concluding observations on the Influenza" and that he had been disappointed in not receiving contributions from the continent: a sadly typical ending. There were other contributions to the 'Journal' on "This Proteus Disease", independent of Beddoes' correspondence, but what comes out of this incident is that it was "the indefatigable Dr Beddoes" who had the prestige to command a response to such a survey.

In 1804 there was an opportunity for an additional medical assistant at the Preventive Medical Institution. Beddoes considered it would be interesting to "medical students who have gone through their elementary instruction and are not immediately disposed to settle ... the situation would be highly advantageous, especially as to ample experience it would add opportunities of human and comparative anatomy, physiological researches, etc." Beddoes was concerned that doctors should have a sound professional education. He himself had a leisurely advance from school to qualifying: five years at Oxford, two and a half in London, two in Edinburgh and a useful summer tour on the continent but he was well aware that many were much more rushed. In the last year of his life, Beddoes expressed formally his plan of medical education. As he was never a teacher in a medical school, his 'Letter to the Right Hon. Sir Joseph Banks Bart. P.R.S.' can be thought of as a contribution to his efforts to promote good health.

The letter was occasioned by a proposal made in 1806 for a reform of the medical profession to prevent unqualified persons from practising. Beddoes wrote partly to support the advocacy of a five year course, but more because, as always, he wished for a public discussion of the whole matter. Once he had dealt point by point with the Edinburgh medical faculty's arguments in favour of a three year course and praised the Oxford and Cambridge system of general study followed by a medical course, he gave in considerable detail his own scheme. The student was to move steadily from anatomical to clinical work and each year was to provide a period for supportive studies. Beddoes insisted on chemistry, botany, medical jurisprudence and physiology. If at all possible, the student should have a period of clinical work abroad [39] at the Paris medical school or in Vienna. Above all, there should be plenty of time for practice in anatomy; Beddoes described vividly the .ush of the Edinburgh course and the need to turn to "grinders" and remarked that the 'Soho faculty' thought 24 the right age to graduate. He emphasised the need for time because "There is no royal road to genuine anatomy; nor can it be acquired but by the united help of the senses" − scalpel in hand. Examiners needed to be trained so that this work in anatomy could be thoroughly tested, in the same manner, Beddoes hoped, as in the Paris school. There students were required to perform an autopsy as well as write a

dissertation. Beddoes considered the teaching of anatomy in England inadequate and writing on the subject "so scanty, so jejune and so stiff that
it hides from the learner that which most requires to be taught — the free
play of nature" and he wished that some benefactor would do for medicine
what the Duke of Bedford and Sir Joseph Banks had done for agriculture and
natural history and found an anatomical school in London.

Beddoes saw that the best way of realising his scheme would be by Act of
Parliament and he gave the outline of a proposed Bill. All this effort would
be wasted, he suggested, if quack medicines were not prohibited and he
proposed that they should be made illegal. In another essay of the same year,
Beddoes proposed that progress in medicine should be encouraged by reward
from taxation:

Let the happy combinations of genius (for example Hunter or Erasmus
Darwin) be honoured by admiration. But for every practical and extensive
[advance] in medicine, there should be a proportionate levy on the public
. . . . What family in fact, but will own that deliverance from the smallpox
is a privilege to be valued in specie; and if collectors had gone round from
house to house as with a brief, I fancy there are few who would have refused
a mite in return for protection against the dangers of typhus and scarlet
fever.

Since Beddoes had also suggested that there should be public contributions
to 'Preventive Medicine Institutions', it would seem that almost casually
he was suggesting some governmental sanction and support for his dream of
Preventive Medicine. It must be admitted that such suggestions are thrown
out in an unpremeditated manner and it cannot be supposed that Beddoes
had any 'Public Health' programme in mind.

The discussion of medical education in the two publications of 1808,
the 'Letter to Sir Joseph Banks' and 'Researches . . . concerning Fever', fill
in the picture of what Beddoes hoped for medicine. Beyond the immediate
details of the practical advice he had to offer, Beddoes had a two-fold programme of Preventive Medicine. He looked forward to a people instructed
in such a way that they understood what was needed for good health and
as far as was in their power could live to keep themselves well. The counterpart was to be doctors who saw beyond their work for individual patients
and accepted a more general responsibility. Beddoes gave his description
at the end of his letter to Sir Joseph Banks:

There exists no set of men more meritorious than the body of our military
practitioners of medicine. Their ardour seems as much to have exceeded that

of their brethren in civil life, as their situation has been more uncomfortable and their rewards more scanty .... [they] employ the hours due to repose in noting down their observations.

Beddoes had already described his own habit of collecting facts and his conviction that

No striking fact can be accurately stated, in conjunction with its antecedent and concomitant circumstances, without improving our acquaintance with human nature. Our acquisitions in this most important branch of knowledge, may be compared to a number of broken series, of which we have not always more than one or two members. But every new accession bids fair to fill up some deficiency; and a large supply would contribute towards connecting series apparently independent, and working up the whole into one grand all-comprehending chain.

Assiduous observation of the daily state of the human microcosm will be the unfailing consequence of attention to its striking phaenomena. Such is the progress of curiosity. Such the origin of all the sciences.[40]

CHAPTER 11

RELIGIO MEDICI

The Institution for the Sick and Drooping poor, with its 'Rules', and *Hygeia* are the twin pillars of Beddoes' gateway to health. The first is the practical embodiment of his ideas; the second shows how the advice he has to give is grounded in his understanding of the nature of man and society. Understandably therefore *Hygeia* aims at instructing "the middle and more opulent classes" because only these have the leisure for study and through these the poorer can benefit. With the change in readers there is a change in tone of his writing. All Beddoes' rules are illustrated from and applied to middle class life. The need for good diet, cleanliness, suitable exercise; sensible clothing; a temperate way of life and great moderation in the use of alcohol, all reappear. He spared no feelings as he castigated the way of life of the well-to-do Englishman and pointed to its consequences: "If we reckon from the middle state upwards, it would ... be more just to assert that the unhealthiness of families is in direct, than that it is in the inverse proportion of their wealth". Faced with such an unhealthy mode of life, doctors can be of no help; Beddoes makes the startling admission that they are paid for "bustle" which only consoles but has no real use.

In dealing with particular illnesses Beddoes had naturally to be concerned with those that attacked all classes of society: scrofula, consumption and fevers, particularly infectious fevers. He extended his list to discuss others which might be considered of little interest to the poor. Among these were diseases of affluence such as disorders of the stomach for "a strong stomach is an infallible remedy against blue devils"; and the vague indisposition of the fashionable; "No mortal would ever have thought of making apathy the mode but a worn out beau". Even more interestingly Beddoes includes a group of conditions which are mental as well as physical, epilepsy, nervous disorders and insanity. Though in *Hygeia* Beddoes paid more attention to the principles from which his advice was developed, he still, by vivid descriptions, brought his readers to face those details of their way of life which were the cause of ill health. He expected of the well-to-do, no less than of the poor, self-examination and reform. Indeed, as the rich had more power to make changes, even more was asked of them than of the poorer readers; where in "The Rules for ... the Sick and Drooping Poor", he had been kindly in pointing out to the poor where they needed to try to make changes, he was often stern in describing to the comfortable their follies and self-indulgence.

The straightforward and practical advice on matters of health contained in *Hygeia* would naturally have commended the essays to their original readers and for many this would have been their main value. Viewed in this light they appear as a major contribution to Beddoes' work for a positive rather than negative attitude to health care. At the same time there is an underlying philosophy and a scientific theory which Beddoes clearly intended to bring home to his affluent readers. These were the people among whom he now lived, who were his fashionable patients and whose children he attended either at home or in their boarding schools. He saw in their habits and attitudes much that led to their own ill-health and to suffering for the whole nation. His previous writings, apart from his political essays, had almost all been addressed to scientists and medical men or to poor people whom he wished to help. In *Hygeia* he hoped to educate not just by explaining practical matters of health, but by bringing about a change of thinking. He intermingled these matters so closely that it would have been impossible for his readers to take one without the other. In each essay Beddoes dashes from homely topics of family health to attacks on luxury and then to his schemes for reform; or he endeavours to concentrate attention on the connection between man's physical and mental well-being; always he hurries his readers along with him. Beddoes himself uses the metaphor of a dance: "polished life makes an ever moving picture . . . presented by certain dances in which all the parties join hands and tread a ring; then dispersing, each in his several line, towards different points of the compass speedily rejoin and pace their round once more".[1] This same metaphor could well apply to *Hygeia* where ideas mingle, meet and pass each other; but *Hygeia* is different from the earlier writings where thoughts are merely tumbled together, for each one of the essays is designed to advance his main argument. There is admittedly much repetition; as in every essay, Beddoes pursued his aim of leading his affluent readers to pay critical attention to the way in which their lives were conducted and to see there the need for the practical measures which he was advocating. In every essay he encouraged them to study the unifying principles he was explaining and he understood well the usefulness of repeating his basic ideas in different contexts. The interest of his material and his vivid descriptions would have carried along the readers who came to *Hygeia* part by part; modern readers cannot help but be more aware of the repetitions. They will also regret that Beddoes found it appropriate to adopt

the 'grand manner'; to abandon the ease of *Isaac Jenkins* and the plain speech
of his more objective writings, for a high-flown, "hortatory" style. Probably
Beddoes felt that to approach in the guise of a medical adviser was the
best means of spreading his ideas among those he had selected as his most
influential audience. The titles of the essays suggest a plan. The first two
essays are directly addressed to those who had most power to influence the
well-being and the thinking of others: 'Heads of Families' and 'Ministers of
the Gospel'. Then come two essays on schools, where Beddoes saw much
need for reform. There are, besides the essays treating particular conditions,
two essays on health in general; one on 'Care of Health' which deals with the
individual and one, the last essay of all, 'On Infection' where Beddoes' concern is for the health of the community.

Beddoes' thesis is clear.

Illness and suffering do not, any more than the phaenomena which we behold
with less emotion, spring from malignant supernatural agency, against which
there is no defence; but may in most instances be satisfactorily traced to the
want of appropriate intellectual culture, and its genuine offspring, blind
temerity.[2]

Man need no longer be at a loss to know how to remedy his condition, for the
basic principles have been ascertained. Beddoes points first to the "pleasure
principle"; men should select what is pleasant to them:

the individual in fact need only to be taught to sympathise with himself, if
that term may be applied to the feelings excited by different possible states
brought strongly into contrast. Consciousness of health, thus contemplated,
will become just as much a source of pleasure as the consciousness of virtue.

The choices made following this principle are to be reinforced by praise and
disapproval, for feelings are more fundamental than understanding. Again and
again in a variety of connections, Beddoes stresses the importance of the feelings as determing action and physical state. Next comes knowledge, both
in private life, where the individual should have a "knowledge of the human
structure and functions"; and in public life where improvement needs to be
founded, not on supposition but on carefully collected and organised information. In his view, "To accelerate the progress of civilization is probably our
only practicable and effectual plan". The belief that it is the function of
government to bring about good conditions for the people, which in the
'Essay on the Public Merits of Mr Pitt' was stated so baldly still lay behind
much that Beddoes proposed in *Hygeia*. In these essays it is presented more
persuasively, woven into the argument and often related to specific proposals.

In this manner it was probably less alarming to his readers. This was a utilitarian view, as Beddoes himself realised, and he makes a point of disclaiming the intention of requiring people to "live by rule". Underlying his advice, in contrast to his theorising, is not a sense of rules and regulations but a humane plea for personal moderation and discipline in the habits of daily life. This was the change which would lead to a society more temperate, more pleased with sufficiency than with luxury, and more ready to enjoy simplicity than sophistication. With the optimism of his day, Beddoes took for granted that these reforms would consequently lead to a society that was more just and equitable.

His experience as a family doctor gave Beddoes the insight to write very intimately to the "Heads of Families" and nowhere is his kindliness better seen than in this essay. He knew the disappointments that resulted from the consequences of illness even where there seemed to have been a cure, giving the vivid example of the difficulties met by those scarred by smallpox. After such a reminder it was natural to draw parents' attention to the need to give their children the right education and to the unkindness of directing their studies entirely to serve wordly success. He instanced the teaching of accounts for the sake of success in business while leaving children in ignorance of how to take care of their health. The unhappiness that children suffer in the privacy of family life, from lack of sympathy, from teasing or from insensitive mockery, had all been observed by him and as a preliminary to all his advice he emphasised that "Unkindness is the natural provocative of malignity; nor will it often be in the power of Fortune to counteract, by any favours the early union of painful feelings with a large mass of ideas. So important is it that nothing should intercept those sympathies of voice and countenance which are to children a source of such exquisite enjoyment, and a portion by no means contemptible, of the discipline of humanity."[3] The same warm sympathy can be found in many details, in his indignation at the anxiety and misery caused to sick people by their being unable to free themselves from superstitious beliefs and folklore; in the way in which he urged parents to consider the individual capabilities of their children and not to impose a uniform Spartan discipline on all; and in his warning against the dangerous consequences of physical attacks on children. He would have the possible tragic results of rough behaviour made clear to children and he reminded those in authority that by choosing beating as a means of discipline they are setting an example for future bullies. Though he gave much practical advice and information, his chief concern as he addressed parents and teachers was to foster this humane spirit so different from the

harsh authoritarianism of the time. He regretted an attitude towards education "where the desire of knowledge began at the remotest extremity of
visible nature and in proceeding from the planetary masses to the inanimate
bodies of our earth, fixes at present, with ardent and diffusive zeal, upon
plants and inferior animals, but passes by the majestic species of man as
unworthy to be studied but for the ignoble end of gain".[4] The objective and
tolerant view which he had first expressed in his 'Letter to a Lady on Early
Instruction' is again voiced but with a greater dignity: we "should first teach
[man's] duty as a human animal; the interest or duty of that vocation which
makes him a human being of peculiar habits ought to be but a secondary
object with his preceptor".

This same humanist view informs the argument of the essay addressed to
Ministers of Religion. Beddoes saw them endeavouring to improve the well-
being of their people and it was with man as an inhabitant of this world that
Beddoes asked them to be concerned. It seems that Beddoes had come to
recognise that a parish priest who took the trouble to equip himself with
basic medical knowledge might legitimately have power to influence and help
his parishioners, particularly the poor. He hoped for help from ministers in
exposing misinformation about medicines, and about the miseries caused
by the tyranny of fashion and the pursuit of luxury. More particularly,
possibly thinking of the part played by the Scottish ministers in contributing
to Sir John Sinclair's Survey of Scotland, he hoped to interest ministers in
his plan for "a minute *probing* examination of the different classes that act
and re-act upon each other within the sphere of the British Isles".[5] He found
common ground and a shared moral aim with them only in his belief that "To
excite a dread of the consequences produced by personal mismanagement
appears the only thing belonging to this world that can be opposed with
effect to inordinate vanity".[6] Beddoes and the clergy could agree, "Our
foresight, whether it be confined to an hour or stretch into eternity, is fixed
upon enjoyment and suffering. To be masters of the sources of both is
still the object of our care"[7] and so they should cooperate in teaching
"The sentiments of self veneration and the practice of self management
[which] as founded on the essential properties of the object to be reverenced
and preserved can never suffer from any versatility of taste."[8] The vehement
attack on religion found in his earlier writings had been laid aside. Beddoes
himself had become more civilised and moderate; but he held to his old
convictions. His views might have found acceptance in liberal circles, but
they were greatly at variance with Anglican evangelical ways of thinking and
with the tenets of many non-conformist sects.

Three essays, the first and the two on schools, gather together the educational ideas that ever since 1792 Beddoes had been putting forward in scattered writings. The principle that knowledge and understanding rest ultimately on sense-experiences and the power of the association of ideas, is kept before his readers. Beddoes does not develop this as an abstract argument but recurs to the theme in connection with practical situations. Sometimes he shows how these ideas are of general application by introducing them in contexts which are not educational in the limited sense of the term. In the midst of a discussion of diet, for example, Beddoes encourages his readers to have confidence that they can re-educate themselves "by fixing attention steadfastly upon feelings, connecting the circumstances upon which they depend, strongly together in the mind. Ideas stamped upon the memory with great distinctness, have undoubted powers in deciding the will, and very frequently they prove capable of resisting the seductive tendency of impressions made by present objects".[9] As we should expect from such a view, Beddoes rejected the authoritarian imposing of programmes of learning and the setting of arbitrary standards; in twentieth-century terms, his method is 'child centred'. As we should expect, it is in activity that learning is grounded, especially in early childhood: "the faculties should have full proportionate play during the first ten years of life. This implies an almost uninterrupted and an immediate intercourse with the powers of Nature, and such occasional communications from seniors, as shall render that intercourse safe, wholesome, and instructive". He stressed that children should not be depressed by having to do tasks which are too difficult for them, but should be led by graded experience to overcome difficulties. Beddoes saw this as laying the foundation for a livelier and healthier adult life, especially as he was convinced that mental depression in childhood, the result of unsuitable studies, was often the cause of physical illness. He must have observed his child patients with considerable sympathy and understanding and he stressed he had had many such patients and had observed them carefully.

All this is familiar ground but the past ten years had brought Beddoes greater practical knowledge and closer experience of children. His most comprehensive account of the studies he would like to see appears at the end of Essay IV, 'On Schools for Boys'. What we should term 'human biology' is, as would be expected, given an important place. So too is botany which should be "connected with practical instruction in plant physiology — a science always fascinating and now from year to year enriched by most curious discoveries" — the approach he equally advised for girls, as Beddoes did not approve of treating botany as no more than a genteel occupation. Just as girls were to have the same teaching in science, experiments not lectures, so they were to learn the same "mechanical arts" as boys, and like them to use the lathe. This would of course have been a lathe for wood-working, operated by means of a treadle, and the suggestion is the more understandable in view of Beddoes' belief that exercise was best provided through interesting occupation rather than sport. He saw disadvantages in sport since it was often discontinued in after-life and often led to drunken-ness. Botanical and mineralogical expeditions were commended on the grounds that they "cannot fail to inspire a taste for the charms of external nature; and the acquisition of such a taste is among the chief means of maintaining the constitutional vigour to the longest allotted term and of raising it to the highest pitch" — 'Qui fait aimer les champs,' he quoted, 'fait aimer la vertu.' In this passage we find the suggestion that children should watch the birth of farm animals and see the dissection of reproductive organs and of a spawning frog. Best known of all Beddoes' remarks on education, these suggestions for dissection have been taken out of context and the possible horror of the experience made much of; so far as the frog is concerned, Beddoes was remembering the experiments of Spallanzani which had so impressed him, and he is particular to emphasise that the frog should first be killed by the quickest method, immersion in boiling water. The whole fits into his scheme of learning through observation and is the more understandable in relation to the problem of 'sex education' which Beddoes had just been discussing. Beddoes was mainly concerned with boys and their problems and his treatment of the subject is unusually clumsy and apologetic. He did in fact feel that it might be the ideal to keep the young in a state of ignorant innocence but realised that this was not possible. There must, he emphasised, be accurate information from responsible people, not from servants: "the understanding must restrain and guide the impulses of the appetite ... no one can regulate the imagination of another except by the communication of ideas and feelings".

These passages suggest that Beddoes' own passion for observation and his somewhat naive belief that knowledge of facts led to right behaviour, limited his ability to enter into the feelings of others and made him forget the natural tenderheartedness of children. The proposal that boys should be taken on visits to hospitals to see patients suffering from the consequences of sexual indiscretions seems particularly horrific. In putting forward these practical schemes, Beddoes must have had chiefly in mind the arrangement which he considered the best for children, that two or three private families should have their children brought together for teaching. By this plan children would have the advantage of meeting others of their own age and at the same time of not being cut off from adults. They could then be taught in the groups of five or six which Beddoes thought was all a teacher could do justice to. In this way the parents could observe the children who would be "at liberty to act, but their every action noted". He must have had in mind the educational efforts of the Wedgwoods and of Edgeworth. This would certainly have made it possible for parents and tutors to adapt experiments and observations to the needs and temperament of their pupils. There is a certain grimness about the idea of observing the children of a family so closely; it does not seem to have been an entirely happy system in the Wedgwood family but Maria and Richard Lovell Edgeworth, who claimed that "To make any progress in the art of education it must be patiently reduced to an experimental science", produced a very humane account of their observations in 1798. Their *"Practical Education"* was certainly known to Beddoes and he was to a minor extent a collaborator.

Beddoes had to be realistic and admit that after the age of ten many children would be sent to boarding school and he wrote forcefully about the changes he would like to see. It becomes clear that he had been a physician in girls' schools and had collected much information in spite of the difficulties raised by "the lady abbesses of our temporary nunneries".[10] He saw the girls suffering from a diet that, besides not providing sufficient food, was intentionally low. He understood the misery of cold rooms, too long sitting at lessons and too long intervals between meals; when fashionable clothing with its pinching stays and badly designed shoes were added, the girls had much to endure. His sympathy is shown by the way he writes not only of serious conditions but of the supposedly trivial matters that the girls had constantly to bear, the "artificial malady" of chilblains; the distress arising from never being sufficiently warm and the "lack of ease and cheerfulness" at meals. Beddoes did not seem to feel that there was deliberate harshness in girls' schools, but rather mistaken ideas and lack of understanding. He drew up a set of recommendations which are very exact and sensible and which show how thoroughly practical he could be. Boys' schools appeared to him by contrast to have a regime that was deliberately planned to be harsh and to tolerate, even encourage, horseplay and bullying. He levies against boys' schools all the criticisms that he had raised against the girls' but added to them a vigorous attack on the lack of care and the insufficient food which led boys to form a habit of getting into debt. Fashion at the public schools encouraged in quite young boys the habit of drinking wine.

The boys' curriculum, too, came in for more criticism as being not appropriate for later life and not stimulating, the reason why boys who have failed at school do so much better in later life. As well as the reforms that he proposed, Beddoes advised that there should be a physician appointed to each school, not a doctor simply called in on occasion, and a system of inspection. Such provision was worthwhile for, in his view, a sound system of education was as important to the health of society as to the individual:

Mathematical, philosophical, chemical, botanical and technological instruction judiciously intermixed but at the same time carried on in conformity with a system clearly made out in the mind of the preceptor, would keep the physical and moral faculties of the children in perpetual and proportional progress. The thoughts would never stagnate; the heart never prey upon itself and intemperance of every kind would be abhorred according to the nature of its consequences. [11]

The essays which follow are directed more precisely to matters of health and are to some extent arranged to correspond to the different times in a man's life. The essay on the 'Care of Health' relates mainly to the needs of babies and small children. There is a typical irrelevance when in warning about the dangers of sudden cold, Beddoes remembered ices and cannot restrain his thoughts from turning to ambassadors who are in danger of eating too many. He advised having warm drinks at hand as an antidote and commented, "Nor would it be amiss, if sovereigns, in selecting their representatives were a little mindful of the power of the stomach as well as of the head". Essays on Scrophula and Consumption follow and then another general topic, on 'The Physical power of Enjoyment' which is concerned with maintaining good health in the later part of life. Appropriately here Beddoes was much concerned with temperance and with a good digestion, for "A strong stomach is an infallible amulet against blue devils". These essays on physical health elaborate the explanations and advice that had appeared in earlier medical writings but in *Hygeia* Beddoes paid much more attention to making clear the mind/body relationship. He devoted the two essays, 'On Nervous Disorders' and 'On Insanity' to the subject which we would now call psychological medicine. Essay IX, though entitled 'Nervous Disorders', deals principally with epilepsy which Beddoes saw as the central condition. As with physical diseases, he lacked the information which would have enabled him to diagnose more accurately and considered that neglect of simpler troubles such as convulsions or hysteria might lead to epilepsy while mental deterioration, depression and even madness might in their turn be produced by this condition.

With his usual common sense, Beddoes pointed out that the discontinuity caused by frequent epileptic seizures is in itself sufficient to "interrupt the memory" and lower the quality of normal life. Beddoes had identified the very brief seizures that take the form of a momentary loss of consciousness and he believed that more people suffer from such nervous disorders than was commonly supposed. Recognising this should make us able, he suggested, "to understand why in polite circles it is requisite to change the topic of conversation every second minute". Yet he wished to make clear the underlying connection between all these forms of mental breakdown. Beddoes considered that strictly nervous complaints were those in which the nerves themselves had been interrupted. But he realised that this was not always the cause. He saw that the old idea of a "malignant spirit who had possession of the patient, and managed his members as the master of a puppet show does those of his diminutive representatives of man" was an attempt at an explanation. He regretted that there were some people still haunted by the same "pernicious beliefs" and wished "that some luminous and elegant writer should take up the subject of the human mind".

Using his own observations and referring closely to Darwin's *Zoonomia*, Beddoes gave a clear explanation of the nervous system. This included an account of the recent discoveries of the part played by electrical impulses. He emphasised to his readers the difference between damage to the nerves themselves and to the organs. Such a distinction enabled him to explain why a boy who could not speak while he was in a cataleptic fit was able later to answer questions which had been put to him during the seizure. "It is evident," Beddoes explained, "that the ear was duly affected by its appropriate impressions, but that a suspension took part in some part of the series of animal actions between the affection of the external sense and the motion of the vocal organs".[12] Both in this essay and in writing on insanity Beddoes stressed that seizures and madness may come from physical damage to the brain and that this could as well arise from an accident as from other causes. In his anxiety to make his readers understand this he began his study of madness with a lengthy description of the post mortem examination of the brains of a number of 'the melancholic dead', referring largely to the researches of William Hunter. Whether such detail was appropriate in a work for general readers is very much open to question. He was obviously very anxious to make clear the facts that had so recently been discovered and to convince his readers that by following the path of scientific investigation further progress would be made:

It should be considered that the dissecting knife, though it be the instrument to which we owe our superiority over the antients in deducing inward lesions from external signs and from particular sensations, and which therefore may be considered as the glass that shews the state of the viscera through the otherwise opake walls of skin and flesh, can be by no means adequate to lay open *all* the effects of unhealthy processes. There may be changes of parts a great deal too subtle to shew themselves upon the sides of a fresh cut. We should not therefore permit ourselves to reason negatively from dissections, which are often cursorily and ignorantly made — possibly chemical analysis may, one day, add to our acquaintance with the alterations of the living substance as much as the best *morbid* anatomist knows beyond one, who has never seen a dead body opened.[13]

As well as hoping to instil a rational attitude to nervous illness and to madness, Beddoes was leading in the direction of greater humanity and sympathy. His approach made it possible to focus attention not on differences but on similarities and reminded his readers that "Mad is one of those words which mean almost everything and nothing". He challenged them to point to a clear dividing line:

The insane have the same muscles with the sane. In both they perform the same general office. Few need be told that in all men, discourse, look, and gesticulation, depend on fixed or alternating muscular contractions.[14] These alone are the outward tokens of the state within.

... Let us attend to the countenance only. If we had a series of drawings, ascending by the closest gradations from the face of a sleeping infant (a being incapable of insanity), to that of the most furious inhabitant of Bedlam, who would undertake to point out the last among the sane, and the first among the insane?

Beddoes saw that there was a gap in our knowledge of the mind/body relationship and that the symptoms of nervous complaints could occur when there was no physical change. So in dealing with this group, epilepsy, hysteria, catalepsy, he was thrown back on describing the power of the feelings to control our physical state. He saw in building up of a balanced style of life the best means of preventing nervous disorders:

study to still the vibrations any impulse of the day may have thrown [the] nerves. A simple natural life affords many resources for the purpose. From spring to autumn the occupations attached to the cultivation, or at least to the superintendance of a garden will preserve the mind tranquil or help to quiet its commotions. After a light early supper a walk will diffuse that complacency over the system which is the best preparation for sleep in all cases and which will by degrees bring the nervous valetudinarian to wake with the lark and with somewhat of the lark's alacrity, the dejected often feel as if the fresh air ventilated the mind.[15]

He reiterated the advice about associating pleasure with learning that he had already emphasised in writing on education. The vivid and pathetic description that he gave of the sufferings of a patient in a serious epileptic seizure should have been warning enough to have his advice taken to heart. Beddoes' descriptions of a health-giving life, sometimes almost lyrical in their tone, have all the quality of owing as much to experience and observation as to theory, but the doctrine of association which underlay them is avowedly the basis of any treatment which might patently be helpful. To him "the associating power of the feelings ... is the most neglected, and perhaps at the same time the most pregnant topic in the doctrine of the mind."[16]

This doctrine is profusely illustrated from individual histories of mental breakdown and epilepsy, from Dean Swift's tragic life to a newly published 'History of seven years Epilepsy'. Beddoes gave many examples of how seizures could be prevented or controlled: the porter who found he could control his seizure when carrying his load; the calming effect of repeating something that has been learned by heart; in general, "whatever brings back the sensations of a healthier period". He had observed that "those things which tend to strengthen the associations formed in the progress of life appear to have power in preventing epileptic paroxysms" and saw this as the explanation of epileptic seizures and convulsions being more frequent in early life. There is an intensity and a sense of urgency in these accounts and Beddoes has no sense that his readers might find them overwhelming. Although it appears nowhere else in his works, this problem obviously interested Beddoes deeply, and he saw it as a challenge; he had chosen pneumatic medicine but he saw that nervous diseases were another group waiting scientific study. The explanation of the many 'case studies' is that he mistrusted general descriptions and observations made on large numbers of patients. He hoped for progress from careful study of the accounts that individuals could give of their attacks, especially those who had warning of their onset. "From these only can we hope to determine that irregularity in the springs and wheels of the animated machinery which answers to the violent disturbances in the movements of the external parts."

In his account of insanity too, Beddoes emphasised the importance of studying individuals. There are however fewer detailed histories; his aim in this essay was more to help his readers enter into the experience of the insane. He recognised the classification into mania and the melancholia but had been led to make his own working definition: "To me it seems self evident that the very belief in things unseen as seen, must be followed by a mental disease". He described the gradual development of such a condition:

Madness perpetually realises the wonders of Ariosto's magic, and prepares palaces or dungeons for those, whom it possesses. A person under disappointment or chagrin, sets about to imagine by what possibilities he may be delivered from his perplexities. They return to his thoughts by day and by night, gathering more or less force according to the intensity of his feelings. Then, after a long struggle, the imagination obtains the mastery compleatly, and ever afterwards uses calamity as its hobby horse.[17]

Nevertheless he warned against being insensitive to the feelings of the insane:

There is, I think, one grand mistake which we perpetually commit in judging of the moral nature of one another. It extends to the insane and the sane alike. We conclude respecting the existence or non-existence of sensibility, according as it speaks the same language, and keeps the same company with our own, or the reverse. Each makes his particular inclinations the standard for the world.[18]

Beddoes realised that he had little — nothing — to offer as a cure and for prevention he relied on the same general precepts for living he had given so often, adding the conventional warning about the connection of insanity with habitual drunkenness and with the excessive use of mercury. He had observed how some fevers and abscesses were followed by insanity and his supposition was that the brain had suffered damage. There is no advice about any violent form of insanity but only on the melancholic and those suffering from delusion, presumably because these were the patients Beddoes' readers were most likely to need to help in their own homes. He did have some practical advice to give. Convinced that early treatment might prevent the development of insanity, Beddoes warned that those who are over-anxious or who have disturbing dreams might be showing early signs of mental illness and needed special care. He also emphasised that it was right to be sensitive to the inner turmoil of those who were ill:

It is, we see, therefore very much with maniacs as with those that dream. The whole picture with its finest shadings remains impressed — or only the outlines of some of the figures are preserved — or else the whole is obliterated; nor would it be known but for indubitable effects, that manifest themselves to the bystanders, that any striking images, or any at all, had presented themselves to the fancy. Indeed, during intervals of reason and after recovery,

those who may be supposed best able to give an account of their own feelings, compare their phrenzy fits to dreams. And they agree with those who
have been occupied by any interesting exhibition, in being surprised at
the length of time that has elapsed since their attention was drawn to
themselves. In describing the state of their intellect, they speak of an un-
controulable hurry and whirl of ideas, by the rapidity of which every
endeavour to fix upon any one subject of thought in particular has been
rendered abortive.[19]

This sensitiveness led him to give a reminder that the insane do keep the
ability to paint and make models and that such occupations should be
arranged, as they were a relief to feelings that had been suppressed.

This essay gave Beddoes the occasion to explore mental activity more
generally. Imagination, dreams, the mysterious activity of the mind during
sleep and its power to influence our waking actions, as in the way something
learned before going to sleep is well remembered, all are introduced into the
discussion. The power of pleasurable associations to act on the mind is
emphasised as much as it was in the essays on education and here it is
amplified by a warning that it is possible to be deceived by sensations which
are in essence physical — "to believe the glow after a full meal to be a particular inspriation of the Holy Ghost". In all this Beddoes' authority is Erasmus
Darwin. He referred frequently to his *Zoonomia* which he had discussed with
Darwin before its publication. He quoted his explanations of delirium and of
madness, that the first "consists in convulsions of the fibres which [Darwin]
ingeniously supposes to exist in the organs of sense" and insanity arose from
"the excess of the sensorial powers of volition", and gave an account of his
writings on sleep.

Essays IX and X are the most surprising and interesting of the *Hygeia*
series. The conditions described as Nervous Disorders — epilepsy, convulsions,
hysteria — Beddoes could have met in his general practice and the essay on
Insanity suggests that possibly Beddoes as a doctor had met many families
endeavouring to care for the mentally sick at home; but he went far beyond
the practical advice they might have expected from him.

If we gather together all the diverse material in these essays, in the writings
dealing with the nurture and education of children, and his description of the
process of maturing and ageing in Essay VIII, 'On the Physical Powers of
Enjoyment', it becomes clear that Beddoes had an appreciation of the psy-
chology of individuals and of its significance in education and medicine.
In this field as in pneumatic medicine he was among the advanced thinkers
of his time. Some of his insights and suggestions anticipate what has only

become commonplace in recent times. Much needed to be done before the elements of his theory could be combined, as he well knew.

Impressions, ideas, pleasure and pain mix and alternate in an endless diversity of ways, as everyone must be sensible in himself and in others, though the order that presides over these phaenomena, has not been by any means completely investigated. Indeed a compleat investigation would not only comprehend the whole history of human nature in all its states, but disclose the means of maintaining the system in its most desirable known condition, and perhaps, of exalting it to a higher.[20]

He rejected metaphysics, abstractions and systems and asked for knowledge resulting from "refined tested and interesting" observation. Yet sadly, he feared that this approach was just what his readers might reject.

Beddoes was not all objective observer. He saw that he had come to the point where science and poetry met. The importance of feelings in bringing ideas together was, he suggested, both "the chief secret for unriddling the inconsistencies of dreams and the key to the boldest flights of lyric and dithyrambic poetry". In its turn reading of the right kind could be beneficial:

Feelings of every kind obtain a settlement in the breast by being associated with harmonious language and strong images.[21]

At the end of the essay Beddoes made yet a further digression, to discuss the poet himself. We may regret that he did not attempt a study of mental activity as coherently planned as his account of pneumatic medicine but any attempt to collect his main ideas from the discursive chaos of *Hygeia* shows that Beddoes was among those who in his day were advocating a more humane attitude to the mentally sick. His own humanity and his sense of the mystery of such illness is clear and almost his last comment on epilepsy is typical:

It is with the state of the soul in the fit as after death. No one has come back to make us a report concerning the one, nor has anyone recalled the ideas he has had during seizure as sometimes happens with regard to a dream, which on waking we know we have had, though we cannot bring to mind the particulars, 'till some associated ideas call them all up at once.[22]

It becomes clear that sanity as surely as physical health depends on a good environment and on experiences which associate pleasure with right ways of living; the connection becomes obvious between Preventive Medicine and Beddoes' conception of what society should be. In his introduction to the series of essays, he admitted that his plan of Preventive Medicine could not be complete because he had "not learned from experience the unfavourable effect of all the variety of human callings upon the human frame" and could not "point out with sufficient certainty the means of counteracting each unfavourable effect". A complete plan, he explained, would need a knowledge of society so extensive that it could only be collected by a co-operative effort. Even so, in each of the essays he found it possible, alongside the advice that he gave as a family doctor, to describe limited changes, some practical, others involving a new attitude of mind, which would bring about improvement in the state of society. He would have liked to see, for example, the sweeping away of superstitious beliefs and he would have welcomed a change in fashion which would make possible "a plan of social intercourse independent of the pleasures of the bottle". In this he had some encouragement, for he was able to give the Evening Discourses of the newly-formed Royal Institution as an example of social occasions without alcohol, wishing at the same time that these were open to young people. Besides the insistence on reforms in individual habits reiterated in every essay, Beddoes' last essay, 'On Prophylactic Medicine', was particularly concerned with matters of public health. Among the practical changes which he suggested were the inspection of schools and the setting up of Medical Boards. These would have power to collect information for the use of doctors and to insist on the isolation of patients suffering from infectious diseases. Money for such local hospitals should be raised by subscription. As the whole community would benefit from this, it would be a worthwhile expense. Another reform was to demand a higher standard of cleanliness in the great public hospitals. The most beneficial change of all would be for individuals to cease to worry about unavoidable catastrophes and to aim instead at reforming those habits which are the cause of disease. In a less guarded passage, where his old political views were allowed to reappear, he let slip a delightful proposal:

Should mankind ever cease to be content with slaughter, famine and pestilence, as a recompense for their senseless admiration of political adventurers and should the power of society ever be employed for the greatest good of the members ... edifices by their destination putting to shame the monuments of ancient art would rise under every inhospitable sky; and these conservatories of old age ... would afford complete shelter against the inclemencies and dangers of the season.[23]

Such reforms relating to particular circumstances are not all that Beddoes had in mind. He directed attention to matters of industrial health and detailed the risks inherent in the working practices of various occupations — not only in heavy manufactures and cottage industries which he had attacked in earlier writings but among the increasing numbers of workers in cramped offices and counting houses. Recognising that for changes to be soundly based it was necessary to collect a mass of facts, the 'social survey' which he proposed in his opening essay was much more sophisticated than any he had considered earlier. He wished to be able to compare the circumstances and conditions, the mental attitudes of various parallel groups in the community. This "minute, probing examination of the different classes that act and re-act upon each other within the compass of the British Isles" was to be a series of studies following through a whole lifetime with the purpose of showing, "how much pleasure or pain each derives from the abundance or want of spirits; from the sense of health, and from the gratification of appetite; from vigour of constitution or from infirmity".[24] Over a hundred years passed before anyone attempted the sort of comprehensive 'longitudinal study' that he envisaged.

By the time he wrote *Hygeia* Beddoes had arrived at a clear conception of the function of government. This he had already outlined in his 'Essay on the Public Merits of Mr Pitt', but in *Hygeia* he is more concerned with social problems than with political aspects. He saw that men cannot isolate themselves, either as individuals or in classes rigidly divided within society. He rejected, though he showed a sympathetic understanding of it, Rousseau's view that the ideal society is that of the primitive savage. Rousseau would, in his opinion, have done better to have been led by the miseries he had suffered and the evils he observed around him, not to reject but to reform society. To Beddoes the value of Rousseau's work lay in its exposure of evil, and the remedy proposed was both wrong and impracticable. Unlike Rousseau, he did not advise a "resort to filth and barbarity" but reminded that "the going of a clock depends neither upon the paint of its cover nor on the brightness of its face, but upon the perfection of each piece in the interior and the nice correspondence in the movements of all".[25] Men must use their powers of reflection and reasoning to remedy the ills of society and to rid it of the evils that come from the remains of barbarism. It was this effort of will and intellect that he stressed in his essay, 'To Heads of Families', and returned to in the second essay, addressed to Ministers of the Gospel. Here he again emphasised the defects of primitive society, but while he dismissed the dream of a virtuous simple life he vigorously drew the attention of the clergy to the worthlessness of the society around them where all classes of people, "whatever be the distance between them are inseparably linked by the chain of destructive vanity".[26] The luxury of the rich destroys their health and their power to enjoy life, while those who produce the luxuries are, "fixed down beside machines whose eternal rotation produces no greater variety of sounds than the rattling of the turnkey's bunch of keys or the creaking of the prison doors". Later, when he returned to the subject of tuberculosis in the essay 'On Consumption' and described the conditions which render people most liable to suffer from it, his criticism of society appeared as a positive philosophy combining the views of the doctor with those of the democrat. Beddoes had come to the conclusion, perhaps more easily assimilated by his readers in the context of a study of disease, that "The principle of civilization is none other than the spirit of equity. It is the abolition of the jus fortioris".[27] Children, women and slaves need to be given equality — and he made clear that among slaves, he included those condemned to labour to produce luxuries.

This is "the method in the madness" of *Hygeia* and, since much of the medical advice it contains had already been given elsewhere in his writings, the nugget to reward the labour of exploring its convolutions. For ten years Beddoes had been struggling to find how the hopes for a better society encouraged first by the American, then by the French, Revolution could find expression. In 1791 he must have been examining the reformist/conservative search for a renewal of the traditional freedoms of the English Constitution. In January 1792, he wrote from Oxford to Davies Giddy a long letter trying to find in the history of England some ground for the traditionalist view. His effort to work out the growth of the Constitution only led him to the conclusion that it would be "in vain to trace back the course of ages to discover a good form of government"[28] and he denied that the people, however much power they may on occasion have wrested from the government, have ever achieved truly representative government. He broke off to ask Giddy to enter into a discussion of his analysis; not surprisingly, for he had allowed himself to be led perilously near support for revolution. During the following years, while others such as Godwin and Paine published their 'systems', Beddoes, mostly responding to events, had made piecemeal efforts to gain a hearing for his own ideas. In the 'Essay on The Public Merits of Mr Pitt' he had come very near to Paine's "Civil government does not consist in executions; but in making such provision for the instruction of youth and the support of age, as to exclude, as much as possible, profligacy from the one and despair from the other". The description, in the last chapter of the '*Rights of Man*', of government conceived as actively bringing about the conditions of well-being which are the right of the people, must have been very convincing to Beddoes. Paine's advocacy is purely political, or perhaps political and moral, but in *Hygeia* Beddoes has made a new synthesis.

How did he come to make this leap? Clearly most important for his thinking was Erasmus Darwin's *Zoonomia* and his discussions with Darwin himself and with Davies Giddy. Here was the extension of his own work on chemistry and medicine. At the same time he knew, though he rejected, the challenge to this system from Kant's metaphysic. In the heart of *Hygeia*, in the discussion of mental states he wrote,

I hate metaphysics too; that is, the school of learning of old and modern Kantism. But the knowledge of the human understanding resulting from the most refined and interesting species of observation and capable of useful application I prefer to all other kinds of knowledge. Every day I lament not being a greater proficient in it.[29]

In Bristol his thinking had become deeper and had ranged more widely. He had seen more and more closely the sufferings of the sick, of children and of the poor. He had gathered at his home a brilliant circle; the young poets among them, Davy as much poet as chemist and Coleridge, cannot help but have turned discussion to the nature of imagination. This, as much as the frustration of political repression, may have given the impetus to make Beddoes bring his ideas into a more philosophical synthesis. As in so many things, he was too precipitate to achieve anything of lasting effect. The Essays were successful at the time; Coleridge admired them all except, significantly, the Essay 'On Insanity'.[30] The method which Beddoes devised to lead his readers to take in his thoughts on society too easily encouraged his excitability and love of 'galloping off in all directions'. There was always in him a tension between reasoned argument and an urgent desire for change. His experience as a physician faced with the suffering of his patients, both rich and poor, and the insoluble problems of disease led him to form his solution and in *Hygeia* to express it in his own eccentric form.

CHAPTER 12

BEHIND THE PRINT

With the publication of *Hygeia* and the organisation and establishment of the Institution for the Sick and Drooping Poor, Beddoes brought to an end his efforts to describe systematically his beliefs and their practical applications. Perhaps he had hoped, when he wrote to Davies Giddy in January 1801,

It can, I think do no good to have people linger under easily avoidable disease — and to excite a relish for the study of human nature is the most likely way to find the outlet from the dire dilemma if one exists — I have a very definite idea of the principle on which this outlet will be discovered . . .

that he could produce an organised treatise. But by the time he had finished *Hygeia* he had to admit his lack of success:

I take my leave of an engagement, which nothing but the perpetual spectacle of the fatal effects, arising from domestic errors in every part of the country, and the hope of paving the way for more salutary practices could have induced me to undertake. From my previous professional avocations, I had no right to calculate upon the frequent and continued interruptions to which I have actually been subject. From circumstances, many of which I could not controul, the author and the printer set to work nearly at the same instant. Hence the month, previous to the publication of each number, was the utmost space allotted for preparing it; and from this, in almost every instance, distant journeys have subtracted a fourth. Such is the cause of much want of arrangement, of frequent obscurities, and many inelegancies.[1]

Since this was written in 1803 it seems probable that the changes brought about by Davy's leaving the Pneumatic Institute and the financial difficulties referred to by Southey formed at least some of the circumstances which Beddoes could not control. His fact-finding activities however continued right up to the end of his life. He was still in 1808 working at his plan to use military information about the causes of death among soldiers and wrote to the Secretary of State to ask that the regular surgeons be empowered to give information. There are among his papers rough notes for a general survey, and almost his last work 'Good Advice for the Husbandman', is more a social than a medical study. This little pamphlet is entirely concerned with warning against excessive drinking of strong beer and cider at harvest time. He described how he determined to see for himself and discovered that it was "easy to mix early among harvest people . . . in different districts".[2] His findings make an astonishing and instructive study of the drinking habits of field workers in different parts of England and Scotland and incidentally reveal that in many parts of the country the corn was cut by women. He observed that where the harvesters "either drank no strong drink at all, or not above a quart a day [they were] cool and pleasant as the fields around them after being refreshed by the dews of the night. But for your six and eight quart men — they often cut as sorry a figure as any which the rising sun had to shine upon." Like many another, he warned that cider was particularly intoxicating: when it is plentiful "you met people reeling from hedge to hedge. The turnpike road was too narrow to hold them". For good measure, he included an account of the immense quantities of beer drunk by London coal-heavers, whose employers, significantly, were linked with the brewers. It was a pamphlet directed to the particular need of the moment, but it is typically Beddoes, bringing in his old suggestion that other crops should be substituted for barley and coming down to a recipe for a fizzy drink: ½ drachm carbonate of soda dissolved in ½ pint of the coldest water, and added to this 2½ drachms of dilute vitriolic acid — to be drunk quickly.

During these years his interest in statistics took a new form as he came to realise that historical as well as contemporary information was needed for full understanding of disease. In 1808 he not only opened his paper on fevers with a history of ideas about the subject but took the further step of asking the help of the bibliophile Francis Douce.[3] He wrote to Douce that he had been making a study of the way diseases increase or decrease in prevalence at different periods, with the purpose of discovering how this might be associated with changes in habits and diets. He explained that an understanding of these related changes and of the meaning of medical terms as they were used in the past, could be of great importance and asked Douce to direct him to books he might have overlooked. Such a study, he realised, was both valuable and difficult:

The connection between the state of health and the manners of respective periods will scarce receive material illustration till some clear union take place between antiquarian and medical information. To make a beginning some experienced and learned practitioner of medicine must be assisted by the antiquary; or some antiquary assisted by a physician, must extract and digest the scattered notices of ages.

Beddoes was on his own admission an obsessive collector of items of information that might 'come in useful one day' and "whenever [he had been] able to obtain certain information concerning any useful appearance in animal nature, it [had] been his custom to preserve it".

While these theoretical interests continued, Beddoes now abandoned
medical experimenting and his political activities; he wrote to Davies Giddy
in 1807, "I have given up all idea of writing on public affairs or men",[4] and
he concentrated on his practice as a family doctor. He had a considerable
success and was admired and liked by those who trusted him and perhaps he
could afford to ignore those who still regarded him with vehement antipathy.
He was not a man to ingratiate himself with all and sundry. From his early
days in Bristol Beddoes has left us records of his work as a medical scientist
or writer but we have little to show him as Dr Thomas Beddoes, family
physician. Fortunately, for a year, from March 1798, there are letters from
him to Davies Giddy about his patient, Miss Baines.[5] Beddoes' distress that
in the end 'airs' had not saved his young patient and his regret that new
treatment, probably with digitalis, had not been available show real concern.
He had encouraged moderate hope of improvement from a sea voyage,
though at the same time being careful not to go beyond hope, and suggested
that she should take no medicine until the voyage was over. His prescription
of half a grain of opium a day, and figs and 'plumbs' at the same time, is
much less drastic than the three grains Dr Darwin, in desperation, advised for
James Watt's daughter. Lydia Baines returned to try not only 'airs', but
Beddoes' 'cow house' treatment. This is another treatment for which Beddoes
was, and still is, ridiculed, but he was not alone in using it and, as has been
noticed, treatment of this kind may have been part of folk medicine. Beddoes
did not in fact suggest what we imagine from the expression 'stabling with
cows', but the partitioning of the patient's bedroom with the cow's head
thrust through curtains, and he had sympathy with his patient's disgust. He
realised that there were limits to using distasteful treatment. For Giddy, who
knew twenty year old Lydia Baines well,[6] the failure was even more poignant.
These are sad letters, but a young doctor [7] in Devon was able to give an
account of Beddoes' treatment just as warm in his appreciation of his kindness but with a happy result. When he was in despair and 'worn to a skeleton'
Dr Allen of Collerton wrote to Dr Beddoes who seems to have had a long
correspondence with him. This time he did advise digitalis but the two letters
Dr Allen published when he wrote the history of his cure to *The Medical
and Physical Journal*, show that Beddoes understood how quite small, homely
details would help a patient, even instructions about how to arrange his
bedding when he was sleepless, keeping his feet warm and the upper part of
his body cool. Dr Allen wished to pay a tribute to Beddoes' "disinterestedness
and humanity" as well as his skill and wrote of the "many, many letters [he
has] been at the trouble of writing".

In 1806–7 Southey wrote to advise his friend Grosvenor Bedford, "For God's sake, take care of yourself and if you go to Bath, pray, pray, go to Dr Beddoes at Bristol" [8] and in January 1807 he reinforced his advice: "I do not like him (i.e., Beddoes) — but I have the highest regard for his professional talents, and would, in any case of illness, resign myself into his hand with the utmost confidence· that he would do for me whatever could be done by medical skill." Southey felt he needed to reassure his friend that Beddoes did not experiment in his treatment, a reminder of the accusations made at the time of the Pneumatic Institute, but there seem to have been no lack of patients who had confidence in him. In 1807 he was called to treat Miss Estcourt,[9] daughter of J. Estcourt of Tetbury, and the series of letters Beddoes wrote to her father gives a lively account of his methods and of his practice; and it even includes some of his prescriptions. The preliminaries are to arrange to visit, either for a day or to stay overnight to observe the patient in the evening and morning, since Beddoes is "loathe to attempt anything without a personal interview". At the same early stage fees have to be arranged. Beddoes' letter on the subject fits Stock's comment that the doctor was not grasping but, on the contrary, generous to his poor patients and realistic in what he asked the well-off to pay. He seems to have put his fashionable patients in the position of deciding for themselves what to pay in accordance with their own estimate of their position in society:

I make it a rule never to have the smallest difference with any one on the subject of fees — And I never fail to find persons in a certain situation in life sufficiently willing to make the physician adequate compensation for his time and attention — I have never the least objection to be as explicit as any body can wish — and that no dissatisfaction may take place on either side, I state in answer to your question that, independently of travelling expenses which are easily computed, I have received, on account of a visit occupying about three days, from twenty to thirty five guineas.

Beddoes appears to have acted on the advice given him by Erasmus Darwin in 1796, "I think you should have 10 guineas a day, and a sum for what should be conceived to be chaise expenses." This was much the same as Darwin himself expected to receive. These matters settled, Beddoes was accepted as the family doctor and he advised on everything, from Mr Estcourt's bowel troubles to a tumour developed by the maid. He saw that Mrs Estcourt was surprised that he did not chat with her about the patient and explained to Mr Estcourt that he did not follow the habit of fashionable physicians and flatter the nurse with long discussions but confined himself to "dry, short plain directions". So he himself avoided the rather spurious consultations which he had condemned so vigorously in *Hygeia*. The children had whooping cough and measles and Beddoes had his own ideas about the treatment of both, particularly warning against the use of hemlock in whooping cough. Diet, his system of warm and cold bathing, the desirability of keeping at an even temperature, all appear in the letters. It seems on the whole that he advised a rather gentler system of drugs than some of his contemporaries but he advised that if Mr Estcourt stayed indoors for his cold he should keep *cool*. Another piece of advice directed against indulgence was that the "smallest quantity of food which will support life will be best for you" and, after a tactfully worded explanation of the foolishness of over-eating, Beddoes laid the responsibility on his patient, "You must settle for yourself whether it be within your power to confine yourself to such an allowance of food." The letters come to include other topics, Beddoes' hope in January 1807 that there was "prospect of an end to Buonaparte's career of mischief" and his expectation that in spite of the attention of his fashionable doctors Davy as a young man would recover "by benefit of nature". On a number of occasions Mr Estcourt gave Beddoes franks for his letters and undertook to forward a letter to Douce. One member of the Estcourt family needed especial care, the son Edmund, and Beddoes arranged to treat him in his own home at Clifton. Edmund seems to have stayed for several months in the summer of 1807 and Beddoes wrote kindly and encouraging letters to the parents.

These letters help us fit together other disconnected hints about Beddoes' practice. It would seem, from the visit of Edmund Estcourt, that members of the family and intimate friends were not alone in being taken into Beddoes' home when they needed treatment. Davies Giddy and Anna's bother Sneyd [10] both were nursed there, Sneyd moving from Dr King's home because he found that "Dr B's uninterrupted placidity [was] more suited to an invalid than Dr K's perpetual bustle and volubility". An unexpected comment this, for Beddoes could be brusque; but Anna too, in encouraging Tom Wedgwood to stay when he was ill, promised "The Dr, you know, is a most peaceable being, and could not disturb you".[11] Beddoes had assured Mr Estcourt that he often travelled further than Tetbury to visit patients and in the course of the correspondence wrote of visiting invalids in Herefordshire and Monmouthshire. Like many doctors of his time he had patients who were attracted by his reputation. Some, like Mrs Clarkson, wife of the campaigner against the slave trade, would go to Clifton to be near him. Others, he would travel long distances to visit, to Yorkshire and Wales as well as to the nearer counties; once, we know, he went to Bury St Edmunds, presumably to visit the Clarksons. Even as late as September 1808 he travelled to Aberystwith,[12] taking with him — rather unusually it would seem — Anna and "little Anna and Henry [who were] good and happy little travellers". Those who appreciated him must have been glad to have so "modern" a doctor; he was, for instance, using a clinical thermometer [13] and must have been one of the few doctors to attempt this accurate way of finding a patient's temperature, for such thermometers did not come into common use until the nineteenth century. Beddoes realised that these needed to be made in a way that was not tiresome for the patient. The bulb, "should be as large as a garden pea — at least — the bore of the tube very fine. The sooner it acquires the heat of the body the better otherwise people grown out of patience and you do not get what you want at last." Even more, we can imagine that it must have been a very sympathetic doctor who could write, "Let me advert to the necessity of soothing attentions . . . a blow on the mind will produce contusions as much as a blow on the body". By 1806 Beddoes had reached a point of success where he felt he could move to London.[14] He was wondering how quickly he could sell out in Bristol and calculating that he would need to plan for a year without income in London. The move, he hoped, would cheer Anna who was despondent and unwell, so unwell that he even thought of the possibility of her death and was anxious about what would become of the children. In the event, it was Beddoes' own ill-health that frustrated his hopes. He had been suffering from rheumatism and a fever of the mucous membrane, and in June 1806, only a month after he had been weighing up the possibilities of the move in a letter to Giddy, he was so ill that his re-covery appeared impossible. Recover he did, and took up all his old concerns.

The patients who stayed with Beddoes for treatment must have found themselves in a kindly and happy household. Not perhaps so romantic in his view of marriage as were his friends the young Pantisocrats with their dream "commune" by the Susquehanna, Beddoes showed more true sensitivity for his future wife. As we know, he entered in imagination into Anna's feelings on leaving home and marrying a man so much older than herself and he took care to ensure that she was made to feel at home in his family. Perhaps it was to hide deeper feelings that he claimed he married Anna because she could sew and because she had none of the artificiality that often came from a boarding school education. He was a penetrating observer of the artificiality of society and described it often enough but he wished in his own life to avoid it and expressed the hope that he and Anna would "escape the indif-ference of most married couples". In the early years the house must have been lively enough. There were the two sons of Mr W. H. Lambton, the wealthy Durham landowner whose legacy [15] Beddoes had used to help him

set up the Pneumatic Institute. They had come [16] to Beddoes in 1797 when
their father died in Italy and the eldest boy, John George, was a lively five
year old. Both Anna and her husband enjoyed having these two in their house
until it was time for them to go to Eton.[17] Nor can the group that gathered
round Beddoes from the time of the protest against the 'gagging acts' have
been entirely devoted to high seriousness in discussing politics, Beddoes'
medical/chemical plans and hopes, and Davy's experiments. There was much
fun and enjoyment too and Anna Beddoes' charm would not have been so
warmly noted, had she not joined in. Davy was from the time of his arrival
an agreeable member of the family and made a vivid impression of intense
life and energy on all who met him. As a romantic poet he had much to
share with Anna. Understanding the chemical work might have been hard
but Davy was well able to convey the excitement of breathing nitrous oxide:

> Not in the ideal dreams of wild desire
> Have I beheld the rapture waking form
> My bosom burns with no unhallowed fire
> Yet is my cheek with rosy blushes warm
> Yet are my eyes with sparkling lustre filled
> Yet is my mouth implite with murmuring sound
> Yet are my limbs with inward transports filled
> And clad with new-born mightiness around.[18]

There were light occasional verses too; and the poems describing the Cornish
scene which Southey admired and published in the 'Annual Anthology'.
Writing and discussing poetry went together among these friends.

To the end of his life, Beddoes enjoyed classical literature,[19] for he never
lost the ability to read Latin and Greek; and he enjoyed what was current in
both English and German literature. His own writings show that he knew well
the great writers of his own century, Dr Johnson and Swift, Fielding and
Sterne, his especial favourite. He was quite an acute critic and the loss of
Davy and Coleridge from the Clifton circle can hardly have meant an end to
literary talk. From passages in *Hygeia*,[20] Essay X, it would seem that Beddoes
had read Wordsworth's 1802 Preface to the *Lyrical Ballads* and been struck
by its description of the power of "harmonious language". though he does
not appear to have given Wordsworth's thought sustained attention. Beddoes
enjoyed Shakespeare's plays with an imaginative intensity that was in tune
with the new feeling that was replacing the attitude of Garrick which in his
schooldays had repulsed Beddoes. He wrote of the plays with understanding,
describing particularly how Shakespeare used madness to carry "terror and
compassion to a height which they cannot perhaps be made to match by any

other means". This was not solitary enjoyment in his study, for Beddoes loved conversation and exchange of ideas, particularly, we are told, with "educated women".

The scientific work of the laboratory turned to light-hearted pleasure on occasions. The literally hilarious sessions of breathing nitrous oxide may have been bad for Beddoes' scientific reputation and a source of anxiety for him, but for those without responsibility, considerable fun while they lasted. Beddoes himself must surely have contributed to the gaiety of the group. He could show quite a whimsical humour at times, as in his unconventional welcome to Dr Frank.[21] When this eminent physician from Vienna first called at Dowry Squre, Beddoes came into the hall to greet his visitor carrying an armful of books, all written by "Dr Frank". Which of these Dr Franks was his visitor, Beddoes wanted to know. He obviously gave no offence and the two got on well together, Dr Frank remarking, "Dr Beddoes reads German as well as he does English and is intimately acquainted with all our best authors". Beddoes' wit was sometimes less appropriately exercised, finding its way even into his serious writings. The ironic letter from "Jeremiah Morrison" to "Dr Daniel Renshaw" which opens the last part of 'Considerations' and the attack on pneumatic medicine in heroic couplets, also supposed to be written by "Dr Morrison", are both well done. Coming as they do in the paper which was planned as the decisive conclusion to the series preparing the way for the Pneumatic Institute, they are a striking example of the way Beddoes allowed himself to be provoked. Like the mocking verses and dramatic episodes in the 'Essay on . . . Mr Pitt' they must have been entertaining for his supporters and they certainly show both skill and inventiveness but they would not have won over his enemies. As Daniel in the lions' den, attacked by the Jeremiahs of the medical and political establishment, Beddoes felt keenly the prejudice against him and his wit is often sharp. There is bitterness in Beddoes' verses too, except for the 'Recipe for a Legendary Tale' and 'Domiciliary Verses' — but he is undeniably funny. The enthusiasm and high spirits of these days inevitably diminished when the Pneumatic Institute came to an end. The scattering of Beddoes' group would of itself have brought this about but there was a change of a different kind. In 1801 Anna Beddoes' first child, a girl 'little Anna', was born.

Anna and the three younger children appear briefly in the letters of both their mother and their father. There are the happy details that all parents write when family life is going well and the corresponding expressions of anxiety at times of illness. In a letter[22] to his uncle Mr Whitehall, written in 1806 and dealing mainly with financial matters connected with his father's estate, Beddoes expressed the hope that his "mother and sister will see the oldest two children this year. They are very healthy and fine children. The youngest is particularly pretty — is now somewhat above three months old — scarcely cries and thrives apace". The ex-baby Henry appears in a letter from Anna Beddoes to Davies Giddy, written in 1808 when she accompanied her husband to Aberystwith. Young Thomas Lovell Beddoes went off to stay with his Edgeworth grandparents and his mother described how, "Tom went to Ireland with his uncle Henry, where he has been very kindly received and where time passes merrily away — in Ireland the youth passes for a wit. Your little god-daughter, the particularly pretty baby, improves in person and in mind and also Henry is all life and vivacity." [23] The visit must have been a success, for Maria Edgeworth, writing after Dr Beddoes' death, considered it was fortunate that "little Tom" had visited for it had "raised feelings of kindness towards the whole family in Dr B's breast". Stock in his *Memoir*, suggests that Beddoes grew calmer as the years passed and in the privacy of his own home was relaxed and happy. He particularly enjoyed the company of young people and in Stock's phrase, was "attentive" [24] to his children. He took great delight in the first baby, Anna, going to see her as soon as he came into the house and picking her up to carry about in the room in a way that seems more twentieth than eighteenth century. An almost idyllic description tells how Beddoes, like Joseph Priestley, worked in the family sitting room. His little ginger kitten would sit on his paper while he was writing and Beddoes waited to turn over. In his family life we can see the warm-heartedness that Beddoes often hid behind a cold, even harsh, manner.

Things were seen a little differently by others. Her mother admitted that little Anna was a 'wild creature' and to her young uncle Henry Edgeworth she seemed, "a very fine child but . . . a little capricious" and he was in no doubt that this liveliness "arises from Dr Beddoes having and continuing . . . to spoil her when she is with him".[25] Anna at the time was only four years old; perhaps her father was merely putting into practice the conviction expressed in all his writings on education, that young children should be encouraged and their feelings respected. All that Stock tells us about the way the Lambton boys were educated while in Beddoes' care suggests that the advice he gave in his writings was the result of his own experience. He took the boys with him when he journeyed and encouraged their questions and curiosity, just as in his 'Letter to a Lady' he advised that the teachers should make opportunities for conversation with their pupils. Just as he advised in *Hygeia*, he observed their individual characters and drew attention [26] to the tempestuous nature of the elder boy, John George, with his combination of great ability and intensity of feeling, Beddoes' method was to use "the optative mood rather than the imperative", but he was not, all the same, too soft. The boys were free to choose their amusements but they were always to be active; the freedom to "stand and stare" does not seem to have entered Beddoes' philosophy. He talked to them as an equal and by his own enthusiasm he encouraged their powers of observation. Beddoes saw that conventional subjects were not neglected and the boys were up to standard when they went to Eton but Stock passes over this quickly in favour of details about lessons in science and mechanics. With James Sadler's son to teach chemistry and Dr King anatomy, these must have been thoroughly practical. Beddoes engaged artists to make models and working toys. We are able to imagine these toys from a description given in Maria Edgeworth's 'Practical Education', for she noted that the details there were suggested by Dr Beddoes. The models were made to come apart and the "names of different parts [were to] be written or stamped on them; by this means the names will be associated with realities" and be more easily remembered. Beddoes' own skill in making apparatus and models and his first hand knowledge of the manufacturing world probably reinforced his educational theories. Sensible [27] toys fitted with his general view that education should take account of new industrial needs and he saw that with such toys children "will neither learn by rote technical terms nor will they be retarded in their progress in mechanical invention by the want of language". Beddoes and Miss Edgeworth both appreciated the difficulties of finding suitable toys, and Beddoes besides having them made for his own pupils, tried to arrange that such toys should be produced for sale.

As early as 1793, Beddoes had advocated the use of models in teaching geometry and when, in 1796, the Bristol mathematician Benjamin Donne [28] published a description of his mathematical models, Beddoes recognised that the proposals fitted well with his own ideas and wrote in support. His own scheme, with the toys he envisaged, was not confined to the teaching of mathematics; he also seems to have recognised that opportunities to enjoy the sense of touch were important for a child's development. Though the plan had good support in "progressive" circles, Beddoes was not able to put it into practice. He tried again in 1800, having found a man capable of making the toys. Even then, outside Beddoes' own family, nothing came of it and the Rational Toys live only in Maria Edgeworth's 'Practical Education'. Donne seems not to have been as good a teacher as his schemes for mathematical toys makes him appear, if we are to believe Davies Giddy's assertion that the time he spent at the Mathematical Academy in Bristol was quite thrown away.

Beddoes seems certainly to have enjoyed teaching. Stock has preserved for us a series of letters[29] written to the younger sister of the lady who had such an unfortunate experience in breathing nitrous oxide. These are full of encouragement and designed to give confidence that she is quite capable of serious study. The young student is not to spend too much time on lady-like activities such as botanical drawing but to read the modern authorities, Withering and Erasmus Darwin, and to see in Botany more than classification and identification. He appears to have had no hesitation in directing his young friend to "The Botanic Garden" itself, despite the poem's exuberant descriptions of sexuality. Beddoes saw Botany, the subject which was accepted as suitable for young ladies, as a starting point. Good in itself, it could lead to an education that would be less restricted and less confined to what was conventional. Accordingly he insisted that she was quite capable of moving on from Arithmetic to Mathematics and of learning Latin and Greek – if she used the right method. Like himself, she needed to persist and then she would find writing easy and enjoyable and like the French prisoners-of-war she would discover that activity dispels gloom. The letters show his sympathy with women. He wrote as a friendly tutor but, as in his 'Lectures to a Female Audience', he made an opportunity to help the cause of women's education. He encouraged his young friend with the thought that education for women was indeed improving and gave this added importance by impressing on her that such improvement would bring about a change of tone in society as a whole. Despite this feminism, Beddoes adapted his programme a little by widening it to include literature, particularly Shakespeare's plays. He was sure she would enjoy them and learn from them just as he had come to do, for she would find that Shakespeare "throws out in his fits of playfulness observations upon life that would set up half a dozen moral philosophers".

These letters show none of the insensitivity that appears in Hygeia when he was advocating animal experiments to familiarise young people with biological facts. In the Hygeia essays he did not give any warning that such lessons and the observation of mating animals, needed to be introduced with any particular care. Perhaps, as he always advised that children should be taught in small groups, he assumed that observation and instruction would be adapted to the particular pupils. In his work, certainly, Beddoes had carried out what he described as "cruel experiments" – and he had been glad to dis-continue them. We know too, that he and Davy used frogs for experiments on Galvanism, at the time he had the two boys in his care. But exactly how far he went in carrying out these experiments at home with his pupils we do not know, or even whether little Thomas Lovell Beddoes saw anything of them.

Apart from this 'blind spot' produced by his scientific enthusiasm, Beddoes was a keen observer of people. Whatever his topic, medical, political or educational, his descriptions convince the reader that he campaigned for reform because he knew from close experience the distresses of his time and their effect on individuals. As a young doctor he had seen, and makes us see, the mother nursing her child in the neglected cottage, going out to fetch water in her black crock from the spout by the hedge; the father snatching up the crushed boy from beneath the horses' hooves as the colliers rush off to hide in the pits. Ten years later, writing as an experienced physician who knew them at first hand, he confronted the affluent readers of *Hygeia* with the miseries young people were made to suffer — the young ladies tormented by tight stays[30] even when the fashion was for loose. flowing dresses; the boy clerks, given too little time for their meals and forced to hurry back to the counting house and lean their full stomachs against their uncomfortable desks. The distortion of the economy that resulted from the demand for luxuries was brought from the realm of theory and given reality in his 'Letter on ... Scarcity', where he pressed Pitt to see that, "the manufacture of fringes and varnished tables might give way to that of worsted stockings and flannel jackets".[31] He saw the decline in trade brought about by disturbed political conditions, not as a detached economist, but as one who knew the men Reynolds was obliged to dismiss from the Coalbrookdale iron works;[32] "the poor men," he wrote to Davies Giddy, "have been remonstrating; they ask, 'Why should I be discharged rather than another', and then mutter out an oath that they would not starve." His sister's letters[33] reveal how the scarcity which led him to write on measures to combat food shortages, affected his own family. Rosamond Beddoes described how her friends were serving cold boiled potatoes, instead of bread, at meals and how her mother found that the best way of using potatoes to stretch out the bread flour was to boil and mash them first. In view of the high price of sugar Rosamond and her mother were also considering keeping bees. Farms near Beddoes' home in Shropshire were being broken into and, according to Rosamond, Mrs Beddoes had oiled all the door locks and proposed to have locks put on the shutters; she also planned to set a watch at night. Food shortages,[34] often little short of famine, were widespread and frequent from 1792 throughout the whole period of the Revolutionary and Napoleonic wars. In 1795 conditions were particularly severe in the West Country. At Nether Stowey, Thomas Poole, later one of Davy's "subjects" in the nitrous oxide experiments, had to deal with the near-starving workmen who tramped from Cornwall. The Poole family too, experimented in bread-making;

they were even short of potatoes and tried using turnips. In Bristol the shortage of flour was so severe that mobs threatened the bake-house. Beddoes must have seen for himself the effect on the poor, especially in the later years when so many were treated at his Institution. Unfortunately, no letter from Anna Beddoes reveals her household arrangements; we do not know whether he husband persuaded her to try out a pressure-cooker for broth making. He could easily have given her, just as he had given William Pitt, a clear and practical account of how such a cooker worked.

Beddoes' most commonly observed characteristic was a brusqueness of manner and aloofness but equally most people came to see that beneath there was kindliness and warmth. Stock considered this a trait which went right back to his boyhood and which combined with 'habitual shyness'; "but when his features relaxed into a smile, the character of benevolence was so strongly impressed upon his countenance that the most careless observer ... felt himself irresistibly attracted to it." With this kindliness, it must be recognised, Beddoes still observed with disgust the indifference of the fashionable world to the sufferings of the poor and he developed a thoroughly down to earth attitude. In his writings he admitted that he preferred an unfashionable directness and determination to make unpleasant facts plain. He was less harsh when he had poorer readers in view. His advice on how to save money by moderating expenditure on ale and tea was gently given but when he turned from readers for whom these were luxuries to rich parents who could afford, yet grudged, £20 spent on toys which would both teach and give lasting enjoyment, he sharply accused them of a perverse sense of values. His personal dealings were just as direct. During his engagement to Anna when he wrote to ask his father what it would be proper to settle on her, he insisted that his mother and sister should also be consulted. Unfortunately his efforts with his family did not always meet with a good response, especially after his father's death.

Happily he was more successful with his friends. He was very clear and definite in his proposals when he asked Humphry Davy and the Irish experimenter, Mr Boyd, to join him in Bristol, just as he was when he first became Mr Estcourt's family physician. He left no-one in doubt, stated his terms bluntly and was obviously prepared for an equally outspoken reply. His unreserved acceptance followed agreement, as Davy found when he was immediately, to his surprise as well as gratification, received with warm friendship into Beddoes' home. In the letter which described to Davies Giddy the circumstances of his leaving Oxford, Beddoes was equally plain and realistic about himself; "Last June I went to the Vice Chancellor, this you see was before any political ferment took place, I said I had no desire to read any more lectures; but that as I had given no chemical lectures that year and as I thought it unjustifiable to keep possession of the Elaby without reading any lectures and as I could not with convenience move my specimens until the summer, I would not formally resign until this summer." He then told how the Vice Chancellor persuaded him to write the memorial asking for a salary for the chemical chair and continued, "I went into the country, became eminently and much beyond my importance odious to Pitt and his gang as I know from a hundred curious facts – the memorial from this and other causes, was forgotten or destroyed." [35] A plainer statement could hardly have been made. With such success at Oxford, right to the end of his career there, Beddoes cannot entirely have lacked social graces, though absorption with his work and with the exciting development of science may have resulted in his not seeing any need to soften his natural directness. There is another possibility, suggested by Stock's comment that Beddoes' manner changed towards the end of his life. It is conceivable that as a democrat [36] and out of disgust, first with this sudden rejection and later with the artificialities of society, he deliberately adopted this blunt manner.

Beddoes' brusqueness was often held against him but there were many who appreciated his qualities and found him a good friend. Before the unhappiness over their father's estate, Rosamond and her brother got on well. She wrote him a lively letter to describe the workmen's "attack" on William Reynolds' home in August 1791, with just the details he would wish to know. [37] Rosamond admired Mrs Reynolds' courage in staying at home and watching from an upstairs window and she knew her brother would be interested to hear that it was Sadler who went to fetch arms from Birmingham. But her tone was not hostile to the workmen. Rosamond shared, too, her brother's horror at the revelations in the evidence submitted to the Commons by the Society for the Abolition of the Slave Trade. At the time of the petitioning in support of the Bill for abolition, Beddoes was at home [38] and reading a copy of the Report, and they must have talked about descriptions of conditions on the slave ships which it contained. These with their shocking details had forced the public to face what was going on. After that, perhaps, Beddoes had his sister's help in the petition he organised at Shifnal. How much influence Thomas Beddoes had with her once he was settled in Bristol, we do not know. We do know that Anna visited Hopesay Farm with her children and that Tom hoped to go again: 'Harry' (Charles Henry) was a baby at the teething stage: "a nice little creature like his papa only blue eyes." This cannot have been before 1805. In 1794–6 Rosamond was still on the same side as her brother politically. While most of her friends were prepared to pay Pitt's tax on hair powder, Rosamond and just one other lady determined not to take out a licence.[39] The purchase of a white wig was the way out for another friend. There is no record of the hair bandeau invented by Miss Edgeworth reaching Shropshire. Rosamond Beddoes' decision to order "2 or 3 dozen pairs of shoes from poor Hardy" after the end of the 1794 treason trial is even stronger proof of where her sympathies lay. Two or three dozen seems an astonishing number; but Hardy was indeed to be pitied. Though he was drawn in triumph through the streets from the Old Bailey to Piccadilly, after his acquittal, his trial ruined his shoe-making business and caused the death of his wife. Thomas Beddoes too, as might be expected, contributed to help Hardy and, later, gave assistance to another victim of the treason trials. This was Thelwall who came to Bristol hoping to join Coleridge but who had to journey further on to hide in rural Wales, where Beddoes gave him support.[40] To help these known opponents of the government certainly shows Beddoes' political sincerity, for it could not fail to be noted. Even generous Thomas Poole could not take the risk of encouraging Thelwall to stay in Somerset — the suspect Wordsworths had proved embarrassing enough.

Beddoes was always ready to help his friends. Coleridge and Davy are the two names that come first to mind, but the help he gave them, though it had its practical aspects, came from a meeting of minds and was quite exceptional. There are other less well remembered histories that show how thoughtful his friendship could be. Sadler's is one. Beddoes appreciated his skill as an instrument maker and took a keen interest in his experiments to improve the efficiency of steam engines. During 1791 and 1792 he was in touch with Reynolds about the progress of the engines and with Giddy about Sadler's work and possible uses for the engines in Cornwall. In August 1791 Sadler's engine was among a variety of topics discussed by Joshua Reynolds[41] in a long letter to Beddoes. He was "convinced that it will answer no good purposes for large engines but . . . convinced its advantageous form and easy application to lathes will make it a useful machine for manufactures of bright work when polishing is by lap burnishing." The engines and Sadler's attempts to improve them, appear in Beddoes' letters during the early part of 1792 and in August Sadler sent his own account of the engine which had been tried and found superior to Boulton's: "It works 19 pounds to the square inch"[42] and would cost half as much as Boulton and Watt's to build. This engine, Beddoes was able to tell Giddy in November, met with Reynolds' approval. Coalbrookdale and Cornwall, where Giddy was keenly interested in the development of steam engines for use in the mines, were the natural places to try out Sadler's engines. When he was replaced by James Watt as the designer of apparatus for the pneumatic experiment and became Barrack Master at Portsmouth his association with Beddoes did not harm him.[43] The work at the dockyard and as chemist to the Naval Board may have suited him better and while he was there he did make an apparatus for preparing oxygen. In spite of his interesting work, when the office of chemist was brought to an end in 1805 Sadler was shabbily treated. Once more the connection with Beddoes was useful. Davies Giddy, by then MP for Bodmin, was able to write of Sadler's work for the 'Professor' of Chemistry at Oxford and was among those whose efforts, in the end, obtained some compensation for him. Beddoes had, meanwhile employed his son as tutor in chemistry in his educational programme for the Lambton boys.

Another young man who, as Beddoes' assistant, found that he was befriended and his talents were soon noticed and encouraged, was the Swiss refugee Johann Koenig.[44] In 1799 Beddoes recommended John King – as he anglicised his name – to Thomas Wedgwood as medical companion on a voyage to Jamaica. On his return, King joined the group at the Pneumatic Institute. Beddoes may at first have chosen him as a skilful experimenter, but King was also an accomplished engraver and Beddoes soon found an opportunity to make use of his artistic talents. He planned that they should collaborate in an illustrated work on physiology and a series of drawings was made. The work was never produced but Beddoes encouraged and made known King's researches on physical education and on the relation between dress and health. Dr King stayed to take charge of the physiological investigations at the Dispensary and he had his own medical practice in the town. Anna Beddoes' sister, Emmeline, came to visit in Bristol and in 1802 she and John King were married. The marriage was not welcomed by the Edgeworth family and Emmeline's father was not generous in the financial settlement he made for his daughter. King was a popular doctor among the poor as he did much to help them but he had difficulty in being accepted in Bristol. A Quixotic attempt to help a reformed prostitute by employing her as a nurse after his wife's confinement not surprisingly gave rise to gossip and his atheistic views were held against him. He was constantly struggling with financial worries and Beddoes consulted Davies Giddy about how they could arrange for him to move to a small town where things would be easier. The trouble seems to have been exactly the situation Beddoes described in his Letter to Sir Joseph Banks. King had no friends among the fashionable practitioners to send him rich patients. He stayed in Bristol, always feeling unfairly treated in the medical world but active and well regarded in the city's artistic circles.

Friendship flowed towards Beddoes as well as from him, and never more than from the friends he made at the beginning of his career. Beddoes was of a stature to win the support of men who were themselves original thinkers and active in public life; Joseph Black, Erasmus Darwin, James Watt. Among his contemporaries, William Reynolds, Thomas Wedgwood and the younger James Watt were probably at least as valuable as partners for the exchange of ideas as for the financial support they themselves gave to the Pneumatic Institute and their energy in raising subscriptions from others. Reynolds, sharing in the early days many of Beddoes' political convictions, gave him a retreat, a time to think and help in planning his future after the shock of finding himself on the list of treasonable persons in 1792. Of all such friends, it was Davies Giddy who was to prove the most faithful and devoted. At first, Beddoes, as a senior member of Oxford University, took the lead.

It was Beddoes who in 1791 introduced Giddy to Sir Joseph Banks and then supported his application to be admitted Fellow of the Royal Society. Giddy was a good mathematician but in 1791 to be F.R.S. conferred social even more than academic standing so evidently Beddoes was at that time acceptable in "established circles". In the last years of Beddoes' life it was Giddy, by then an MP and a respected public figure with very different ideas of how to solve the problems of poverty and hunger, who was giving support. The concerns that Beddoes and Davies Giddy shared at the beginning of their friendship came to an end one by one. Giddy soon accepted the public duties expected of him as a landowner and the realisation that, as Sheriff of Cornwall, he was under an obligation to put aside his own feelings in carrying out his duties, brought to an end the possibility of their sharing an enthusiasm for the French Revolution. Instead, Beddoes had to sympathise [45] with his friend's inner conflict when in May 1792 it was his duty to present the County's loyal address to the King. They were drawn together first by Beddoes' geological interests which led to the tour to the South West. When concentration on the plans for pneumatic medicine obliged Beddoes to give up formal geological work, so that he declined to help Mr Polwhele [46] with a mineralogical Survey of Devonshire, he could still appeal to Giddy whenever he needed mineral specimens. Giddy entered into the new schemes. He was not, in the early 1790s, so nationally influential as others of the group who helped Beddoes, and he was unable to inspire in Cornwall support for the Institute, but he was the friend to whom Beddoes was able to write intimately. They discussed by letter the scientific ideas behind the plan and the relevance of Darwin's *Zoonomia*. Beddoes later wrote to Giddy about the details of his cases. Above all it was Giddy with his intimate knowledge of Beddoes and of his work who recognized Davy as the right assistant. Giddy was the friend to whom Beddoes wrote about the day to day efforts that lay behind the published series of 'Considerations'. All the time, Giddy moved more and more into the world of political action where ideas had to be embodied in legislation or administration. Beddoes was glad to see Giddy in Parliament and gave him advice about toughening himself before an election: he was to take a little ginger or sal volatile; sponge his body with brine and then take a cold bath or shower.[47] At the same time, Beddoes himself remained on the outside, without public responsibility and the challenge that it gives to put theory into practice.

Beddoes expressed to Giddy all his conflicting hopes and anxieties. His published references to contemporary affairs show his anger both at events and at the political and social conditions in the country; it is only in the letters to Giddy that we see the feeling beneath his polemics. His distress at mob violence; his horror at the slave trade; the intense swings of mood from hope to horrified despair that he felt as he watched events in France, all appear and show not a political agitator but an anguished man tossed about and trying to come to a balance. In writing to Giddy, Beddoes could argue out his views on government, coming even to admit, in spite of his distress at the terror, that he could find the republican form consistent with honesty and commonsense. Most outspoken of all, he recorded instances of the government oppression in England, the surveillance of himself and the unfair trials after the Birmingham riots. A notably ill-written letter from Shifnal on 19th November 1792, anxious about reactionaries and about the way he himself had been under surveillance, marks the depth of his depression. In spite of the divergence of their views, the old intimacy remained and in March 1795 Beddoes wrote frankly of his pessimism about the future with England's reputation lost; her government corrupt and France's power on the continent apparently unassailable. He saw danger at home from the obstinacy of the aristocracy, which in his view was no less than that of the French "noblesse". It must have been an alarming letter to receive, especially as Beddoes indicated that he had some political publication in mind. He had complete confidence in Giddy, who in turn could say to Beddoes: "I write to you with a freedom I should not use to any other person."

After this, political topics appear less often and gradually disappear. The letters are concerned with the Pneumatic Institute, Beddoes' lectures and family affairs; such comparatively trivial concerns as the barouche with elbow

room for four which Beddoes wished to buy and the books from the Leipzig
Fair he had ordered for himself and Giddy, to matters as important as
Beddoes' investments. Beddoes came to entrust his financial affairs entirely to
Giddy. Sometimes there were delicate and unhappy problems. When his sister
Philippa was planning an unsuitable marriage Giddy confided to Beddoes his
desperate anxiety and sought his opinion as doctor. It is understandable that
Giddy was distressed at his sister's choice for Mr Guillemard had a mistress and
children in Liverpool; but to seek Beddoes' confirmation that he was insane
was less well judged. We must hope Giddy did not reveal all his anxieties
to his sister, for he failed to prevent what proved to be a happy marriage.
Similarly, when Beddoes thought it best to sell the tanyard in Shifnal, which,
understandably, he felt he could not oversee from Bristol he turned to Davies
Giddy for help in dealing with the difficulties Rosamond and Mrs Beddoes
were making. He hoped that Giddy would go to Shropshire to see the problem
"on the ground". More intimate still, and comparable with the letters about
his feelings for Anna just before their marriage, is the letter Beddoes wrote at
the time of his father's death in 1803. It was a time when he was overcome
with melancholy, partly because he had to remember the unhappiness his
father caused and partly because the return to Shifnal after a long absence
brought home the sad, inevitable changes in human life. The "Hamlet" mood
must sometimes come, but he hoped that Giddy could tell him the 'secret' of
keeping it at bay.

It was Anna Beddoes who put the friendship to the test.[48] Charming and
delightful as she seemed to all who knew her, from her efforts to 'improve'
herself, it would seem that she felt left behind among her husband's lively
and intellectual friends. When, in 1800, Giddy feared he had developed
consumption and came to put himself under Beddoes' care, Anna Beddoes
fell madly in love with him. Though only seven years younger than Beddoes,
Giddy was not burdened by illness and disappointment. He was already a
figure in public life in Cornwall and, unlike the "short fat little doctor", was
tall and good looking. But he had too, a rigid code of right behaviour learned
in his childhood and developed by his reading and experience into a stoic
sense of rectitude. So at first Giddy reasoned and calmed; when Anna went
so far as to suggest that they become lovers he resisted with determined
firmness. Twice she pursued him, in 1804 to Oxford and in 1806 to London,
though part of her purpose in going to stay with her friend Philippa Giddy,
now Mrs Guillemard, may have been to meet her brother, Henry. Giddy had
to endure her despairing cry that but for little Anna she would kill herself.
Giddy seems to have felt that he could teach Anna to accept a brother/sister

affection. This was the mode of relationship with women that seemed easiest to him. Anna could not — or would not — change and it was left to Davies Giddy to bring about the end. In 1806 he began his plans to marry the wealthy Miss Ann Mary Gilbert. They were engaged in 1807 and married in April 1808; a cool, rational, from a worldly point of view, suitable marriage. For the whole year Anna was in a turmoil, first for herself, then fearing for Giddy's own happiness in a too-hasty marriage. Kind and firm as he was, the discipline he imposed on himself has hidden for ever his innermost feelings. What is not in doubt is the sincerity and depth of his friendship for Thomas Beddoes, soon to be most poignantly tested.

Of Beddoes' feelings during this time we have not the slightest hint. A doctor, trained to observe, it would seem he could not have been totally unaware. Perhaps, remembering their age difference and his ill-health, he generously could not blame Anna; and he took lightly an attempt she made to explain her feelings. More probably he so relied on Giddy's loyalty that he really was unable to see what he would never imagine. He was happy that Anna enjoyed her stay at Tredrea where she went to recuperate after the birth of her first child and hoped that she would be able to holiday with Miss Giddy again. He must, in spite of his own reservations about religious observances, have agreed at Christmas 1807 that Giddy's father should baptise 'little Anna', Tom and Charles Henry, and have approved that Mary Giddy should be godmother to the baby girl born in 1808. Beddoes could write in 1803 of the "constant serenity I have enjoyed at home and which is little likely to be disturbed by any internal cause." And so he would have wished it to continue.[49]

This serenity Beddoes valued above all for himself and it was the quality he aimed at teaching others. His hope was to find the way that would defend man against the "nihility" of life and save him from the destructive mood in which "the mind perpetually changes into a huge vault in which the world and everything it contains lies entombed".

Three periods of Beddoes' life would appear to promise success of a kind that might have won him lasting reputation: his Oxford Readership, from 1787 to 1792; the time of intense medical and chemical investigation when the Pneumatic Institute was first in being, 1799–1801; and the organisation of the Preventive Medicine Institution after 1802. The years in Oxford, with his original lectures and his grasp of new scientific theories and of their significance for both medicine and industry, had the appearance of the beginning of a notable scientific career. His own work on alkalis and electricity suggests the possibility he might have made further advances in chemistry and might even have become the pioneer of modern chemistry teaching in the University. Circumstances, the conservative atmosphere and declining numbers at Oxford, even without his own political indiscretions, were against him. At the Pneumatic Institute, the significance of what was achieved by Davy's work, expecially his observation of the pain-killing effect of nitrous oxide, is now so clear that his failure to grasp it stands as the tragic mistake of Beddoes' life. Yet it is far from certain that, even if he had recognised it, he had the means to seize the opportunity. The other aspect of the Institute, the work for preventive medicine, Beddoes did succeed in developing soundly, and it was in its own right original and far-sighted. It was a pioneering work, deserving to become a model for similar projects. This time neither circumstances nor over-hastiness stood in the way of success but, except in Bristol, his Institution went unnoticed and Beddoes felt keenly the absence of interest. Only Dr Frank, on his visit in 1803, seems to have appreciated what Beddoes was doing and was so impressed that he gave up other plans in order to stay longer with Beddoes and study his work. Beddoes had to face much prejudice. The essay on Pitt which appeared in 1796 was the last of his political writings but he was always outspoken and memories are long. He was unreserved, too, in criticism of the medical profession, exposing the greed and charlatanism of fashionable doctors and drawing attention to the reality that medicine could provide no remedies once illness was established. Some among the medical profession supported his investigation of gases and he found some doctors to collaborate in the collection of medical information. But the revenge of the medical establishment was to ignore what he had to say. In the Medical Institution Beddoes at last set his work on a sound foundation. Rules for the patients were worked out and published; the medical work was apportioned among a group of assistants; and in time new premises were found. Anna Beddoes wrote of the early days:

Dr B is going on successfully with the Institute. Crowds of poor people flock there twice a week 300 patients have attended for this last month. By this means new remedies are tried and the results stated in a book kept for that purpose so that good and bad is fairly detailed. This is regularly seen by an Irish medical man by name of Scully, who executes Dr B's plans as punctually as he would himself or more so.[50]

In preventive medicine alone, in the administration of the clinic and in the surveying of conditions relevant to disease, Beddoes had a programme possible in terms of the knowledge available to him. The frustration of his work in this field is consequently even saddler than his earlier disappointments which are now understood as having been inevitable. There are practical, even amusing details which make clear how much in advance of his time were his dreams. He clearly loved technical devices; not only was he an early user of the clinical thermometer, he gave careful thought to the details of its design. In considering the health problems of women he saw that mechanical inventions might be of more help to them than medicine. The pressure cooker he probably thought of as something to use for community kitchens but it was for the individual mother of a family that he dreamed of so modern an apparatus as a washing machine, rightly thinking how much her health would benefit. "Among the hardships incident to married poor women who have a family, I have noticed one as particularly severe ... This is neither more or less than their frequent dabbling in water to wash," he wrote to James Watt in 1808,[51] only a month before his death. "Could no good Genius invent a machine, by which opulent neighbours, attentive to the welfare of these hard fated human creatures, shall be able to redeem some of them from this destructive drudgery?" But for domestic use a washing machine driven by steam was not much more practical than Darwin's steam powered aeroplane.[52]

During the last years of his life Beddoes worked on in face of ever more frequently occurring attacks of serious illness. It may be for this reason that he settled for a more orthodox and even fashionable practice as a doctor. He must have been successful and respected, since, at the time of Beddoes' death, Richard Lovell Edgeworth thought that Henry, newly qualified as a doctor, might consider taking over his brother-in-law's practice. Maria Edgeworth described to her brother Sneyd how Anna "wrote herself to Henry the day or two before Dr B. died to beg he would go to her" adding "I suppose he is of course there by this time. My father has written to Henry ... a most kind and sensible letter recommending him to try to get into Dr B's place at Bristol ... if he got into good practice there he can remove to Dublin with his Bristol reputation whenever he pleases."[53]

Early in 1808 Beddoes was relatively well, travelling to visit his patients, but in the autumn he fell ill. Anna wrote of him as very weak and deaf — 'reduced'[54] — Mr Estcourt sent him grapes and a hare which he 'drank as soup'. In October Davies Giddy and his wife visited, staying first with Emmeline and John King and then with Beddoes and Anna. Still in November he would write about collecting medical records but his struggles with illness and with the treatment his doctors prescribed took over his letters to Giddy. Then came Anna's, "I have not the power of mind to make him take what is proper for him."[55] On Christmas Eve 1808 she wrote imploring Giddy to come and on 28th December she sent an unsigned note announcing her husband's death.

Illness brought down the barrier between Anna and her husband. She wrote, "In illness Dr B appears to advantage in many respects: his mind by being weakened dwells more upon individuals than upon general objects. Of course he expresses much more feeling and has less appearance of selfishness than when his mind is occupied upon larger objects." Her immediate thought was to avoid, "anything that would hurt his fame" in the writing of a memoir. Though Davy could not – as Anna Beddoes hoped he might – write a "Life", he and Coleridge together paid tribute that reaches back to the warm friendships of the creative time at Clifton. Coleridge learned of his friend's death from Davy who wrote, "Alas, poor Beddoes is dead! ... He is gone at the moment when his mind was purified and exalted for noble affections and great work ... My heart is heavy ... I am interrupted by very melancholy feelings ..." [56] Coleridge on reading the letter was overwhelmed by a violent fit of weeping. A month later he put down his own feelings. "I felt that more hope had been taken out of my life by this than by any former event. For B was good and beneficient to all men, but to me he had always been kind and affectionate and latterly I had become attached to him by a personal tenderness." [57] If the tributes of these young men to whom Beddoes meant so much at the beginning of their careers may seem too partial, we may turn to the letter that Beddoes asked Estcourt to forward to Douce on 8th March 1808. It has survived, with Douce's inscription: "The excellent and amiable writer of this letter did not live quite 9 months after the above date leaving me to lament the loss of great personal kindness and advantage." [58]

There is no letter from Giddy: he was with Anna at Dr Beddoes' funeral and his tribute to his friend took a different form. Beddoes had once been anxious about what would happen to his children if their mother died and in his last illness he made the best provision for them that he could by appointing Giddy as his trustee and their guardian. Anna was left with a comfortable income and provision for the upbringing of her four children. Davies Giddy carried out his responsibilities with more than efficiency. He showed generosity and friendship to Beddoes' two sons and understanding kindness to Anna. He had played a part in the setting up of a pioneer institute for medical research and started on his career Humphry Davy who was to end as President of the Royal Society – and in an odd closing of the circle he succeeded him in that office. But in any account of Thomas Beddoes, Davies Giddy must above all be remembered for the loyalty to the friendship begun when he was a student in Oxford.

Contemporary notices at the time of Beddoes' death tend to describe him as a kindly, eccentric doctor, to make little reference to his scientific achievements and none to his politics. The Edinburgh Medical Society's account does more justice to his work. In England the earliest attempt to record Beddoes' originality as a chemist and medical philosopher was Dr Grane's contribution to the *'Gentleman's Magazine'* [59] in February 1809. His very succinct account of Beddoes' achievements could well have stood until Anna Beddoes could find a more practised biographer than Dr Stock, had it not been for the jogging couplets in which it was expressed.

> True Genius kindles fires, whose piercing light
> Reveals all Nature's secrets to the sight
> Of him, whose turn for observation draws
> Conclusions founded on unerring laws.
> Beddoes in Chemistry thus gathered fame
> And few contested the Professor's claim.

Dr Crane continued with a catalogue of Beddoes' medical writings until, faced with the need to refer to the Pneumatic Institute he avoided making a judgement, calling it a 'Visionary dream', and ended with a rebuke to Beddoes' detractors:

> But hold, my Muse, and let no strictures here
> Attempt to pluck the laurels from his bier.
> To worth departed act a kinder part
> And recollect the goodness of his heart.

Beddoes' enemies lacked this generosity. 'Amicus' had already written a spiteful account attacking him for atheism, impudence in expressing his political opinions and wrong judgement in medical experiments.

Neglect, silence, worse enemies to reputation than hostility, followed. Dr Beddoes had no intellectual heirs. His work and his ideas had to be taken up afresh. In chemistry he never claimed any continuity between Davy's work and his own early, inconclusive experiments on alkalis and on electricity as a means of analysis, but wrote on 17th November 1808, "Davy has just solved one of the greatest problems of chemistry by decomposing the fixed alkalis." [60] In medicine, neither the use of nitrous oxide gas as an anaesthetic nor the movement to establish dispensaries for poor people can with absolute certainty be traced back without a break to his original work.

Beddoes did think more than once of leaving the country. He had an idealised picture of the American countryside and wished he could settle, "where the Ohio meets the Muskingham". [61] In a mood of despondency, he thought that Davies Giddy too would do well to leave England; with his

experience of the problems of the countryside, he would be more appreciated
in Philadephia. Yet he realised that property makes it difficult for a man to
move; perhaps, as well as thinking of Giddy's Cornish estate, Beddoes had in
mind his own family's farms in Shropshire. Later, when Beddoes thought of
the West Indies where he had patients, his conscience would not let him go to
the land of slavery. There was much in Bristol to remind him of conditions in
those islands which were so closely linked by trade, particularly the sugar
trade, with the city. Slaves themselves could no longer be bought and sold in
the city — it was far back in history that Bristol had had its own slave market
— but in Beddoes' day there were many people of African/West Indian origin
living there and in auction rooms at home the wealthy could buy West Indian
estates complete with their slaves. The experience of Joseph Priestley and
Thomas Paine suggests that Beddoes too would not have found that his work
and ideas were more easily accepted in America. Both had at first been
welcomed as refugees from a country hostile to supporters of liberty and
after a while both met in America a spirit of intolerance which crushed them,
Beddoes wished for reform rather than for the overthrow of the government
in Britain. As the war with France progressed, overt opposition was less
attractive even to those who had ardently hoped for change. Many who had
opposed war with France and had tried to show that it would only drive the
revolution to become more violent and extreme retreated from agitation after
1795, once there was no longer hope of a peace treaty. Coleridge's *'Fears
in Solitude'* and his *'France: An Ode'*, both written in 1797/8 while he
was still in Somerset, express a conflict that must have been widely felt,
disillusionment at the betrayal of freedom in France and distress that honest
criticism of wrongs at home must now conflict with newly realised love of
Britain. Beddoes found work he could do in this country, as a doctor and in
advocating a positive attitude to health. He did not need to go into exile, to
lie low in the remote countryside like Thelwall in Wales or, like Wordsworth,
to spend his old age covering up his early expressions of revolutionary hope.
He worked on, with astounding mental and even physical energy as ill-health
increased. It is hard to imagine what it meant, in Beddoes' time, to suffer
from shortness of breath, the pain of rheumatism; the misery of intestinal
disorders and, near the end, the frustration of deafness — not in old age but
in the years when we now expect to be active and healthy. The eighteenth
century had two resources — alcohol and opium. The first Beddoes rejected,
for himself as he advised others; the second he regarded as a medicine. The
worst affliction was loneliness; and he speaks his own sad epitaph: "I lack the
consolation of a friend."

CHAPTER 13

FAMILY AND REPUTATION

The thought that Henry Edgeworth might take over Beddoes' practice in Bristol and share Anna's house led nowhere. No. 3 Rodney Place was sold in February 1809 and Beddoes' fine collection of medical and scientific books and journals went to London to be auctioned by Sotheby's. Anna Beddoes settled first in Malvern and then moved to Bath, where she remained while her children were at school. Thomas Lovell and Henry went to the Grammar school and Anna was sent to boarding school at Warwick. After all that Dr Beddoes had written about girls' boarding schools, this plan for Anna seems surprising but Miss Byerley, sister of Wedgwood's partner Thomas Byerley, might have met with his approval. Anna's friendship with Philippa Giddy continued. Mrs Guillemard, as she became after her marriage in 1805, visited [1] Anna in Bath and later, when Henry Beddoes decided to join the navy, she was able to interest her friends Captain and Miss Willoughby who could be helpful to him. In 1821 Anna wrote to her sister Emmeline that she was too busy fitting out Henry for his ship to be able to go to Clifton; she wished Emmeline to go to her instead. Henry was just sixteen. In 1823, the time when she was anxious about Thomas publishing his poetry, Anna described herself as "extremely happy on Henry's account". He had been a rated Midshipman for two years and had gone to the West Indies on board the *Tribune*. In this letter, written from Brussels, Anna again referred to help from Mrs Guillemard's friends. It would seem that they continued to interest themselves in Henry's career, for he named his eldest son Thomas Henry Willoughby [Beddoes]. Once her responsibilities to her sons were over, Anna lived abroad. After her visit to Brussels, where she was shocked by the ostentatious extravagance of the English, she settled for a while in Tours. Then in 1824 she moved to Italy, visiting Rome, where her younger daughter, Mary, kept a record of the places visited. Anna and her daughter were only recently settled in Florence when in May 1824 she died. Thomas Lovell, on his way to visit her, had left England before receiving the news of his mother's death.

When in 1811 Anna Beddoes spent the summer at her father's home in Edgeworthstown, Davy met her and found her "greatly improved in bodily and mental health"[2] but it seemed to him she had not yet recovered the

sparkle that he remembered so well. Her husband's death must have brought
a turmoil of emotion as well as sadness, coming when she had not subdued
her feelings for Davies Giddy and so soon after his marriage. Anna came
to feel the sadness of realising that she had not appreciated her husband's
goodness. Davies Giddy showed her understanding and sympathy far beyond
the mere management of her financial affairs and this she learned to value.
From Brussels, when her home in Bath was being given up, Anna wrote to
Emmeline King:

Mr Gilbert [i.e. Giddy] is so gentle in his manner of writing to me now, that
it is very soothing to me, as he with reason might find fault with my want of
foresight and bad management.

Giddy, as well as being Trustee, was the children's guardian [3] and particularly
took care of the boys' education, leaving Anna to care for her two daughters.
While Thomas and Henry were still at school in Bath he arranged for them to
visit at Cheney Longville. To Dr Beddoes' cousin (in the extended sense of
the word) Giddy reported with some pride Tom's success at Charterhouse,
where both he and Henry were sent in 1817. He wrote with satisfaction
again of Tom at Oxford, at Pembroke, his father's and Giddy's own College:
"I believe our young friend goes on very well at College." This was in 1822
in Thomas Lovell's second year but by 1824 he had decided to leave without
taking his degree, and go to Germany. By then he was twenty one and Giddy's
guardianship came to an end. He had managed the finances of Beddoes'
children with skill. Sometimes when there were uncomfortable difficulties
over the rents from the Hopesay farm, he had himself to make advances.
All this Thomas Lovell must have seen most clearly when his finances were
handed over to him; he remembered with affection Davies Giddy's generosity
and warm care. The friends with whom Beddoes had shared his scientific
ideas and who had understood and supported his hopes for pneumatic medicine had all died within a few years during the first decade of the nineteenth
century: Erasmus Darwin in 1802, Thomas Wedgwood in 1805 and William
Reynolds, perhaps the closest of all his friends, in 1803. Only James Watt
remained. With such a succession of losses, no wonder Beddoes felt alone.
He had turned always to Giddy for practical advice and his trust in looking
to him to care for Anna and the four children proved more than justified.
    Thomas Beddoes' life is curiousy paralleled [4] in his son's. Very naturally,
the young Thomas chose to follow his father in studying medicine and like
him he became dissatisfied with Oxford and with the 'hauteur' [5] of English
society; he too set himself to learn German, still not a very popular study.

Revolution and war had prevented Dr Beddoes from visiting Germany but Thomas Lovell was free to choose for his medicial studies Göttingen, the University so well known to his father, possibly at the suggestion of Dr King. An earlier judgement is brought to mind by his comment on the German university: "There is an appetite for learning, a spirit of diligence and withal a good natured fellow feeling wholly unparalleled in our old apoplectic and paralytic Almae Matres." At first, the medical course, especially the anatomy school; the lectures of Blumenbach; the literary culture, were absorbing enough but soon Thomas was involved in nationalist movements and became like his father "obnoxious" to authority. Possibly it is to be expected that two men with lively temperaments and high intelligence should react in similar ways, one to the events of the 1790s and the younger to the uprisings which were part of the nationalist movements of the 1830s and 1840s. What he came to know of his father from reading, in all probability *Alexander's Expedition* with its essays and *Hygeia*, would have confirmed rather than discouraged Thomas Lovell's choices. He was expelled from Bavaria and obliged to leave Zürich as a result of his political activities. He died in Switzerland in 1849; his family found it hard to accept the fact of his suicide which was concealed by his friend and first biographer, T. F. Kelsall.

Thomas Lovell Beddoes differed in one respect from his father. He left a definite body of work which could stand alone: his poems. It has come to be recognised for its originality and its poetic achievement [6] and was admired by a few even in his lifetime. The intensity and technical skill of the lyrics and the quality of the blank verse have found admirers ever since the first careful edition of his poetry was published by Kelsall in 1851. Like his contemporaries, Beddoes clearly enjoyed the exuberant style of the late Elizabethan and Jacobean poets, but H. W. Donner, pointing to some of the descriptive passages [7] of *Alexander's Journey*, thinks that in these there may be found another source of Lovell Beddoes' richness; he considers "the exotic poetry of the son" may spring from his father's "bold and luxuriant descriptions". A copy in a childish hand of passages from *Alexander's Journey*, surviving among Anna Beddoes' letters to Davies Giddy, lends support to this suggestion. Among his father's writings, both in the poems and elsewhere, Thomas Lovell would have found other passages "romantic" in feeling and descriptive power. Unfortunately T.L.B's macabre imagery and concentration on the theme of death is still unacceptable to some. *Death's Jest Book*, the title of the play that he worked over and over again, has come to stand for all Beddoes' work and through the emphasis on this characteristic we are brought back to the original Thomas. It has been suggested that the dissections and anatomy [8] lessons that were part of the educational programme at Clifton, seen by Thomas Lovell when he was too young and too impressionable, were the origin of his obsession with death. Perhaps: but this gives too little weight to Thomas Lovell Beddoes' own experience as an anatomist and to his own particular felt understanding of death as "an area which stops just short of annihilation, a suspended area of confusion of identity, dissolution without re-creation, where this world and the next interpenetrate reluctantly but unavoidably".[9] If Thomas Beddoes' methods of education are to be held to have influenced the next generation, John Lambton, who, with his brother, lived for seven years at Rodney Place, should also come into the picture alongside Thomas Lovell Beddoes. He was thirteen when he left in 1805 to go to Eton, old enough to have been influenced by Beddoes' methods. Indeed, Beddoes had insisted on keeping the boys with him long enough for his 'plan' to succeed. John Lambton, who became 1st Earl of Durham and the first Governor General of Canada, is most remembered for the report named after him, which recommended responsible government for Canada and "inspired all subsequent British Colonial policy". This was the climax of his career; but from the beginning of his public life as Whig MP for Durham he had shown his radical views, denouncing the government for the Peterloo massacre in 1819; advocating parliamentary reform and being one of the most vigorous supporters of the first successful measure to extend the franchise, the Reform Act of 1832. "Radical Jack" he was, too, in his sympathy for the miners in his Durham coal-pits and his early introduction of Davy's miner's safety lamp. Coleridge and Davy were the most brilliant of the young men to come under Beddoes' influence but his personality and ideas also lived in the poetry of his son and the public life of John Lambton.

By the end of his life, Thomas Lovell Beddoes was, among all his family, most closely in touch with his cousin Zoë King. Thomas stayed with Zoë's parents in Bristol and he had a real affection for them. It is clear that Emmeline and Dr King [10] were very much the link with the days in Clifton and "little Anna's" message on the 1811 visit to Edgeworthstown: "I have an aunt in Clifton I shall never forget" foreshadows this. Anna certainly wrote constantly to her sister with news of herself and her two daughters, sometimes commiserating with Emmeline when things were difficult for her and hoping the time would soon come when she would no longer need to fear the postman's knock. It was from Emmeline that Anna learned that there had been damage to Dr Beddoes' tomb, and Dr King who was asked to put things right. If he was unable to do so he was to ask Dr Clayfield, for the sake of his old friendship. Dr King, working on through his financial vicissitudes, earning a £3.0.0 fee where Beddoes would have asked £10.0.0, attended various members of his wife's family when they were ill, in his own home in Clifton or in Edgeworthstown. Mary Beddoes, the baby born only a few months before Dr Beddoes' death, was in his care during the long and distressing illness which led to her death in 1830.

One incident shows most vividly Dr King's [11] loyalty to Dr Beddoes, whose work he had shared and whom he must have known so intimately in family life. From it we learn very clearly how Beddoes was appreciated by his colleagues. In September 1836 Dr King, taking part in the meeting of the British Association for the Advancement of Science in Bristol, attended a lecture which he dismissed as a "geological farce". To his disgust he found the lecturer first gave an inaccurate description of Dr Beddoes' scientific work and then repeated an old story which had been current in Bristol after a mishap to a consignment of frogs sent to Beddoes for laboratory work. King, in the letter he wrote to the Bristol Journal, told the story well:

Dr Beddoes was [next] represented to have been a devoted admirer of the fair sex, and in his solicitude for the health of the ladies, to have discovered that their liability to nervous diseases, and frequent deaths from pulmonary consumption, originated in their intemperate use of tea, and in order to convince them of that error, he devised a scheme of stocking two ponds, one containing an infusion of green tea, the other pure water, each with an equal number of live frogs; that he bespoke a cargo of these animals from a friend in Shropshire, who sent him ten thousand frogs in a huge hogshead, perforated with ten thousand air holes, similar to those afterwards bored by Davy in the floor of the House of Lords; the cask duly directed to Dr Beddoes, at Clifton, being landed on the quay at Bristol, fell from the crane and broke, so that ten thousand frogs over-ran the quay, and many hopped into the Frome, an event which threatened the most dire consequences to Dr Beddoes – the burning of his house, and perhaps that of the city, because the common people being at that time very loyal, and Dr B suspected of sinister political designs, imagined that the frogs were intended as provision for some French jacobins concealed in the doctor's cellar, would have proceeded to patriotic acts of revenge, had they not been timely restrained, and brought to their senses by the harangue of a benevolent physician, arriving by chance on the spot etc. etc.

Now for the sober truth of this matter, for every tale, however absurd and incredible, has some real foundation, let truth have a small chance with fame, said to be *tam ficti pravique tenax quam mentia viri*. Though he also had his facetious moments Dr Beddoes' conversation was always terse, and epigrammatic, like his respiration, not long-winded, and I do not believe that

even in answer to some silly enquiries about the destination of the frogs, he
would have said he meant to poison them with green tea.

The fact was, that for some of Davy's galvanic, as well as other physiological experiments, carried on at that time, at Dr Beddoes' desire, by the
writer of this letter, a supply of live cold-blooded animals was required. The
toads of Clifton Down had been nearly extirpated by the experimentor,
zealously aided by two fine boys, the pupils of Dr Beddoes, the eldest of
whom is now an illustrious statesman, the representative of his soveriegn at
a foreign court. The Doctor wrote for a supply, and a cask containing about
300 frogs was sent. Through the carelessness or curiosity of the men engaged
in landing it, the cask was opened, and many of the frogs were liberated,
to die in the deadly waters of the Frome. The rest were saved, to be sacrificed
in galvanic and physiological experiments. [12]

Dr King's account of the facts is a useful reminder of the work at Clifton,
particularly of the physiological studies he himself was engaged in with
Beddoes. This incident, too, is very revealing in showing how prejudice and
rumour could be combined in the standard way and then applied to Dr
Beddoes. Frogs were indeed used by Beddoes to explore the effects of tea
drinking; he described this in *Hygeia* when warning against too much tea
drinking in girls' boarding schools with the caveat that there was no certainty
that humans would be affected in the same way as the frogs. A house and a
town were, it is true, burned, when a mob suspected a scientist of subversive
ideas — Priestley's house and Birmingham; and there was an occasion when
a 'benevolent physician' intervened to bring a riot to an end — Dr Long Fox
at the time of the Bristol Bridge riots. The elements of ill-verified truth were
so combined to make a fiction harmful to Beddoes' reputation that Dr King
felt more was required than setting the frog story right. Early in his letter
he gave a concise but accurate account of Beddoes' writings, laying emphasis
on the scientific papers and ended:

In justice to the memory of Beddoes, I may be permitted to add a few words.
He died at the age of 48 — his whole life was marked by an ardent love of
science, in the most extensive sense of that word. Had he lived to the present
time, he would have been a distinguished member of the British Association,
for he was unwearied and unsparing of his pecuniary means, and of those of
wealthy friends, in his endeavours for the advancement of science. He did
not live long enough to have the luck of making any important discoveries;
some fortunate combination of circumstances beyond his control, might
have led him, like many others, to the honourable distinction of a discoverer.

For the "mass of facts" that showed why Beddoes "enjoys so large and well
deserved a fame in the sphere of science" Dr King referred the lecturer to
the biography "that had stood stock still during 25 years, on the shelves of

his friends". Dr King was not the most polished of men, but his description of a man who had died less than a year before "as a literary undertaker" reveals how deeply he felt about the inadequacy of Stock's account.

Dr Buckland,[13] who gave the geology lecture, had no excuse. In 1836 there was an easily accessible and full account of Beddoes' work in the 1824 supplement to the Encyclopaedia Britannica. From this the lecturer could quickly have learnt that what he had put together was not consistent with either the character or the scientific work [14] of Dr Beddoes. The article was contributed by Dr Peter Roget who at nineteen, newly graduated from the Edinburgh medical school, had spent a year with Beddoes at Clifton, the first exciting year of the Pneumatic Institute's existence. He took part in all the work of the Institute, scientific as well as medical, and in the laboratory performed his own chemical and physiological experiments. He stayed only a year at the Institute but he contributed further to Beddoes' work on pulmonary tuberculosis by sending him for his "Causes ... of tuberculosis" reports on the diet and life of the fishing communities of the east coast of Scotland. By 1824 Roget was well established in medical and scientific life in London, a Fellow of the Royal Society since 1814, participating in many learned societies and respected as a lecturer on a wide variety of subjects in medicine and science. He was fully able to give a well-judged account of Beddoes' work. Like Dr King, he stressed the importance of Beddoes' early scientific work and publications, particularly the study of Mayow. Roget recognised Beddoes' success at Oxford and found it difficult to understand why he decided to "relinquish all these advantages". He was quite clear that the article on the French clergy only hastened and did not cause Beddoes' decision to leave Oxford. In his description of the Pneumatic Institute Roget did justice to both the scientific and the medical purposes of the work; to him, the biggest thing the Institute accomplished was the investigation of the effects of nitrous oxide on the human system. Roget's is unusual among early accounts of Beddoes' work in appreciating his concern for preventive medicine:

The object which Dr Beddoes had ever most at heart was to excite a lively and general attention to the means of preserving health and repelling the first inroads of disease, by the diffusion of medical knowledge throughout all ranks of the community.

His summing up makes an interesting complement to Dr King's:

Very justly characterised as a pioneer in the road to discovery [Dr Beddoes was] more active in exciting the labours of others, than in labouring himself

in the field of experiment. He had the imagination of a poet and could paint
in the most vivid colours the sufferings entailed by disease.

It is a warm account, especially when we remember that Roget had not been
at ease during his year at Clifton. He foresaw that the work there would
run into difficulties and he, or possibly his mother on his behalf, was anxious
over political difficulties. Disappointed, he left Beddoes in the summer of
1799 and in September was invited to stay at Bowood where his uncle,
Samuel Romilly, was a member of Lord Lansdowne's circle.

Roget's own experience might well have led him to appreciate the developments in medicine for which Beddoes hoped. His public career as a doctor
began in earnest in 1804, with his appointment as physician at the Public
Infirmary at Manchester. There he worked under Dr Ferriar who was fighting
to reform the evil conditions in the cotton mills and in the lodging houses
where the pauper apprentices lived. With him, Roget wrote a simple guide
to health, 'Advice to the Poor'. During his five years there, Roget lectured
at the Manchester Literary and Philosophical Society and took part in setting
up the Manchester Medical School. Once established in medical practice in
London, he continued his interest in the Dispensary movement and helped
to plan the North London Dispensary, which opened in 1810. Roget worked
as its honorary physician for many years. In 1823 he had the shocking task
of trying to halt an epidemic of fever in the Milbank Prison. Nothing seemed
effective. He was particularly disappointed that dispersal, from which he had
hoped much and which at first seemed to succeed, was quickly followed by
recurrence of the fever. At the end, Roget cautiously noted that it seemed
"some injurious influence had been at work over and above the causes to
which the outbreak was originally imputed". Five years later he was appointed
Royal Commissioner to survey the London water supply and his detective
work for this was accepted as having demonstrated scientifically the connection between contamination and disease. The method which Beddoes and a
few of his contemporaries suggested, of trying by finding "exceptions" to
establish the connection between circumstances and disease, had in Roget's
hands much greater success and it is understandable that it was he who drew
attention to this part of Beddoes' work. Roget in the next generation found
that his methods were understood; he had, in contrast to Beddoes, a conservative temperament that accorded with the dominant mood of his time.

One more of the young men who worked at the Pneumatic Institute has
left impressions of Beddoes. He wrote very differently; in spite of the personal
distress he had felt at the time of Beddoes' death, the account Humphry

Davy [15] left in a set of biographical notes on eminent scientists was severely and professionally critical:

Reserved in manner and almost shy; but his countenance was agreeable. He was cold in conversation and apparently much occupied with his own peculiar views or theories. Nothing was a stronger contrast to his apparent coldness in description than his wild and active imagination which was as Darwin's. He was little enlightened by experiment and I may say, little attentive to it. He had great talents and much reading; but had lived too little among superior men. On his death bed he wrote me a most affecting letter regretting his scientific aberrations. I remember one expression "like one who has scattered abroad the Avena fatua of knowledge from which neither brand, nor blossom nor fruit has resulted. I require the consolation of a friend". Beddoes had talents which would have exalted him to the pinnacle of philosophical eminence if they had been applied with discretion.

As a comment on Beddoes' work as a chemist during the sixteen months that Davy worked with him, there may be justice in these observations. If they show Davy himself in a less pleasant light than the letters of the ambitious nineteen year old arriving at Clifton, it is as well to remember that they were written early in February 1829 just before Davy's last illness.

All three accounts put on record how his work appeared to thoughtful contemporaries and their judgement of him as a man. They are in sharp contrast to the traditional stories repeated either for amusement or, in King's view, out of 'party rancour' which ought to have been forgotten. They are the last assessments we have for many years. Beddoes appears incidentally in accounts of other men who overshadowed him, when their biographies begin to be published in the mid-nineteenth century. First in 1847 was Joseph Cottle's lively and first-hand, but irresponsibly inaccurate *'Reminiscences of Coleridge and Southey'*. Cottle was mostly concerned with the less scientific side of the breathing of nitrous oxide. Dr John Davy, whose 'Fragmentary remains of Humphry Davy' appeared in 1858, by publishing Davy's early letters reveals how both Anna and Dr Beddoes impressed the ambitious, but at the same time anxious, young Humphry. Other letters,[16] where Beddoes appears as a friend and doctor, were made available when the correspondence of Southey and Coleridge began to be published. But these are only glimpses; Dr Beddoes as someone whose life was interesting for its own sake, disappeared. T. F. Kelsall was the intimate friend of Beddoes' son and had opportunity to talk to Zoë King but in the Memoir he wrote for his edition of Thomas Lovell Beddoes' poems he had no more to say of Dr. Beddoes than that he was "a man of vigorous and accomplished mind and large philosophic views".[17]

So it remained until in the twentieth century, Beddoes began again to be noticed for what he accomplished. First were the chemists.[18] The work he did as a chemist on the alkaline earths had been lost sight of when attention had concentrated on the Pneumatic Institute. When this was recalled, it became possible to estimate more clearly the importance for Humphry Davy of the period spent at the Pneumatic Institute. The range of Beddoes' own previous work and of the interests he wished to continue can be set against the falling off of experimental rigour which he accepted, whether from pressure of work or from a desire to satisfy the Managers of the Institute with a quick report and by producing results. Overlooked quite naturally because he could be credited with no specific discovery in chemistry, Beddoes has in recent times come to be seen as contributing to the historic development of chemistry by bringing Black's style of teaching to Oxford and as taking a part, however minor, in the phlogiston debate. It was, too, Beddoes' work that gave body to the idea of pneumatic medicine.

There has been a similar development in medicine. The significance of Humphry Davy's note on how breathing nitrous oxide made him insensible to the pain of his wisdom tooth and his observation of the anaesthetic powers of nitrous oxide, recorded at the end of 'Researches', has been re-emphasised. This has led to discussion of the place of the work at the Pneumatic Institute in the history of anaesthesia;[19] to appreciation both of the importance of Beddoes' and Davy's work on the physiological effects of gases and of the value of the apparatus Watt and, following him, Clayfield, designed. It is difficult to accept that the use of nitrous oxide as an anaesthetic arose from a fresh start when in 1844 Horace Wells rediscovered its power as giving relief from pain and inhaled it before having a tooth extracted. Though it is not possible to establish any record showing that there was a deliberate intention of taking up this particular suggestion made by Davy, or of reconstructing the apparatus used at the Pneumatic Institute, research has shown a very strong continuity [20] in knowledge of Davy's work; Medical and Chemical lectures and even the popular use of 'laughing gas' for entertainment have been traced so increasing the possibility of a true link between the work of Dr Henry Hill Hickman (1800—1830) and Davy. Hickman was in medical practice in Ludlow in 1820 and by 1824 was settled in Shifnal, Beddoes' birthplace. Hickman used carbon dioxide to anaesthetise animals and he modestly but unambiguously proposed the trial of gases in human surgery. Hickman hoped that his work might be brought to the notice of Sir Humphry Davy but he was unable to find support either in England or in France. We now know that in Edinburgh, Hickman attended Professor T. C.

Hope's lectures and that these included at least a mention of the experiences of Beddoes and Davy. Hickman, as a member of the Edinburgh Royal Medical Society, very probably heard a lecture 'On Asphyxia' given by a fellow member, Henry Goldwyer, who was completing studies begun at the Bristol Infirmary. There, it is interesting to note, he had attended lectures given by Dr J. E. Stock. This exploration of the history of nitrous oxide by Dr W. D. A. Smith in the 1960s has brought us close to the end of the trial; we cannot know how, if ever, it may be traced right home. While some American[21] accounts have been more prepared to ignore the gap and hail the Pneumatic Institute as the origin of anaesthetics, we can find among them an interesting history of a different line of descent. Joseph Priestley,[22] by then settled in Pennsylvania, was particularly anxious to receive reports of Davy's work at the Pneumatic Institute and at the end of October 1801 wrote to him a long and generous letter. He expressed his satisfaction that he would "leave so able a fellow-labourer of my own country in the great fields of experimental philosophy". There is a touch of sadness in the wording of Priestley's request for news of Davy's work:

My son, for whom you express a friendship, and which he warmly returns, encourages me to think that it may not be disagreeable to you to give me information occasionally of what is passing in the philosophical world, now that I am at so great a distance from it, and interested, as you may suppose, in what passes in it. Indeed, I shall take it as a great favour. But you must not expect anything in return. I am here perfectly insulated, and this country furnishes but few fellow-labourers, and these are so scattered, that we can have but little communication with each other, and they are equally in want of information with myself.

Beddoes was not alone in being impressed by Davy's early work for that, as well as what was being done in Clifton, had impressed Priestley too. Through him, details of Davy's experiments would have reached Benjamin Rush, who had hoped Priestley would become Professor of Chemistry at Philadelphia. Though Priestley felt unable to accept, Woodhouse who was appointed must have included Davy's work in his teaching, for his students are known to have breathed nitrous oxide. One of them, William P. C. Bartram, in a paper published in 1809, suggested the medical use of the gas. So it would seem there was a direct link from Davy at the Pneumatic Institute through Joseph Priestley to American science.

Cartwright drew attention to the pioneering work in preventive medicine which of course included Beddoes' efforts to promote health education. This, as we have seen, was only part of Beddoes' interest in the upbringing of children and young people but his place among the educationalists of the late eighteenth century began to attract attention only in 1960.

The nature of Beddoes' friendship with poets and writers has become better understood. The poetic quality which Roget saw; the imaginative sympathy which both he and Coleridge recognised in *Isaac Jenkins* as well as the intellectual excitement of his experiments at Clifton, gave Beddoes a particular 'rapport' with the literary circle of young men. C. A. Weber's German study [23] of Bristol's importance in Anglo-German Romanticism was the first to place Beddoes in a literary context; this theme was broken by the 1939—45 war, not to be taken up again until the time when he came to be seen in a similar context,[24] in studies relating to Coleridge and Wordsworth. Beddoes, neither a genius meriting the title of polymath, nor a specialist, has had to submit to a process of fragmentation. As a member of an 'invisible college', he takes his place among men who found close links between chemistry, geology, industry, medicine, literature and politics. So, the incident of the Regius Professorship of Chemistry; the Pneumatic Institute; the friendships at Clifton all have a different tone and we become more aware of his links with the Lunar Society. Whereas for mid-nineteenth century writers of biography he was a minor character, Beddoes has come to appear as a significant figure in recent [25] accounts of his contemporaries.

In his own context, that of the period of the French Revolution, the war against the new Republic and the arrival of Napoleon, Beddoes was not a chemist interested in politics, a doctor who saw the application of chemistry to medicine, but a man for whom these ideas made a unified whole. The new knowledge of the physical world, growing in chemistry and geology, and the political and social reforms which appeared to be coming into being, were for him dependent each on the other. As a chemist and a geologist he accepted and taught new knowledge. He saw that men would have to use the geology that showed how the world had been formed and abandon the idea of the Flood; that they would need to understand the evolution of different species of living creatures to know the physical needs of men and women and to forget moralising in order to build a new society; but philosophically he looked back to Locke. He rejected Rousseau, but his thoughts on education, especially for women, were not traditional but in accord with progressive ideas of his time. The poet in him had all the exuberance of the romantic but he wrote in poetic forms that were being rejected; even in his response to Shakespeare as supreme in his exploration of human character, Beddoes could in the same appreciation link Shakespeare with Hartley — "Shakespeare first dramatised what Hartley afterwards analysed". In him we can see the intense disillusion that for many succeeded the enthusiastic response to the French Revolution. Underlying the outbursts of anger in his writings on Pitt in 1796, was his attempt at making realistic proposals for reform but he showed no sign of wrestling with the dilemma posed, and always posed, by revolution.

The question that remains is whether Beddoes ever saw the dilemma. For all his effort to construct a scientific/medical/political synthesis, Beddoes remained substantially limited to a materialistic solution. Coleridge alone [26] struggled to find an answer that was not ready-made. Does this suggest that Beddoes was merely able to catch the ideas of his day that surrounded him and, with lively intelligence, to toss them about and bring them together in new ways? Certainly his was not a particularly original mind, but he was alert and not uncritical; there is value in seeing new connections and attempting a new synthesis, as he did in bringing together health, chemistry and reform of society. It does seem that he was in essence a practical man, as when his work on gases led him to decide to devote himself to their possible application in illness and to test in use that application. It is agreed that he arrived at this decision before political consideration affected his position; Beddoes did not choose pneumatic medicine to replace scientific work he was obliged to leave in Oxford. In all his writings from *Isaac Jenkins* onwards we can see sympathy for suffering, an understanding of human needs and a kindliness which was recognised by all his friends. The hurry and bustle of his life, the unfinished work, may well be related to the desire to put his ideas into action. At the end disappointed, he may have found some compensation in his medical practice and writings and in the wide respect that was accorded him as a doctor. Yet it is his early scientific writings and what we know of his teaching of chemistry and his experimental work, that show us where Beddoes had insight and skill. We cannot help wondering, since for Beddoes "Chemistry [was] daily unfolding the profoundest secrets of nature" and since the discoveries of chemistry were the origin of his thinking, what Beddoes' achievement might have been had he remained a chemist.

THOMAS BEDDOES' CONTRIBUTIONS TO
THE MONTHLY REVIEW, 1793–1801

The material given in this Appendix is based on the work of Benjamin C. Nangle and his original scholarship is here gratefully acknowledged. His two Indexes to the Contributors and Articles in the *Monthly Review* [a] together give a history of the Review and biographical accounts of the contributors. In them, the main reviews are listed in alphabetical order of the name of the author of the book reviewed. The Indexes to the Foreign Supplement, which appeared three times a year numbered independently, give only the number of the review article.

To show more clearly the significance of Beddoes' work as a reviewer and its relation to his other work, his writings are here placed in chronological order. It can be seen that once the Pneumatic Institute was in being, Beddoes no longer contributed main articles; his reviewing came to an end when he was busy with *Hygeia* and his attentions turned to Preventive Medicine.

Main Reviews

1793   Vol. 11   Philosophical Transactions of the Royal Society. Anonymous. p. 419.

          Vol. 12   John Abernethy. Surgical and Physiological Essays. p. 48.
Thomas Reide. View of Diseases of the Army. p. 89.
Medical Facts and Observations, Anonymous. p. 94.

1794   Vol. 13   Experiments and Observations Relative to Animal Electricity. Richard Fowler. p. 297.
Transactions of the Royal Irish Academy. Anonymous. pp. 385–388.
Alexander Philip Wilson. Inquiry into Urinary Gravel. p. 166.
Memoirs of the Literary and Philosophical Society of Manchester. pp. 65 and 182.

---

[a] Nangle, B. C.:   1934, Monthly Review, First Series. Indexes of Contributors and Articles (to 1790).

     Nangle, B. C.:   1955, Monthly Review, Second Series. Index of Contributors and Articles (1790–1815).

Vol. 14 Matthew Carey. On Malignant Fever. p. 187.
Medical Facts and Observations. Anonymous. p. 25.
E. Valli. Experiments on Animal Electricity. p. 40.
A Translation of the Table of Chemical Nomenclature.
Proposed by de Guyton, formerly de Morveau, Lavoisier,
Berthollet, and de Fourcroy: with additions and alterations:
to which are prefixed an Explanation of the Terms and some
Observations on the new system of Chemistry. p. 317. [Dr
George Pearson named as author in the review.]

Vol. 15 John Abernethy. Surgical and Physiological Essays. p. 299.
Benjamin Rush. An Account of the Yellow Fever as it
appeared in Philadelphia. p. 161.

1795 Vol. 16 William Crump. An Inquiry into Opium. p. 68.
John Ewart. History of Two Cases of Ulcerated Cancer. p. 308.
George Fordyce. Dissertation on Simple Fever. p. 279.
James Hutton. Dissertation on Natural Philosophy. p. 246.
James Peacock. A Short Account of Filtration. p. 178.

Vol. 17 John Dalton. Meteorological. Observations. p. 178.
Michael Ryan. History and Cure of Asthma. p. 417.
John Hunter, FRS. Treatise on Blood with a memoir by
E. Home. p. 261.
Alexander Gordon, M.D., Physician to the Aberdeen Dis-
pensary. A Treatise on the Epidemic of puerperal fever of
Aberdeen. p. 316.
John Hunter. On Blood. p. 75. Continuation of his previous
article.
Memoirs of the Medical School of London. Anonymous.
p. 193.
Rev. Joseph Townsend, Rector of Pewsey. Guide to Health.
p. 99.

1796 Vol. 19 Colin Chisholm. An Essay on Fevers. p. 62.
Samuel Ferris. General View of the Establishment of Physic
as a Science. p. 320.
E. Peart. The Antiphlogistic Doctrine of Monsieur Lavoisier
Critically Examined and Demonstratively Confuted. p..194.
Benjamin Rush. Medical Inquiries and Observations Part 2.
p. 408.
James Russell. A Practical Essay on Certain Diseases of the
Bones. p. 69.

Vol. 20 Joseph Adams. On Morbid Poisons. p. 57.

Mrs Fulhame. An Essay on Combustion, with a view to a new Art of Dyeing and Painting. Wherein the Phlogistic and Antiphlogistic Hypotheses are proved erroneous. p. 301.

Memoirs of the Literary and Philosophical Society of Manchester. Beddoes. From p. 416 to the end.

James Carmichael Smyth. Descriptions of Jail Distemper.

Minutes of the Society for Philosophical Conversations. p. 284.

1797 Vol. 23 John Crisp. Observations on the Nature and Theory of Vision. p. 64.

E. Peart. On the Composition and Qualities of Water with remarks on the opinions of different reviewers on the author's preceding tract entitled 'The Antiphlogistic Doctrine of Monsieur Lavoisier Critically Examined'. p. 139.

Vol. 24 John Abernethy. Surgical and Physiological Essays. p. 47.

Rev. Joseph Townsend. A Guide to Health Vol. 2. p. 19.

Robert Townson. Travels in Hungary. p. 169. Continued in the next month.

J. G. Schmeisser. System of Mineralogy, formed chiefly on the plan of Cronstedt. p. 26.

1798 Vol. 25 Dr John Rollo, Surgeon General to the Royal Artillery. Two Cases of Diabetes. With William Cruickshank, Chemist to the Ordnance and Surgeon of Artillery. Trials of Various Acids. p. 58.

Foreign Supplement

The page number is the page on which the Supplement begins.

Vol. XII Page 481
1793 Article 7. Journal der Physik, i.e., A Journal of Natural Philosophy. F. A. C. Gren, Professor at Halle.

9. La Médecine éclairée par les sciences physiques. A. F. Fourcroy.

Vol. XIV Page 481
1794 Article 7. Chemische Annalen. Lorenz von Crell.

10. Journal of Nat. Philos. F. A. C. Gren. Cont. from Vol. XII.
11. Experiments on Substances capable of extinguishing Fire. Assessor Aken, Stockholm 1793, and Nils Nystrom, Norrkoeping 1793.

Vol. XVI Page 481
1795 Article 9. Transactions of the College of Physicians, Philadelphia.

Vol. XVIII Page 481
1795 Article 16. Philosophie Chimique. A. F. Fourcroy.
              17. Les Révolutions de France et Genève. d'Ivernois.

Vol. XX Page 481
1796 Article 1. Origine de tous les cultes. M. Dupuis.
              2. Zum ewigen frieden, i.e., To Perpetual Peace. Emmanuel Kant, Koenigsburg 1795.
              3. An inaugural Dissertation on the Chemical and Medical History of Septen, Azote, or Nitrogen. Winthrop Saltonstall, Connecticut.
              4. Account of the Epidemic Yellow Fever as it appeared in New York in 1795. Val. Seaman M. D.
              5. Physiological Observations on Amphibious Animals. Part I on Respiration. Part II on Respiration with a fragment on Absorption. Robert Townson, Göttingen.
             29. de Morbis Vasarum Absorbentium. S. T. Soemerring, Frankfort. A prize Dissertation.
             30. de Corporis Humani Fabrica. S. T. Soemerring, Frankfurt on Main.

Vol. XXI Page 481
1796 Article 9. De generis Humani Varietati Nativa, i.e., On the Native Varieties of the Human Species. 3rd Edition by J. F. Blumenbach F. R. S. To which is prefixed an Epistle to Sir J. Banks, Göttingen.
             10. On the Origin, Causes and Early Practicable Extirpation of the Small-pox and Contagious Disorders. Now first proposed to Ferdinand IV King of both Sicilies and demonstrated by F. M. Scuderi. F. M. Scuderi, Naples.

11. Danish Translation of John Brown's work. Pfaff, Copenhagen.
12. Ideas on the Production of diseases. C. W. Hufeland, Jena.
13. On the Vital Principle. J. D. Brandis, Jena.
14. Archives of Physiology. J. C. Reil, Halle.
15. Insanity general and particular, with a century of cases. V. Chiarugi, Florence.
16. Experiments on the Shining of Phosphorus in Azotic Gas. A. H. Scherer M. D. and C. C. F. Jager M. D. with remarks on M. Goettling's Tract. Weimar.
22. Physical and Political Travels thro' Dacia and Sarmatia. Dr Hacquet, Nuremberg.
23. Anatomy of the Nerves of the Heart. A. Scarpa, Pavia.
26. 'The Hours'. A periodical to which Schiller contributed. Tübingen.
27. The Luciniad or the Art of Midwifery. Sacombe, Paris.
28. Apparatus Medicaminium: Minerals. Prof. Gamelin. Göttingen. Continuation of the late Dr Murray's Materia Medica.
29. Compilation of Dissections, J. C. F. Schlegel, Gotha.

Vol. XXIII Page 481
1797 Article 2. Memoir concerning the fascinating faculty which has been ascribed to the rattlesnake. Benjamin Smith Barton, Pennsylvania.
10. Memoire de Physique et d'histoire Naturelle. J. B. Lamarck.
11. Theorie de la Terre. J. C. Delametherie.
12. Inaugural Dissertation on Dysentery. W. Bay.
Reviewed with W. W. Taylor
3. The Art of Prolonging Human Life. C. W. Hufeland, Jena.
4. Works of General Dumouriez, Vol. I. Portugal.

Vol. XXIV Page 481
1797 Article 2. Experiments on the Irradiated Nervous and Muscular Fibre. F. A. von Humboldt. Posen & Berlin.
8. Annales de Chimie. Paris.
20. Handbuch der Pathology. K. Spreugel.
22. Theorie de la Terre. J. C. Delametherie (cont.)

31. Manuel de Philosophie Pratique. Lausanne. (Includes translations of extracts of 'Poor Richard' by Franklin and of extracts from Evenings at Home.)

Vol. XXV Page 481
1798   Article 7.   Letters on Switzerland and Italy. G. A. Jacobi. Translated from German.
        8.   History of the Yellow Fever in New York. Alexander Hosack, Philadelphia.
        9.   Case of the Manufacturers of Soap and Candles in the City of New York. Published by Association of Tallow Chandlers and Soap Makers, N. York.
      21.   Contributions to the Chemical Knowledge of Mineral Bodies. Prof. Klaproth, Berlin.
      22.   Intelligence concerning French Military Hospitals. G. Wedekind. Physician to the Army of the Rhine. Leipzig.
      28.   Plan of a Natural History of the Human Species. C. F. Ludwig, Leipzig.
      29.   The Present State of Medical Learning in the City of New York.

Vol. XXVI Page 481
1798   Article 6.   An account of the Disease and Death of General Hoche by M. Poussielgue, surgeon, Paris.
        8.   Inaugural Dissertation – in what manner pestilent Vapours acquire their acid qualities ... with letter from Dr Mitchell. A. C. Lent, New York.
      12.   A foundation for a future zoonomia. Anon., Jena.
      17.   Carolia Linne system and vegetabilium. Ed. C. H. Pearsoon, Göttingen.
      18.   On the effect of Mineral Waters. J. E. Wickman, Hanover.
      19.   Outlines of Physical Sciences. F. C. A. Gren, Halle.
      20.   Observations on Gastric Juices. F. Chiarenti, Florence.
      21.   On the Action of Frictions with Saliva. V. L. Brera, M.D., Pavia.

Vol. XXVII Page 481
1798   Article 14.   An Inquiry concerning the origin of diseases. A. Roschlaus, Frankfurt.

15. On the Knowledge and Treatment of Fevers. Part I. J. C. Reil, Halle.
17. William Meister's Apprenticeship. Goethe, Berlin.
21. Diaetophilus Psychological History of his seven years Epilepsy. Part I. Zürich.
22. Annales de Chimie. 1797. Paris.
23. Louisa, a Pastoral. J. H. von Voss, Königsberg.
34. An account of the Plants of Mauritius. P. R. Willemet, Leipzig.

Vol. XXVIII Page 481
1799   Article 11. Annales de Chimie 1798–1799.

Vol. XXIX Page 481
1799   Article 14. Annales de Chimie (cont. from Vol. XXVIII). Paris.
          15. On Perkinism. J. C. Tode, Copenhagen.
          22. Letters of a Physician written at Paris and with the French Army. G. Wardenburg.

Vol. XXX Page 481
1799   Article 1.   Observations on a journey into the southern departments of Russia. P. S. Pallas, Leipzig.
          14. Annales de Chimie.

Vol. XXXI Page 449
1800   Article 11. Annales de Chimie. Paris, 1800.

Vol. XXXII Page 449
1800   Article 11. Annales de Chimie. Paris.

Vol. XXXIII Page 449
1800   Article 11. Annales de Chimie. Paris.

Vol. XXXIV Page 449
1801   Article 7.   Annales de Chimie. Paris.

APPENDIX II

DR JOSEPH PRIESTLEY'S LETTER TO HUMPHRY DAVY, OCT. 31, 1803

(In *Humphry Davy: Poet and Philosopher*, by T. E. Thorpe, 1896, pp. 38–39.)

Thorpe's account runs,

"Among the letters entrusted to me is one from Priestley, which must have been particularly gratifying to the young man. It is as follows:

Northumberland, Oct. 31, 1801.

Sir, – I have read with admiration your excellent publications, and have received much instruction from them. It gives me peculiar satisfaction that, as I am far advanced in life, and cannot expect to do much more, I shall leave so able a fellow-labourer of my own country in the great fields of experimental philosophy. As old an experimenter as I am, I was near forty before I made any experiments on the subject of Air, and then without, in a manner, any previous knowledge of chemistry. This I picked up as I could, and as I found occasion for it, from books. I was also without apparatus, and laboured under many other disadvantages. But my unexpected success induced the friends of science to assist me, and then I wanted for nothing. I rejoice that you are so young a man; and perceiving the ardour with which you begin your career, I have no doubt of your success.

My son, for whom you express a friendship, and which he warmly returns, encourages me to think that it may not be disagreeable to you to give me information occasionally of what is passing in the philosophical world, now that I am at so great a distance from it, and interested, as you may suppose, in what passes in it. Indeed, I shall take it as a great favour. But you must not expect anything in return. I am here perfectly insulated, and this country furnishes but few fellow-labourers, and these are so scattered, that we can have but little communication with each other, and they are equally in want of information with myself. Unfortunately, too, correspondence with England is very slow and uncertain, and with France we have not as yet any intercourse at all, tho we hope to have it soon . . . .

I thank you for the favourable mention you so frequently make of my experiments, and have only to remark that in Mr Nicholson's Journal you say that the conducting power of charcoal was first observed by those who made experiments on the pile of Volta; whereas it was one of the earliest that I made, and gave an account of in my History of Electricity, and in the Philosophical Transactions. And in your treatise on the Nitrous Oxide p. 55 you say, and justly, that I concluded this air to be lighter than that of the atmosphere. This, however, was an error in the printing that I cannot account for. It should have been *alkaline air*, as you will see the experiment necessarily requires.

With the greatest esteem, I am Sir, yours sincerely

J. PRIESTLEY."

261

NOTES AND REFERENCES

Chapter 1. Introduction

[1] Davies Giddy of Tredrea in Cornwall took the surname of his wife, Mary Ann Gilbert, on his marriage in 1808 and was subsequently known as Davies Gilbert. Since he was always known to Thomas Beddoes as Davies Giddy, that is the name I have used throughout.

[2] Anna Beddoes to Davies Giddy, 19 Jan. 1809. Cornwall Record Office, Truro. DD DG 89.

[3] John Edmonds Stock, 1775–1835. Exeter College, Oxford. Studied medicine at Edinburgh University. Physician to the Bristol Royal Infirmary, 1811–1828. His Memoirs of the Life of Thomas Beddoes M.D. appeared in 1811.

[4] Coleridge, S. T. to Humphry Davy, 30 Jan. 1809. Letters of Davy, Royal Institution Library.

[5] See Note 3.

[6] Anna Beddoes to Mrs King (her sister Emmeline), 1823. Bodleian Library, Oxford, Papers of Beddoes, MS Dep. C. 134–137, Box 135.

[7] Monro Smith, G.: 1917, A History of the Bristol Royal Infirmary, pp. 180–182.

[8] Stock, J. E.: 'Life', p. 388.

Chapter 2. Early Life

[1] Stock, J. E.: 1811, Life of Thomas Beddoes, p. 2.

[2] Thomas Paine's pamphlet was published in January 1776 and widely distributed.

[3] Plymley, J.: 1803, A General View of the Agriculture of Shropshire, pp. 82–88.

[4] For a general account of eighteenth century Shropshire, see:
Trinder, B.: 1973, The Industrial Revolution in Shropshire, Phillimore, Chichester.
Trinder, B. and J. Cox: 1980, Yeomen and Colliers in Shropshire. Phillimore, Chichester.

[5] Copy of an affidavit of T. Bishop, R. Pigeon and T. Morris relating to land in possession of Thomas and Richard Beddoes. Sworn 1826. Bodleian Library, Oxford, MS Dep. C. 134.

[6] Rosamond Beddoes to Mrs. Whitehall, n.d. Bodleian Library, Oxford, MS Dep. C. 135.

[7] Plymley, J., op. cit., p. 106 ff.

[8] See Rosamond Beddoes' letters in Bodleian Library, Oxford. MS Dep. C. 134–7.

[9] Telford, T. in J. Plymley, op. cit., Ch. 4, p. 289.

[10] Smith, S.: 1979, A View from the Ironbridge. Ironbridge Gorge Museum Trust. No. 13 has a particular interest in being a woodcut by J. Edmunds who later illustrated T. Beddoes' poem, Alexander's Journey . . . .
Klingender, F. D.: 1968, Art and the Industrial Revolution. Adams and Dart, pp. 75–80; Paladin, St. Albans, 1972.

[11] 'Report of Mr John Smeaton etc.' printed for a Select Committee of Civil Engineers. Reviewed in Nicholson's Journal, April 1768.

[12] Will of R. Beddoes. Shropshire Record Office, Shrewsbury.

[13] Auden, J. E. (ed.): 1909, Shrewsbury School Register from Nov. 16, 1738, pp. 4–5.

[14] T. Beddoes to R. Beddoes, n.d. (but 1777). Bodleian Library, Oxford, MS Dep. C. 135.

[15] Jones, W. (of Nayland): 1809, Life of George Horne, Vol. 1, pp. 135–6.

[16] Green, V. H. H.: 1964, Religion at Oxford and Cambridge, Ch. 10.

[17] The Swedish Chemist, Carl Wilhelm Scheele, published his treatise, On Fire and Air in 1777 but his experiments were made between 1770 and 1773. In the four years before 1777, Priestley had made and published the same discoveries.

[18] Gunther, R. T.: 1923, Early Science in Oxford, Vol. 1. Oxford Historical Society, Oxford, p. 64.

[19] Gunther, R. T.: 1933, The Old Ashmolean, Oxford, p. 4 ff.

[20] Wall, M.: 1783, Dissertations on Select Subjects in Chemistry and Medicine, Preface. I am indebted to the Librarian, Museum of the History of Science, Oxford, Mr A. V. Simcock, M. A., for drawing my attention to this reference to the laboratory; and to the similarity of the laboratory in Oxford to the one at Altdorf as illustrated in his own account of the Ashmolean Chemical Laboratory.

[21] Wheeler, T. S. and Partington, J. R.: 1960, The Life of William Higgins, Higgins and the Atomic Theory, Pergamon, Oxford, pp. 66–67.

[22] Stock, J. E.: 1811, Memoir, p. 9.

[23] Thomas Beddoes to Richard Beddoes. Bodleian Library MS Dep. C. 134–137.

[24] William Hunter, 1718–1783. From 1768 had a famous anatomy theatre in Gréat Windmill St., London, where he lectured and trained other surgeons.

[25] Dr John Sheldon, F. R. S., 1752–1808.

[26] For an account of scientific lectures and patrons in London in the second half of the eighteenth century see Musson, A. E. and Robinson, E., 1969, Science and Technology in the Industrial Revolution, Manchester University Press, Manchester, p. 119 ff. For a contemporary account of Sir Joseph Banks, p. 7 ff., and Dr John Sheldon, p. 45 ff., see Faujas de Saint Fond, B., 1797, Voyage en Angleterre and Geikie, Sir Archibald: 1907, Journey through England, p. 37 ff. For a more extensive account with comments on Faujas de Saint Fond's merits as a reporter see Cameron, H. C.: 1952, Sir Joseph Banks, K.B., F.R.S., Batchworth Press, London, pp. 125–127; 167–172. For an account of Richard Kirwan's London life see M. Donavan's Biographical Account in the paper read to the Royal Irish Society 25.2.1850 Proceedings of the Royal Irish Academy 4 (1850), lxxxi–cxviii.

[27] Wheeler, T. S. and Partington, J. R., op. cit., pp. 2–3.

[28] Smeaton, W. A.: 1967, Louis Bertrand de Morveau, F. R. S., 1737–1816, Notes and Records of the Royal Society, London, p. 118.

[29] Hodgson, J. E.: 1924, History of Aeronautics in Great Britain, Oxford 1924, p. 114 ff.

[30] Lazarus Spallanzani, 1729–1799, was appointed to the Chair of Natural Philosophy at Pavia, 1776. His Dissertations were published in 1776 and 1778; Beddoes' translation in 1784.

[31] Spallanzani, L. Dissertations, Translator's Preface, p. xiv.

[32] Spallanzani, L. Dissertations, Translator's Preface, p. xviii.

[33]   Spallanzani, L. Dissertation III, p. 345.
[34]   Rosamond Beddoes to Richard Beddoes, 1783, from Hopesay. Bodleian MS Dep. C. Box 135.

Chapter 3. Edinburgh Medical School

[1]   Beddoes, T.: 1795, Brown's Elements of Medicine, p. li.
[2]   Beddoes, T.: *op. cit.*, p. li.
[3]   For the Edinburgh Medical School and its personalities see:
Garrison, F. H.: 1929, A History of Medicine. W. B. Saunders, Philadelphia and London.
Grant, Sir A.: 1884, The Story of the University of Edinburgh during its first three hundred years. Vol. I. Longmans, Green.
Lawrence, C.: 1979, 'The Nervous System and Society in the Scottish Enlightenment', in B. Barnes and S. Shapin (eds.), Natural Order: Historical studies of scientific culture. Sage.
[4]   Beddoes, T. to C. B. Trye, MS D303C1/61 Gloucester Record Office, Gloucester.
[5]   Lyson, D.: 1848, Life of Mr Trye, Surgeon, with extracts from his private papers.
[6]   Beddoes, T. to C. B. Trye, D303C1/61 Gloucester Record Office.
[7]   Anderson, Dr J. in Chambers Biographical Dictionary of Eminent Scotsmen, 1871 (reprint).
[8]   Beddoes, T. to C. B. Trye, D303C1/61 Gloucester Record Office.
[9]   Lawrence, C. and M. Neve: 1980, Taxonomy and Practice: medical classification in the eighteenth century. Paper presented to the conference on 'New Perspectives in the History and Sociology of Medical Knowledge' organised by British Society for the History of Science, Bath, March 1980, p. 11.
[10]   Beddoes, T.: 1795, Brown's Elements of Medicine, p. lxxx.
[11]   For an account of Joseph Black, see:
Ramsay, W.: 1918, Life and Letters of Joseph Black, M.D., Constable, London.
Simpson, A. D. C. (ed.): 1982, Joseph Black, 1728–1799. A Commemorative Symposium. Royal Scottish Museum, Edinburgh.
[12]   Stock, Life, p. 14.
[13]   Knight, D. M.: 1976, in C. C. Gillispie (ed.), Dictionary of Scientific Biography, Vol. 5. Chas. Scribner's Sons, New York.
[14]   Beddoes, T. to R. Beddoes. MS Dep. C. 135 Bodleian Library, Oxford.
[15]   Beddoes, T. to C. B. Trye, D303C1/62 Gloucester Record Office.
[16]   Stock, Life, Appendix 1. Of the Sexual System of Vegetables, p. xvi, and Appendix 2. Of the Chain of Being.
[17]   Beddoes, T.: 1808, Letter to the Rt Hon. Sir Joseph Banks, Bart. P. R. S.
[18]   Beddoes, T.: 1793, Letters from Dr Withering, etc. Supplementary to two publications on Asthma, Consumption, Fevers, etc.

Chapter 4. Chemical Reader

[1]   For an account of the problems of chemical nomenclature, see Crosland, M. P.: 1962, Historical Studies in the Language of Chemistry, Dover Publications, New York.

[2] Gibbs, F. W. and W. A. Smeaton, 1961, Ambix 9, 'Thomas Beddoes at Oxford'.

[3] Wheeler, T. S. and J. R. Partington: 1960, The Life and Works of William Higgins, 1763–1825. Pergamon, Oxford, pp. 2–5.

[4] Beddoes, T.: 1787 February, Memorial on the State of the Bodleian Library.

[5] Stock, Life, p. 17.

[6] For a very sympathetic early account of Guyton de Morveau, see Granville, A. B.: 1817, Life of Baron Guyton de Morveau; and for de Morveau's significance for English scientists, see Smeaton, W. A.: 1967, Louis Bernard Guyton de Morveau F. R. S. (1737–1816), Notes and Records of the Royal Society, p. 118.

[7] Papers of Thomas Beddoes, Bodleian Library, Oxford, Dep. C. 134.

[8] Beddoes, T.: 1793, Observations on the Nature and Cure of Calculus, Sea Scurvy, Catarrh and Fever.

[9] Correspondence of J. Black and T. Beddoes, 1787–1792. Edinburgh University Library Gen. 875/111/52–53.

[10] Beddoes, T. to J. Black, 6 Nov. 1787. Edinburgh University Library.

[11] Black, J. to T. Beddoes, 24 Nov. 1787. Edinburgh University Library.

[12] Beddoes, T. to J. Black, 23 Nov. 1788. Edinburgh University Library.

[13] Beddoes to Black, 6 Nov. 1787 and Black to Beddoes, 24 Nov. 1787.

[14] Guyton de Morveau (Dijon) to T. Beddoes, prof. Chymie Oxford, 19 Sept. 1788. Bodleian Library MS Dep. C. 134.

[15] Op. cit. (Note 14). That is, "He is a very pleasant young man and one who will learn everything he puts his mind to; I am sorry that his plans do not include devoting himself to chemistry; but there is time for him to develop a liking for it and your lectures are the right ones to inspire him, he has come to some of our sessions and I have clearly seen that he had been listening to you".

[16] Idem., i.e., "experiments on artifical cold produced by salts . . .".

[17] Idem., i.e., "concerning this, a quite strange thing has happened here having put amonia on silver oxide in a wide necked bottle in order to let it digest cold in a cupboard there was a spontaneous explosion which broke everything in the cupboard, there were more than 4 fingers' depth of liquid on the silver, what is it that can have caused the inflammation?".

[18] Idem., i.e., "would bring everything together as regards theory and nomenclature".

[19] Idem., i.e., "I am giving Mr Smith a letter for M. Berthollet whom I am asking to give him a copy for you".

[20] Beddoes, T. to J. Black, 23 Feb. 1788. Edinburgh University Library.

[21] Hodgson, A. E.: 1924, The History of Aeronautics in Great Britain. Oxford University Press, p. 141 ff.

[22] Sadler, J. to T. Beddoes, 14 Jan. 1791, Bodleian Library, Oxford, MS D. 134.

[23] Beddoes, T. to J. Black, 15 April 1791, Edinburgh University Library.

[24] Stock, J. E., Life, p. 10.

[25] For a modern account of Mayow's experiments and illustrations of the apparatus see Partington, J. R.: 1930, A Text Book of Inorganic Chemistry, 3rd ed., Macmillan. pp. 33–34.

[26] Beddoes, T. to J. Black, 23 Feb. 1788 and further requests on 15 April 1791 and 14 June 1792. Edinburgh University Library.

[27] Beddoes, T. to Sir Joseph Banks, 3 and 27 Jan. 1791.
Dawson, Warren R.: 1958, The Banks Letters. A Calendar of the Manuscript correspondence. British Museum (Natural History).

[28] Beddoes, T.: 1791, On the Affinity between Basaltes and Granite. Philosophical Transactions of the Royal Society, Vol. lxxxi Part I, p. 69.

[29] Op. cit., p. 57: 'conducted' is misleading to a twentieth century reader; in modern terms, Beddoes was saying "according as the temperature is programmed".

[30] Papers of Thomas Beddoes, Bodleian Library, Oxford, MS Dep. C. 134. See above. Beddoes wrote to Black that he was anxious to do justice to Hutton's theory.

[31] King-Hele, D. (ed.): 1981, The Letters of Erasmus Darwin, pp. 173–5.

[32] Beddoes, T. to Davies Giddy, 4 Nov. 1791. Cornwall R. O. DDG 41.

[33] Beddoes, T. to Davies Giddy, 4 Nov. 1791. Cornwall R. O. DDG 41.

[34] On April 21st 1789 Beddoes wrote to Black: "Since I had the pleasure of seeing you in Shropshire I have had an opportunity of making some observations on the agency of heat which have impressed upon my mind a very strong conviction of the truth of Dr Hutton's theory of the earth". If he is referring to Black's visit to Oxford and Birmingham this would be the work he had done in preparation for his paper on 'The Affinity between Basaltes . . .'. Edinburgh University Library.

[35] Ramsay, W.: 1918, The Life and Letters of Joseph Black, M.D. Constable.

[36] Memorial on the Present State of the Bodleian Library. For an account of the works missing or defective, see p. 10 ff.

[37] Swift, Jonathan: Polite Conversation (by Simon Wagstaff esq.): "A collection of genteel & ingenious conversation according to the most polite mode & method, now used at court & in the best companies of England." In several dialogues.

These are a set of conversation pieces in fashionable settings, satirising the cliché-ridden, dull conversation of the 'best set', preceded by an ironic essay about the collecting of witty sayings, and would have been purchased by the Bodleian for their Tory stance.

[38] See below, Chapter 5. Levere, T. H.: 1981, 'Dr Thomas Beddoes at Oxford . . .', Ambix, Vol 28, Pt. 2 (July 1981), p. 62 et seq. gives a detailed history.

[39] Gunther, R. W. T.: 1923, Early Science in Cambridge. Oxford University Press, p. 227.

[40] See Lefebvre, G., trans. R. R. Palmer: 1947, Paperback edition 1967, Princeton University Press.

[41] Goodwin, A.: 1979, The Friends of Liberty, Hutchinson, p. 108.

[42] King-Hele, D. (ed.): op. cit. Erasmus Darwin to James Watt, 19 Jan, 1790, p. 200.

[43] Quoted in Porter, R.: 1982, English Society in the Eighteenth Century, Penguin Books, Harmondsworth, Middx., p. 368.

[44] Goodwin, A., op. cit., pp. 111–113.

[45] Club des Jacobins, Bibliothèque Historique de la Révolution. See account in Goodwin, A., op. cit., pp. 201–203.

[46] Goodwin, A.: op. cit., pp. 122–124.

[47] Reinhard, M.: 1970, Le Voyage de Pétion à Londres, 24 Octobre–11 Novembre 1791. Revue d'Histoire Diplomatique, Jan.–June 1970, pp. 1–60. See account in Goodwin, A., op. cit., pp. 186–7.

[48] Lefebvre, G.: trans. R. R. Palmer: 1947, The Coming of the French Revolution. Princeton University Press, Princeton, N.J., p. 205.

[49] Goodwin, A.: op. cit., pp. 106–110, where Dr Price's sermon is extensively quoted.

[50] Paine, T.: 1791, Rights of Man; ed. Collins, H.: 1969, Penguin Books, Harmondsworth, Middx., p. 87.

[51] Goodwin, A.: op. cit., pp. 66–68.

[52] Lefebvre, G.: op. cit., p. 148.

53 Goodwin, A.: *op. cit.*, pp. 67 ff. and p. 264.
54 See Aspinall, A.: 1949, Politics and the Press, Harvester Press, Chapters 3, 7, 9, 11.

Chapter 5. The Midland Circle

1 Stock, J. E.: 1811, Memoirs of the Life of Thomas Beddoes M.D., Murray, p. 12.
2 Trinder, B.: 1973, The Industrial Revolution in Shropshire, Phillimore, Chichester, pp. 198–200. For the industrial developments in Shropshire and the Reynolds family, see: Raistrick, A.: 1953, Dynasty of Iron Founders. Longmans, Green, and Trinder, B., *op. cit.*
3 Plymley, J.: 1803, A General View of the Agricultural History of Shropshire.
4 Beddoes, T.: 1791, An account of some appearances attending the conversion of cast into malleable iron. Philosophic Transactions of the Royal Society, London. Vol. 81, Part 2, pp. 173–181.
5 Crell's Chemical Journal, Vol. 2 1792, Reprint of Beddoes' 1791 Phil. Trans. Paper. Vol. 3. 1793, p. 290.
6 Sadler to Beddoes, 14th January 1791: Bodleian Library, Oxford. MS. Dep. C. 134–137.
7 Sketch Book of William Reynolds, Science Museum Library, London. For an account of the book, see Dickinson, H. W., in Transactions of the Newcomen Society, Vol. II, 1921–22. For further details of the 'battle of patents' in which Sadler was involved see: Torrens, H. S.: 1982, 'New Light on the Hornblower and Winwood Compound Steam Engine', Journal of the Trevithick Society, No. 9.
8 Rose Beddoes to Thomas Beddoes, August 1791, Cornwall R. O., Truro, MS DG41/31. Reynolds, J. to T. Beddoes, Ketley, 26 Aug. 1791, Cornwall R. O., Truro, MS DG41/30.
9 Darwin, E. to R. L. Edgeworth, Feb. 1788, in King-Hele, D. (ed.): 1981, Letters of Erasmus Darwin. University Press, Cambridge.
10 Beddoes, T. to Davies Giddy, Shifnal, 21 Nov. 1791. Cornwall R. O., Truro, MS DR41.
11 Stock, J. E.: *op. cit.*, Appendix 6, pp. xxxvi–xxxviii, quoted in The Letters of Erasmus Darwin. D. King-Hele (ed.), 1981, Cambridge, pp. 174–176.
12 T. Beddoes to D. Giddy, undated. Cornwall R. O. Truro MS DG41.
13 W. Reynolds to T. Beddoes, 1789. Bodleian Library, Oxford. MS Dep. C. 134–7.
14 Beddoes, T.: 1792, Alexander's Expedition down the Hydaspes and the Indus to the Indian Ocean, printed privately, Madeley, p. 4.
Torrens, H. S.: 1982, The Reynolds-Anstice Shropshire Geological Collection in Archives of Natural History (1982) 10(3); pp. 429–441. See p. 433.
15 Bibliotheca Parriana, Catalogue of the Library of the late Reverend and learned Dr Parr: 1827, J. Bohn and J. Mawman, London. For an account of S. Parr, see: Derry, W.: 1966, Dr Parr, A Portrait of the Whig Dr Johnson. Clarendon Press, Oxford.
16 Beddoes, T.: Alexander's Expedition, p. iv.
17 Beddoes' titles were:

    Chapter I.   Observations on Hindoo austerities and on ceremonious devotion, pp. 49–58.
    Chapter II.   On the manufactures of the Hindoos, pp. 58–63.

Chapter IV. On the manufactures of the Hindoos, pp. 68–77.

Chapter VI. On the possessions of the British in Hindoostan, pp. 82–90.

18 Beddoes, T., *op. cit.*, Chapter IV, p. 45.

19 Beddoes, T., *op. cit.*, Chapter VI, p. 89.

20 Nehru, J.: 1946, The Discovery of India, Signet Press, Calcutta, pp. 333–338.

21 Beddoes, T. *op. cit.*, Chapter VI, p. 85; Chapter I, p. 58.

22 Plymley, K.: (i) April 1792, (ii) February 1792, Diary, Book 9, Shropshire R. O., Shrewsbury.

23 T. Beddoes to D. Giddy, 1792/3, Cornwall R. O., Truro, DG41.

24 See Derry, W., *op. cit.* (Note 15).

25 T. Beddoes to D. Giddy: from Shifnal, 3 April 1792, 1 December 1791; from Bath 4 November 1791; from Shifnal 3 April 1792, Cornwall R. O. DG41.

26 Parr, S.: 1792, A letter from Irenopolis to the Inhabitants of Eleutheropolis.

27 T. Beddoes to W. Reynolds, 19 November 1792, Cornwall R. O. DG41.

28 Plymley, K.: Diary, Book 4, Oct. 1791–Feb. 1792, printed leaflet in cover. Book 5, Feb. 10 1792–Feb. 23 1792. Book 6, Feb. 24 1792–Mar. 5 1792, Shropshire R. O. Shrewsbury.

29 T. Beddoes to D. Giddy, 2 February 1792; August 1792, Cornwall R. O., Truro, DG41.

30 Annual Register, 1792: Chronicle, p. 36 September. History of Europe, pp. 53–54 December; Appendix, p. 76.

31 Plymley, K.: Diary, Book 17, May 19–August 17, 1793, Shropshire R. O., Shrewsbury.

32 Jackson's Oxford Journal, June–December 1792.

33 T. Beddoes to D. Giddy, undated, Cornwall R. O. Truro, DG41.

34 Wedgwood, J. to Hester, J. T., 1793 (written by his secretary Chisholm), Wedgwood Archive, University of Keele, 1707–93.

35 Nepean, E. to Hawkins Browne, I., 1 November 1793, Public Record Office, HO 42/22 ERE 9144.

36 T. Beddoes to D. Giddy, undated (1792/3) Cornwall R. O., Truro, DG41.

37 T. Beddoes to D. Giddy, 11 May, 1792; undated letter; 5 July, 1792, Cornwall R. O., Truro, DD41.

38 See B. Trinder, 1973, Industrial Revolution in Shropshire, Phillimore Chichester, p. 73 ff.; p. 128; p. 201.

39 Stock, J. E.: 1811, Life of Thomas Beddoes M.D., pp. 88–89.

Chapter 6. Revolutionary and Educationalist

1 Wordsworth, W.: 1804, The French Revolution, 1.9 and ll. 35–40.

2 Beddoes, T.: 1792, Alexander's Expedition, p. iv–vi.

3 Rathbone, H. M.: 1852, Life and Letters of Richard Reynolds with a Memoir of his Life, p. 33.

4 Whitney, J.: 1947, Elizabeth Fry. Harrap. Guild Books Edition, p. 56. Mrs Trimmer ran a school for poor children in Ealing. Thomas Day's (1748–1789) Sandford and Merton was published in 1783, 1789. See previous chapter. Mrs Barbauld – best known for her poems and for her stories for children published as 'Evenings at Home', 1793.

[5] Beddoes, T.: 1792, Letter to a Lady on Early Instruction, Particularly that of the Poor: subsequently, Letter to a Lady, pp. 4–5.
[6] Clarke, M. L.: 1977, Mme de Genlis and Louis Philippe, History Today, Vol. 28, No. 10, October 1977, pp. 673–8. Mme de Genlis was certainly in England at this time but I have been unable to find any evidence that she and Dr Beddoes met.
[7] Beddoes, T.: Letter to a Lady, p. 2.
[8] *Ibid.*, p. 4.
[9] *Ibid.*, p. 11.
[10] *Ibid.*, pp. 16, 17.
[11] *Ibid.*, p. 7.
[12] *Ibid.*, p. 8.
[13] *Ibid.*, p. 9.
[14] *Ibid.*, pp. 20–24.
[15] Carswell, J. P.: 1950, The Prospector: being the Life and times of R. E. Raspe 1737–1794.
[16] Curran, J. P.: July 1790, Speech on the Right of Election of Lord Mayor of Dublin.
[17] Beddoes, T.: 1792, Isaac Jenkins and Sarah his wife.
[18] *Ibid.*, p. 7.
[19] *Ibid.*, p. 16.
[20] *Ibid.*, p. 9.
[21] *Ibid.*, p. 12.
[22] *Ibid.*, pp. 28–29.
[23] *Ibid.*, p. 38.
[24] *Ibid.*, p. 25.
[25] *Ibid.*, p. 30.
[26] Plymley, K.: Journal, Book 32, Feb.–April, 1795, MS SRO 567, Shropshire R. O. Shrewbury.
[27] King-Hele, D.: 1982, Letters of Erasmus Darwin, p. 255.
[28] Beddoes, T.: Observations, p. 15.
[29] *Ibid.*, p. 31.
[30] *Ibid.*, pp. 60–61.
[31] *Ibid.*, p. 89, *et seq.*
[32] Keele University Library, Letter, T. Beddoes to T. Wedgwood, after 1796, Wedgwood accumulation deposited by Messrs Josiah Wedgwood and Sons Ltd., Barlaston, Stoke-on-Trent.

Chapter 7. Bristol: Reviewing for *The Monthly Review*

[1] Stock, J. E.: *op. cit.*, pp. 90–91.
[2] Felton, J. (Editor of the Picture of London): Guide to the Watering Places, p. 93 ff.
[3] Goodwin, A.: *op. cit.*, pp. 281–282.
[4] Thelwall, K.: 1837, Life of John Thelwall, p. 59, and Cestre, C.: 1906: John Thelwall, A pioneer of Democracy in England. Swan Sonnenschein.
[5] Locke, D.: 1979, A Fantasy of Reason, Life of W. Godwin. Routledge and Kegan Paul, p. 80 ff.
[6] Locke, D.: *op. cit.* and Goodwin, A.: *op. cit.*, pp. 76 ff.

[7] Sadler, T. (ed.), 1869, Robinson, H. C.: Diary and Correspondence. Vol. I, pp. 26–27.

[8] Rosamond Beddoes to Mrs Whitehall, December 1st [1794]. Bodleian Library MS Dep. C. 135.

[9] Edgeworth, R. L. and M. Edgeworth, ed. Hunter, 1820: Memoirs of Richard Lovell Edgeworth, Vol. II.

[10] Beddoes T. to D. Giddy, 29th October 1793. Cornwall R. O.

[11] See Appendix.

[12] Beddoes' contributions to the Main and Foreign Sections of the Review are listed in an Appendix where the articles discussed here can be traced. The contributions to the Monthly Catalogue are often very brief: the most relevant are noticed here. For a full list see Nangle, Index, Vol. 2, pp. 248–250.

[13] Beddoes, T.: 1802. Hygeia, Essay I.

[14] Beddoes, T.: 1802. Hygeia, Essay IX.

[15] Bartram, W.: 1791, Philadelphia; 1792, London. Travels through North and South Carolina. van Doren, M. (ed.): 1955, Dover Reprint.

[16] Robberds, J. W.: 1843, Memoir of William Taylor 1765–1836, Murray, London, pp. 84 *passim*; and Nangle, *op. cit.*

[17] Enfield, W., 1741–97, who had been on the staff of the Warrington Academy, was one of the few R. Griffiths consulted about contributors.

[18] Where would be the Harm of a Speedy Peace?, 9th December, 1795, Biggs and Cottle, Bristol.

[19] Beddoes, T.: 1796, Essay on the Public Merits of Mr Pitt, J. J. Johnson, London. For a full account of Beddoes' Essay see next chapter.

[20] Coleridge, S. T.: The Watchman, ed. Patton, L., 1970. Vol. 2 in The Collected Works of Samuel Taylor Coleridge. General editor K. Coburn. Bollingen Series No. 75. 1969– . London and Princeton, N.J.

[21] Coleridge, S. T.: Notebooks 1794–1804. Vol. 1 in The Notebooks of Samuel Taylor Coleridge, ed. Coburn, K. Bollingen Series No. 50 (5 vols.) 1957– . New York, Princeton, N.J. and London. See Entry No. 249.

[22] Birkbeck Hill, G.: 1887, ed J. Boswell, Life of Johnson, 'Apr. 10, 1776 Dinner at Mr Thrale's. He expatiated a little more on the theme (i.e., reviews): "The Monthly Reviewers (said he) are not Deists; but they are Christian with as little Christianity as may be; and are for pulling down all establishments. The Critical Reviewers are for supporting the constitution both in Church and State. The Critical Reviewers, I believe, often review without reading the books through, but lay hold on a topic and write chiefly from their own minds. The Monthly Reviewers are duller men, and are glad to read the books through".' Vol. III, p. 32.

Chapter 8. Arrival of Coleridge/Political and Literary Activities

[1] Beddoes, T.: 1799, Essay on the Causes, Early Signs and Prevention of Pulmonary Consumption, p. 6.

[2] Beddoes, T. (ed.): 1795, Brown's Elements of Medicine, p. cxvii.

[3] Stock, Life, p. 94.

[4] Beddoes, T. to Davies Giddy, 15th June 1793. Cornwall R. O.

[5] Bodleian Library ms. Affidavit made by Lovell Edgeworth, Maria Edgeworth and George Keating, testifying to the marriage of Anna with Thomas Beddoes in April

1794. The affidavit was needed by Dublin solicitors Hamilton and Co. and probably links with financial arrangements after T. Beddoes' death.

6 Correspondence of Thomas Beddoes with his parents Richard and Anne and his sister Rosamund. Bodleian Library MS C. 135. (2 April, 1794; 5 June, 1794; 11 March, 1794; 26 July, 1793.)

7 Anna Beddoes to Mrs Beddoes and Rosamond Beddoes, 26th Oct., 1794; 28th May 1794. Bodleian Library, MS C. 135.

8 Southey's part in bringing Coleridge to Bristol and the relations of the two young men are particularly well described in Georges Lamoine, *La Vie Littéraire de Bath et de Bristol* (doctoral thesis 1975; Paris 1978, Librairie Honoré Champion). See especially, Tome II, Part 3, Chapter 1, pp. 499–560, and for Joan of Arc II, Part 3, Chapter 2, pp. 556–560.

9 Warter, J. (ed.): 1856, Selections from the Letters of R. Southey. R. Southey to T. Southey 11th Feb., 1810, p. 194 and to John May, 19th July 1797, for Southey's summary of the arrangements.

10 Cottle, J.: 1848, Reminiscences of Coleridge and Southey, p. 2 ff.

11 *Op. cit.*, p. 261.

12 Beddoes, T.: 1792, "Extract from a Letter on ... Early Instruction", p. 9. The use of Hessian mercenary troops in the American War and the building of barracks for a standing army caused much alarm. Rumours in Jan. 1794 that Hessian soldiers had landed in the Isle of Wight added to the fear. See Goodwin, *op. cit.*, p. 313.

13 Goodwin, A., *op. cit.*, pp. 385–6.

14 Cottle, J., *op. cit.*, pp. 93 ff.

15 Coleridge, S. T.: Lectures 1795, ed. Patton, L. and P. Mann, 1971, Vol. 1 in The Collected Works of Samuel Taylor Coleridge, *op. cit.* (Ch. 7, Note 20).

16 Stock, J. E., *op. cit.*, p. 127.

17 Griggs, E. L.: 1932, Unpublished Letters of S. T. Coleridge, Constable. Vol. I, Letter 18.

18 For the course of the protest and the details of dates of publication of the pamphlets see Patton, L. and P. Mann, *op. cit.*

19 16th Aug. 1819. An orderly meeting in St Peter's Fields, Manchester, to press for Parliamentary Reform was broken up by a charge of yeomanry and 12 protesters were killed. This blunder by the Magistrate was approved by the Ministry and the incident became known as "The Peterloo Massacre" and symbolic of the oppressive nature of the government.

20 See Hirst, M. E.: 1923, Quakers in Peace and War, p. 469, and Collection of newspaper items etc. in Friends House Library, Euston Road, London.

21 Coleridge, S. T.: The Watchman, ed. Patton, L., 1970. Vol. 2 in The Collected Works of Samuel Taylor Coleridge, *op. cit.* (Ch. 7, Note 20).

22 Beddoes, T. to Davies Giddy, 8 Nov. 1792. Cornwall R. O., DG41.

23 Coleridge, E. H. (ed.): 1912, Poems of Samuel Taylor Coleridge. Oxford Standard Authors, 1960 reprint, p. 138.

24 Stock, J. E., *op. cit.*, Appendix No. 7.

25 Coleridge, E. H., *op. cit.*, p. 74. The variant reading makes clear how much less realistic about country life Coleridge was compared to Beddoes and that "the dell" was to be "where high souled Pantisocracy shall dwell".

26 Southey, R. to H. Davy, 4th May 1799. Papers of H. Davy, Royal Institution Library, London. Box 9, p. 91.

[27] Coleridge, E. H., *op. cit.*, p. 240.

[28] Stock, J. E., *op. cit.*, Appendix no. 7.

[29] Coleridge, E. H., *op. cit.*, p. 216.

[30] Coleridge, S. T.: Notebooks 1794–1804, *op. cit.* (Ch. 7, Note 21).

[31] Potter, S. (ed.): 1934, A Minnow among Tritons: Letters of Sara Coleridge.

[32] Coburn, K., *op. cit.* The entries are as follows:

Considerations	Notebook entry	133
John Brown	" "	389
Edinburgh Medical School	" "	174
Dreams	" "	188
Nitsch	" "	249

[33] See ante: Arrival in Bristol.

[34] Davy, H. to S. T. Coleridge at Nether Stowey. Papers of H. Davy, Royal Institution Library, London, Box 27 (n.d. 1800).

[35] Erdman, D. V.: 1956, 'Coleridge, Wordsworth, and the Wedgwood Fund'. Bulletin of the New York Public Library; Part I, Vol. 60, pp. 425–443; Part II, Vol. 60, pp. 487–507.

[36] Hazlitt, W.: 1823, My First Acquaintance with Poets. William Hazlitt: Selected Writings, ed. R. Blythe, Penguin, Harmondsworth, Middx., p. 51.

[37] The Watchman No. III, March 17, 1796. Bollingen edition, p. 100.

[38] Essay on the Public Merits of Mr Pitt: Commutation tax p. 92ff.; Tiller of the ground p. 156; Furniture of science p. 138; New poisons p. 160 ff.; Yokefellow in adversity p. 92.

[39] *Ibid.*, p. 32.

[40] Monthly Review, New Series, Vol. 20 July 1796, p. 258.

[41] Essay, p. 160 ff.

[42] *Ibid.*, p. 115.

[43] *Ibid.*, p. 198.

[44] *Ibid.*, Chapter I.

Chapter 9. Pneumatic Institute/Humphry Davy

[1] Beddoes, T.: 1793, Letter to Darwin, pp. 4–5.

[2] Beddoes, T.: 1793, Observations on the Nature and Cure of Calculus etc.

[3] *Ibid.*, pp. 6–12.

[4] *Ibid.*, p. 59.

[5] *Ibid.*, p. 52.

[6] *Ibid.*, p. 56.

[7] *Ibid.*, p. 147.

[8] *Ibid.*, p. 253 ff.

[9] Beddoes, T.: 1793, Letter to Dr Erasmus Darwin on a New Method of Treating Pulmonary Consumption, p. 27 ff.

[10] Beddoes' description of his efforts to account for this change during pregnancy is given also in 'Observations on sea scurvy, calculus, etc.', p. 113 ff.

[11] Letter to Darwin, p. 31 ff.

[12] *Ibid.*, p. 43.

[13] *Ibid.*, p. 50 ff.

[14] Darwin, E. to T. Beddoes, 17th Jan. 1793. See King-Hele, D.: 1981. The Letters of Erasmus Darwin, Cambridge U.P., pp. 228–231. This letter was printed by T. Beddoes in A Letter to Erasmus Darwin, M.D., 1793.

[15] Letters from Dr Withering etc., 1793, pp. 3–4.

[16] Letter to Darwin, pp. 4–5.

[17] Beddoes, T.: 1794, Considerations on the Medicinal Uses of Factitious Airs, p. 10.

[18] *Ibid.*, p. 17.

[19] Muirhead, J. P.: 1854, Mechanical Inventions of James Watt, Vol. II, Letters, J. Watt to E. Darwin, 30th June 1794.

[20] A reconstruction of this apparatus, made from Watt's drawings, is on view in the Wellcome History of Medicine section at the Science Museum at South Kensington: No. 19 in a series of tableaux.

[21] Beddoes, T. to Davies Giddy, 19th June 1796, Cornwall R. O.

[22] Beddoes, T.: 1796, Medical Cases and Speculations including Parts IV and V of Considerations, p. xi.

[23] Dawson, W. R.: 1958, The Banks Letters. A Calendar of the Manuscript Correspondence. British Museum: Natural History.

[24] Bolton, H. C.: 1892 (ed.) Scientific Correspondence of Joseph Priestley. Priestley to Sir Joseph Banks, 25th April, 1790, Priestley to Dr Withering, 15th April 1793, New York.

[25] Beddoes, T. to Davies Giddy, 31 July 1796 and 22 Aug. 1796, Cornwall R. O.

[26] Beddoes, T. to Davies Giddy, Mar. 1795, Cornwall R. O.

[27] For details of the collaboration between James Watt and Dr Beddoes I am much indebted to Lord Gibson-Watt who kindly allowed me access to the papers of James Watt in his private collection.

[28] Letters of Thomas Wedgwood, 1794–1801. Keele University Library, Wedgwood accumulation.

[29] King-Hele, D.: 1982, Letters of Erasmus Darwin, p. 259.

[30] Weber, C. A.: 1935, Bristols Bedeutung für die Englische Romantik und die Deutsch-Englischen Beziehungen. Max Niemeyer, Halle (Saale).

[31] Hartley, Sir Harold: 1972, Humphry Davy. E. P. Publishing, Wakefield, pp. 9–19. Davy, J. (ed.): 1856, Fragmentary Remains of Sir Humphry Davy, Bart., where many of Davy's early letters are quoted.

[32] Todd, A. C.: 1967, Beyond the Blaze. D. Bradford Barton, Truro. Chapter VIII.

[33] Davy, J. (ed.): 1856, *op. cit.*, p. 000.

[34] Charlotte Edgeworth to Mrs Ruxton, Oct. 1802. Edgeworth Letters, Bodleian Library – Papers of Mrs C. Colvin.

[35] Davy, H.: Notebook 13e, Davy Archive, Royal Institution, London.

[36] I am indebted to the present occupants of 6, Dowry Square for the opportunity to see the house and for an explanation of the history of the alterations which have been made.

[37] Davy, H. to S. T. Coleridge at Nether Stowey. Box 27, Davy Archive, Royal Institution, London.

[38] Beddoes, T.: 1799, Notice of some Observations etc., p. 6.

[39] *Ibid.*, p. 7 ff.

[40] *Ibid.*, p. 37.

[41] *Ibid.*, p. 6.

[42] Hartley: 1972, *op. cit.*, pp. 26–37. See Thorpe, T. E.: 1896, Humphry Davy Poet and Philosopher, Cassell, p. 40 ff.

[43] Davy, H.: 1800, Researches, Chemical and Philosophical . . . Facsimile reproduction, Butterworths, n.d. (1972). In the account of Research IV Davy names the people who breathed N_2O.

[44] Published by Beddoes in 1799 immediately after Davy's arrival; included Davy's early experiments on Heat and Light which he accepted quite uncritically. See Thorpe, *op. cit.*

[45] Davy, J.: 1800, *op. cit.*

[46] See below, Chapter XIII.

[47] Pain, intense and brutal, was in the 18th century accepted as an inevitable part of everyday life and inflicted for punishment without restraint.

[48] It was Thomas Charles Hope, 1766–1844, successor to Joseph Black as Professor of Chemistry at Edinburgh, who recommended Davy to Rumford. In 1799 Hope visited Beddoes, who had been his fellow student in Edinburgh, and was very impressed by Humphry Davy and the experiments he was doing at the Pneumatic Institute. See Kendall, J., Endeavour, July 1944.

[49] Peruvian bark, from which quinine is obtained, was a common medicine.

[50] Warter, J. W.: 1856, Selections from the Letters of R. Southey. Southey, R. to J. Rickman, 1802.

[51] Jones, T. W.: 1925, Thomas Beddoes, M.D.: A Neglected Chemist, Science Progress, Vol. 19. Jan. 1925, p. 635. See also Stock, J. E.: 1811, Memoir, p. 157 ff.

[52] Nicholson's Journal, 1808, p. 68, Letter from a Correspondent on the late Discovery of Metals in the fixed Alkalis. 8th Aug. 1808.

[53] Southey, R. to Humphry Davy from Exeter, 4 May 1799. Royal Institution Library.

[54] Grigg, E. L.: 1932, Unpublished letters of Samuel Taylor Coleridge, Coleridge, S. T. to H. Davy, 15 July 1800, Constable. Letter 71.

[55] Coburn, K., *op. cit.*, Coleridge Note Books 1794–1804. Entry 1098.

[56] Davy, H. Notebook 13G. Royal Institution Library.

[57] See Sharrock, R.: 1962, The Chemist and the Poet: Sir Humphry Davy and the Preface to the Lyrical Ballads. Notes and Records of the Royal Society, Vol. 17, p. 57.

Chapter 10. Preventive Medicine

[1] Cartwright, F. F.: 1952, The English Pioneers of Anaesthesia, Wright, Bristol, p. 86.

[2] John Wedgwood to Tom Wedgwood, 12 March 1800. Wedgwood Archive, Keele University Library. Quoted by permission of Messrs Josiah Wedgwood & Sons Ltd., Barlaston, Stoke-on-Trent.

[3] Stock, J. E.: *op. cit.*, p. 156.

[4] Sir John Sinclair: 1754–1835, first President of the Board of Agriculture. 1790 Planned a 'Statistical Account of Scotland'. This was published at intervals during the next ten years. See E. Clarke, D. N. B., Vol. 18.

[5] Beddoes, T.: 1797, Reports concerning the effects of the Nitrous Acid in the Venereal Disease.

6 Beddoes, T.: 1799.

7 Beddoes, T.: 1800, Communications Respecting . . . Venereal Disease, p. v ff.

8 Beddoes, T.: 1799, Essay on the Causes Early Signs and Prevention of Pulmonary Consumption for the use of Parents and Preceptors.

9 Beddoes, T.: *op. cit.*, p. 264.

10 Beddoes, T.: *op. cit.*, p. 3 ff.

11 T. Beddoes to Sir John Sinclair, 18 Feb. 1798, quoted in Sinclair, J.: 1837, Memoirs of Sir John Sinclair, Vol. II, p. 27.

12 The inclusion of fifers among a group of trades where the occupational hazard was occasioned by dust seems curious. The rail at the base of the mast of a sailing ship, where running rigging was belayed, was the fife rail. Fifes were used to give the signals for working the rigging, but there seems no record of a fifer being anyone other than the man who gave these signals. I am grateful to Mr A. W. H. Pearson, Historian at the National Maritime Museum, Greenwich, for help in checking the use of the term 'fifer'.

13 Beddoes, T.: *op. cit.*, pp. 34—35.

14 Beddoes, T.: *op. cit.*, pp. 103—107.

15 Beddoes, T.: *op. cit.*, p. 254.

16 Bradley and Willich (eds.): The Medical and Physical Journal, Vol. 10, June—December 1803.

17 Cartwright, F. F.: *op. cit.*, pp. 163—164.

18 Beddoes, T.: 1804, Rules of the Medical Institution for the Relief of the Sick and Drooping Poor, p. 93.

19 Beddoes, T.: *Ibid.*, pp. 158—159.

20 Beddoes, T.: *Ibid.*, p. 179.

21 Beddoes, T.: *Ibid.*, p. 32.

22 Beddoes, T.: *Ibid.*, p. 110.

23 Beddoes, T.: *Ibid.*, p. 20.

24 Medical and Physical Journal, Vol. 8, July—December 1802, p. 7.

25 Beddoes, T.: 1796, Letter to the Rt Hon. William Pitt on the means of Relieving the Present Scarcity.

26 Beddoes, T.: 1796, Essay on the Political Merits of Mr Pitt, p. 11.

27 Stock, J. E.: *op. cit.*, p. 299 ff. This visit is also described in Weber, C. A.: 1935, Bristols Bedeutung für die Englische Romantik und die Deutsch-Englischen Beziehungen. Halle (Saale).

28 See Goodwin, A.: 1979, 'The Friends of Liberty', Hutchinson, London, pp. 403—405; 414.

29 Beddoes, T.: 1797, Introductory Lecture to a course of Popular Instruction, p. 64.

30 T. Beddoes to J. Watt, Jr. Letters in the Boulton papers, Birmingham Central Library, dated 24 Dec. 1797; 2 Jan. 1798; 25 Feb. 1799; March 1798. Cited by permission of the Trustees of the Matthew Boulton Trust.

31 Beddoes, T. to Davies Giddy, June 1798, Cornwall R. O., Truro, DG43.

32 Stock, J. E.: *op. cit.*, p. 136 ff.

33 Beddoes, T. to Davies Giddy 14th April, 1798, Cornwall R. O., Truro, DG42.

34 Beddoes, T. to Davies Giddy, January 21, 1802, Cornwall R. O., Truro, DG42.

35 Medical and Physical Journal, ed. Bradley and Willich, Vol. 9, Jan—Dec. 1803; Vol. 12, June—Dec. 1804.

36 Beddoes, T. to Davies Giddy, 11 June, 1803, Cornwall R. O., Truro, DG42.

[37] Medical and Physical Journal, Vol. 10, June–Dec. 1803, p. 193 ff.
[38] Medical and Physical Journal, Vol. 9, Jan.–June 1803.
[39] Paris would have particularly interested Beddoes for the opportunities it was able to give its students for practice in anatomy. Vienna, perhaps described to him by Dr Frank, was outstanding for clinical medicine.
[40] Beddoes, T.: 1800, Appendix in Davy, H., Researches, Chemical and Philosophical; chiefly concerning Nitrous Oxide, or dephlogisticated nitrous air, and its Respiration. London, J. Johnson, pp. 577–579.

Chapter 11. Religio Medici

Hygeia first appeared as eleven essays, published separately in 1802. The Essays are:

I   Essays on Personal Prudence and Prejudices respecting health To Heads of families inhabitants of the British Isles.
II   Essays on the Means of Avoiding Habitual Sickness and premature mortality to Ministers of the Gospel of Every Denomination.
III   On Individuals composing our Affluent and Easy Classes. Part 1, Of Schools for Girls.
IV   On Individuals composing our Affluent and Easy Classes. Part 2, Treatment of Boys.
V   On Temperature and Hardiness with Remarks on Diet.
VI   On Scrophula.
VII   On Consumption.
VIII   On the Preservation of the Physical Powers of Enjoyment with remarks on food and digestion.
IX   On the Nature and Prevention of some Disorders commonly called nervous. Part 2, On Insanity.
XI   Essays containing Remarks on Miscellaneous topics of Prophylactic Medicine.

The references that follow are to the three volume edition, Vol. I, Essays I–IV, 1802; Vol. 2, Essays V–VIII, 1802; Vol. 3, Essays IX–XI, 1803. The Essays are paged independently and here the Reference is to Essay and page.

[1] *Hygeia*: Essay II, p. 54.
[2] *Idem.*, Essay I, p. 21.
[3] *Idem.*, Essay I, p. 32.
[4] *Idem.*, Essay I, p. 84–86.
[5] *Idem.*, Essay II, p. 77.
[6] *Idem.*, Essay II, p. 83.
[7] *Idem.*, Essay II, p. 52.
[8] *Idem.*, Essay II, p. 90.
[9] *Idem.*, Essay VIII, p. 60.
[10] *Idem.*, Essay III, p. 13.
[11] *Idem.*, Essay IV, p. 82.
[12] *Idem.*, Essay IX, p. 149.

[13] *Idem.*, Essay X, p. 13.

[14] *Idem.*, Essay X, p. 40. Beddoes added the following footnote to *Hygeia*, Essay X, "Whether madness admit of an essential character?", p. 41:

"The appearance of the eye, which is so striking in the maniacal state, is regulated by its muscles. Though hollow when the patient is calm, it will protrude on the commencement of the paroxysm. This arises from the rigidity of certain muscles to be seen on the back of the ball in any set of anatomical plates. The glistening is a similar operation. The dullness of the eye often arises from a sort of corrugation of the coats, though the furrows are not singly visible. But when all the moving fibres become tense, the coats are fully unfolded, and shine. — Sometimes the presence of the keeper of the madhouse shall overawe the raving patient, till his tongue and limbs become in a moment composed. But the eye will retain its characteristic expression."

[15] Essay IX, p. 107.
[16] Essay IX, p. 90 (Note).
[17] Essay X, p. 74.
[18] Essay X, p. 36.
[19] Essay X, p. 19.
[20] Essay IX, pp. 70/71.
[21] Essay X, p. 85.
[22] Essay IX, p. 184.
[23] Essay V, p. 93.
[24] Essay II, pp. 77–78.
[25] Essay IX, p. 203.
[26] Essay II, p. 63.
[27] Essay VII, pp. 99–100.
[28] Beddoes to Giddy, Cornwall R. O., Truro, DG41.
[29] Essay IX, p. 205 (Note).
[30] Griggs, E. L. (ed.): 1932, Letters of Samuel Taylor Coleridge. S. T. C. to R. Southey, Letter 18, March 12, 1803: "with the exception of the Essay on Mania, the Hygeia is a valuable and useful work".

Chapter 12. Behind the Print

[1] Beddoes, T.: 1803, *Hygeia*, Essay XI, p. 96.
[2] Beddoes, T.: 1808, Good Advice to Husbandmen, pp. 8, 19.
[3] Beddoes, T. to F. Douce, March 1808, Bodleian Library, Oxford, MS Douce d 21.d.29.
[4] Beddoes, T. to Davies Giddy, 26th June, 1807, Cornwall R. O.
[5] Beddoes, T. to Davies Giddy. Seven letters during 1797, 1798, 1799, Cornwall R. O., DDDG 41–43.
[6] See Todd, A. C.: 1967, Beyond the Blaze, D. Bradford Barton, Truro, p. 29.
[7] Medical and Physical Journal, Vol. 9 Jan.–June 1803, p. 263 ff.
[8] Warter, J. W.: 1856, Letters of R. Southey. Southey, R. to Grosvenor C. Bedford, Nov. 10 1806, Vol. I, p. 396.
[9] Papers of T. G. G. Sotheron-Estcourt, Esq., of Tetbury, Gloucester R. O.

[10] Maby, M.: Life and Letters of Dr J. King. Unpublished typescript deposited Bristol Central Library (Reference). I am indebted to Miss Maby for permission to quote from her account and for helpful discussions about Dr King.

[11] Anna Beddoes to Tom Wedgwood, n.d. Quoted by C. Collier Abbott, 1942, 'The Parents of Thomas Lovell Beddoes', *Durham University Journal*, Vol. XXXIV, No. 3.

[12] Beddoes, Anna to Davies Gilbert, Sept. 1808. Cornwall R. O.

[13] Beddoes, T. to Davies Giddy, 19th Feb. 1805. Cornwall R. O. Beddoes devised, even if there is no evidence of its being made, an apparatus for measuring muscle tone.

[14] Beddoes, T. to Davies Giddy, 19 May 1806. Cornwall R. O.

[15] Beddoes, T. to Davies Giddy, 20 Aug. 1798. Cornwall R. O.

[16] Stock, *op. cit.*, p. 150 ff.

[17] See Cooper, L.: 1957, Radical Jack.

[18] Davy Notebook 136. Royal Institution Library. Quoted by T. E. Thorpe; 1896, Humphry Davy: Poet and Philosopher, p. 50.

[19] Stock, J. E.: Memoir, p. 409.

[20] Beddoes, T.: *Hygeia*, Essay X, p. 85.

[21] Johann Peter Frank, 1745–1821: in 1795 moved from Pavia to Vienna where he became famous as a clinician. 1799–1819, publication of *System einer vollstandigen Medizinschen Polizey*. Frank considered that public health was at all times the responsibility of the state; was concerned with health in industry and schools; with education. C.f. Weber, C. A.: 1935, Bristols Bedeutung für die Englische Romantik und die Deutsch-Englischen Beziehungen. Max Niemeyer, Halle (Saale).

[22] Beddoes, T. to Mr Whitehall, 18 March 1806. Bodleian Library, Oxford, MS.

[23] Beddoes, Anna to Davies Giddy, Sept. 1808. Cornwall R. O.

[24] Stock, J. E.: Memoir, p. 389 ff.

[25] Edgeworth, Henry to J. King, 1 May 1806. Edgeworth Papers, Bodleian Library, Oxford.

[26] See Cooper, L., *op. cit.*

[27] Edgeworth, M. E. and R. L.: 1798, Practical Education: 'Toys'.

[28] Stock, J. E., *op. cit.*, p. 128, Appendix 8. Todd, A. C.: 1967. Beyond the Blaze, pp. 16–17.

[29] Stock, *op. cit.*, p. 281 ff.

[30] *Hygeia*, Essay III, pp. 50–51; with an interesting reference to Dr King's work.

[31] Beddoes, T.: 1796, Letter to the Rt Hon. W. Pitt . . . Scarcity, p. 21.

[32] Beddoes, T. to Davies Giddy, n.d. (Winter 1972–3). Cornwall R. O. DDG 41.

[33] Correspondence of Rosamond Beddoes. Bodleian Library, Oxford, MS C. 135.

[34] Levere, T. H., private communication.

[35] Beddoes, T. to Davies Giddy, n.d. Cornwall R. O., Beddoes was concerned that his successor should be Stacey, who had a large family, and seems to have regretted the appointment of Dr Bourne.

[36] Thelwall startled his audience by not wearing a wig when he lectured and simply having his long hair tied back.

[37] Beddoes, Rosamond to T. Beddoes, 21 (August 1791). Cornwall R. O.

[38] Beddoes, T. to Davies Giddy, n.d. Cornwall R. O.

[39] Beddoes, Rosamond to Mrs Whitehall, 8 May [?].

[40] Coleridge, S. T. to J. Thelwall, 30 Jan. 1798. E. L. Griggs: 1932, Unpublished Letters of Samuel Taylor Coleridge, Vol. I, Letter 56. Constable, London.

[41] Reynolds, Joshua to T. Beddoes, 26 Aug. 1791. Cornwall R. O.

[42] Sadler, J. to Mr Stacey of Oxford, 10 Aug. 1792. Cornwall R. O. [Stacey was asked to inform Beddoes.] Letters in Gilbert collection concerning engines.

[43] Hodgson, J. E., 'James Sadler of Oxford'. *Newcomen Society Transactions*, Vol. VIII, 1927–28.

[44] See Maby, M., *op. cit.*

[45] Beddoes, T. to Davies Giddy, 11 May, 1792, n.d. and 8 July, 1792. Cornwall R. O.

[46] Beddoes, T. to Davies Giddy, 14 March 1795. Cornwall R. O.

[47] Beddoes, T. to Davies Giddy, 23 Aug. 1805. Cornwall R. O.

[48] See Todd, A. C., *op. cit.*, pp. 147–151; and Todd, A. C.: 1957, 'Anna Maria, Mother of T. L. Beddoes' in *Studia Neophilologica*, Vol. 29.

[49] Beddoes, T. to Davies Giddy, 3 March 1803. Cornwall R. O.

[50] Beddoes, Anna to Davies Giddy, August 1803. Cornwall R. O. DG 89(i).

[51] Beddoes, T. to James Watt, Nov. 1808. In Muirhead, J. P.: 1884, *Mechanical Inventions of James Watt*, Vol. II, *Letters*.

[52] Darwin, E.: 1791, The Botanic Garden, Part I, p. 26: "there seems no probable method of flying conveniently but by the power of steam, or some other explosive material".

[53] Maria Edgeworth to Sneyd Edgeworth, 30 Dec., 1808. Bodleian Library, Oxford, Maria Edgeworth's letters, 1782–1849.

[54] Beddoes, Anna to Davies Giddy, 24 Oct., 1808. Cornwall R. O.

[55] Beddoes, Anna to Davies Giddy, 24 Dec., 1808. Cornwall R. O.

[56] Davy, H. to S. T. Coleridge, 27 Dec., 1808. Quoted in John Davy: 1858, Fragmentary Remains of Sir Humphry Davy, Bart, p. 106.

[57] Coleridge, S. T. to H. Davy, 30 Jan., 1808. Papers of Sir Humphry Davy, Royal Institution Library.

[58] Douce, F.: inscription on letter from Beddoes d/d March 1808. Bodleian Library, Oxford. MS Douce d 21.d.29.

[59] Gentleman's Magazine, 1809, Vol. 1, p. 157.

[60] Beddoes, T. to Mr Estcourt. Papers of T. G. G. Sotheron Estcourt. Gloucester R. O.

[61] Beddoes, T. to Davies Giddy, 10 Jan. 1793. Cornwall R. O.

Chapter 13. Family and Reputation

[1] Letters of A. M. Beddoes to Mrs King (née Emmeline Edgeworth), 1807–1824. Beddoes Papers Box 135 (3), Bodleian Library, Oxford.

[2] Humphry Davy to Mrs. Apreece (his future wife), Nov. 1811. Quoted in Davy, J.: 1855, Fragmentary Remains of Sir Humphry Davy, Bart., p. 149.

[3] Donner, H. W.: 1935, The Browning Box. Oxford University Press. Todd, A. C.: 1952, Thomas Beddoes and his Guardian. Times Literary Supplement, 10th October.

[4] Donner, H. W.: 1935, Thomas Lovell Beddoes, Blackwell, Oxford.

[5] Kelsall, T. F.: 1850–51, Poems, with a Memoir.

[6] Higgens, J. (ed.): 1976, Thomas Lovell Beddoes, Selected Poems, Fyfield Books/ Carcanet Press, Manchester.

[7] See Donner, *op. cit.*, pp. 39–40. Donner's "intellectual exercise" seems to me too absolute a judgement.

[8] See 'The Oviparous Tailor', Poems of T. L. Beddoes, ed. Donner H. W. 1935, Oxford, pp. 113–114.

[9] Higgens, J., op. cit., p. 14.

[10] Letters of Anna Beddoes to Mrs King.

[11] I am indebted to Miss M. Maby for drawing attention to Dr King's letter to the Bristol Journal, 30 Sept. 1836; see Maby, M.: Life and Letters of Dr John King, deposited in Bristol Central Library, Reference.

[12] The frogs appear twice in Hygeia, once as subjects of experiments to test the effects of various substances including opium, laurel and tea and again when they are suggested as demonstration material in biology lessons. The use of frogs in experiments on galvanism was standard.

[13] Rev. W. Buckland, 1784–1856, Prof. of Mineralogy, Oxford 1813, a pioneer of the British Association for the Advancement of Science, founded in 1831. Dr King suggested that the inaccuracies in the lecture and its tone may have arisen from its being an after-dinner speech at the end of an expedition on the River Frome.

[14] Roget, P. M.: 1824, Supplement to the 4th, 5th and 6th editions of the Encyclopaedia Britannica, Vol. 2.

[15] Papers of Humphry Davy, Box 14 i, Royal Institution Library, London.

[16] Southey, C. C.: 1850, Life and Correspondence of Robert Southey; Warter, J. W.: 1856, Selections from the Letters of Robert Southey; Griggs, E. L.: 1932, Unpublished Letters of S. T. Coleridge.

[17] Kelsall, T. F., op. cit.

[18] Robinson, E.: 1955, Thomas Beddoes, M.D. and the reform of science teaching in Oxford, Annals of Science 11; Cartwright, F. F.: 1952, The English Pioneers of Anaesthesia, Wright, Bristol; Ferguson, A. (ed.): 1948, Natural Philosophy through the Eighteenth Century, Supplement to The Philosophical Magazine.

[19] Parker, G.: 1928, The Discovery of the Anaesthetic powers of nitrous oxide, The Lancet, Vol. 114, 7 Jan. 1928, pp. 60–62.

[20] Smith, W. D. A.: 1965, A History of Nitrous oxide and oxygen anaesthesia, Parts I, II and III, British Journal of Anaesthesia, Vol. 37, pp. 795–798; 871–882 and 958–966. 1966, Part IV, ibid., Vol. 38, pp. 58–72. 1970, Part IVa, ibid., Vol. 42, pp. 347–353; Part IVb, pp. 445–458. As Beddoes himself was treating a patient in Ludlow in 1794 it is possible that there may have been some memory of this in Hickman's lifetime.

[21] Miller, A.: 1931, The Pneumatic Institution of Thomas Beddoes at Bristol, Annals of Medical History, New York.
Gottleib, L. S.: 1965, Thomas Beddoes, M.D. and the Pneumatic Institution at Clifton, 1798–1801, Annals of Internal Medicine, Vol. 63, No. 3, pp. 530–533.
Keys, T. E.: 1969, The Early Pneumatic Chemists and Physicians, The Journal of Anaesthesiology (American Society of Anaesthesiologists), Vol. 30, No. 4.

[22] Priestley, Joseph, to Humphry Davy from Northumberland, Pennsylvania, 10 Oct. 1801, quoted in: Thorpe, T. E.: 1896, Humphry Davy – Poet and Philosopher, Cassell, p. 39. The younger Joseph Priestley's experience in breathing nitrous oxide was reported by Davy in his Researches, p. 535. The whole letter is of considerable interest and the full text as quoted by Thorpe (pp. 38–39) is given as an Appendix.

[23] Weber, C. A.: 1935, Bristol's Bedeutung für die Englische Romantik und die Deutsch-Englischen Beziehungen. Max Niemeyer, Halle (Saale).

[24] Pre-eminently the work of Professor Kathleen Coburn and later Georges Lamoine, see

above, Chapter VIII; and more recently T. H. Levere, Poetry Realised in Nature – Samuel Taylor Coleridge and early nineteenth-century science, University Press, Cambridge, 1981.
[25] See for example: Todd, A. C.: 1967, Beyond the Blaze; Emblen, D. L.: 1970, Peter Mark Roget; Hartley, Sir Harold: 1972, Humphry Davy; King-Hele, D.: 1981, Letters of Erasmus Darwin.
[26] William Godwin, who could never rest content with the solution he proposed in Enquiry concerning Political Justice, 1793, has also a claim to be included.

BIBLIOGRAPHY

Manuscript Sources

Correspondence of Thomas Beddoes

Edinburgh:	University Record Office	Correspondence of Joseph Black M.D.
Gloucester:	County Record Office	Estcourt Archive and Letters of T. Beddoes
Keele:	University Library	Wedgwood accumulation
Oxford:	Bodleian Library	Papers of Thomas Beddoes
Truro:	Cornwall Record Office	Gilbert papers

Related material

Bristol	City Record Office	King family records
Litchfield:	Joint Record Office	Will of Thomas Beddoes 1769
London:	Friends House Library	Papers relating to E. L. Fox
	Public Record Office	Home Office Papers
	Royal Institution Library	Papers of Humphry Davy
Oxford:	Bodleian Library	Edgeworth Papers
Shrewsbury:	County Record Office	Legal Documents
		K. Plymley: Diary

Lord Gibson-Watt: Private collection	Papers of James Watt

A List of Doctor Beddoes's Publications

As given by J. E. Stock

1784   Translation of Spallanzani's Dissertations on Natural History. A second Edition in 1790.

1784   Notes to a translation of Bergman's Physical and Chemical Essays.

1785   Translation of Bergman's Essay on Elective Attractions.

1786   Translation of Scheele's Chemical Essays. Edited and corrected by Doctor Beddoes.

1787   An account of some new experiments on the production of artificial cold. Letter from Thomas Beddoes M.D. to Sir Joseph Banks P.R.S.

1790   Chemical Experiments and Opinions extracted from a Work published in the last Century.

1791*   Observations on the Affinity between Basaltes and Granite.

1791* An account of some appearances attending the conversion of cast into malleable Iron.
1792* Second Part of ditto.
* These three papers appeared in the Philosophical Transactions for 1791, and 1792.
Uncertain: Memorial addressed to the Curators of the Bodleian Library.
1792 A Letter to a Lady on the subject of early Instruction, particularly that of the Poor; printed but not published.
1792 Alexander's Expedition to the Indian Ocean; printed but not published.
1792 Observations on the Nature of Demonstrative Evidence, with Reflections on Language.
1792 Observations on the Nature and Cure of Calculus, Sea-Scurvy, Catarrh and Fever.
1793 History of Isaac Jenkins.
1793 A Letter to Doctor Darwin, on a new mode of treating Pulmonary Consumption.
1794 Letters from Doctor Withering, Doctor Ewart, Doctor Thornton, etc.
1794 A Guide for Self-preservation and Parental Affection.
1794 A Proposal for the Improvement of Medicine.
1794 Considerations on the Medicinal Use and on the Production of Factitious Airs, Parts 1 and 2.
1795 Brown's Elements of Medicine, with a Preface and Notes.
1795 Translation, from the Spanish, of Gimbernat's new Method of operating in Femoral Hernia.
1795 Considerations &c. Part 3d.
1795 Outline of a Plan for determining the Medicinal Powers of Factitious Airs.
1795 A Word in Defence of the Bill of Rights against Gagging-bills.
1795 Where would be the harm of a Speedy Peace?
1796 An Essay on the Public Merits of Mr Pitt.
1796 A Letter to Mr Pitt on the Scarcity.
1796 Considerations, &c. Parts 4 and 5.
1797 Alternatives compared; or, What shall the Rich do to be safe?
1797 Suggestions towards setting on foot the projected Establishment for Pneumatic Medicine.
1797 Reports relating to Nitrous Acid.
1797 A Lecture introductory to a popular course of Anatomy.
1798 A suggestion towards an essential improvement in the Bristol Infirmary.
1799 Contributions to Medical and Physical Knowledge from the West of England.
1799 Popular Essay on Consumption.
1799 Notice of some observations made at the Pneumatic Institution.
1799 A second collection of Reports on Nitrous Acid.
1800 A Third ditto.

1801 Essay on the medical and domestic management of the Consumptive, on Digitalis and on Scrophula.

1801–2 Hygeia; or Essays, Moral and Medical, on the causes affecting the personal state of the middling and affluent classes.

1803 Rules of the Institution for the sick and drooping poor. An edition on larger paper was entitled Instruction for People of all Capacities respecting their own Health and that of their Children.

1806 The Manual of Health, or the Invalid conducted safely through the Seasons.

1807 On Fever as connected with Inflammation, an Exercise.

1808 A Letter to Sir Joseph Banks on the prevailing Discontents, Abuses and Imperfections in Medicine.

1808 Good Advice for the Husbandman in Harvest and for all those who labour hard in hot berths as also for others who will take it in warm weather.

In this List are not included a variety of communications to the Medical Facts and Observations, the Monthly Magazine, the Medical and Physical Journal, Nicholson's Journal, &c.

Biography

Stock, J. E.: 1811, Memoirs of the Life of Thomas Beddoes M.D.

Contemporary Works

Correspondence etc. with references to Thomas Beddoes

Southey, C. C.: 1850, Life and Correspondence of Robert Southey.

Muirhead, J. P.: 1854, Mechanical Inventions of James Watt, Vol. II, Letters.

Warter, J. W.: 1956, Selections from the Letters of Robert Southey.

Davy, J. (ed.): 1858, Fragmentary Remains of Sir Humphry Davy, Bart., with a Sketch of his Life.

Griggs, E. L.: 1932, Unpublished Letters of Samuel Taylor Coleridge, Constable.

Patton, L. and P. Mann (eds.): 1971, Lectures 1795, Vol. 1 of the Collected Works of Samuel Taylor Coleridge, General Editor K. Coburn, Bollingen Series 75, 1969–   , London and Princeton, N.J.

Patton, L. (ed.): 1970, The Watchman, Vol. 2 of the Collected Works of Samuel Taylor Coleridge (see above).

Coburn, K. (ed.): 1957–   , The Notebooks of Samuel Taylor Coleridge, Bollingen Series 50 (5 vols.), New York, Princeton, N.J. and London.

Dawson, W. R.: 1958, The Banks Letters, A Calendar of the Manuscript Correspondence, British Museum, Natural History.

King-Hele, D.: 1981, The Letters of Erasmus Darwin, Cambridge University Press, Cambridge.

General

Newte, J. A.: 1791, A Tour in England and Scotland.
Darwin, Erasmus: 1791, The Botanic Garden, The Poetical Works of E. Darwin, 1806.
Club des Jacobins, 4, 5. Bibliotheque Historique de la Révolution Discours de MM. Cooper et Watt 1792.
Paine, T.: 1792, Rights of Man, Pelican Books 1972, Penguin Books Ltd., Harmondsworth.
Plymley, J.: 1803, A General View of the Agriculture of Shropshire.
Davy, H.: 1800, Researches, Chemical and Philosophical; chiefly concerning Nitrous Oxide, or Dephlogisticated Nitrous Air and its Respiration, J. Johnson.
Huceks, J.: 1795, Pedestrian Tour Through North Wales.
Sadler, T. (ed.): 1869, The Diary and Correspondence of H. C. Robinson (Crabbe Robinson), Macmillan.
de Saint Fond, Faujas: 1797, Voyage en Angleterre, H. J. Jansen, Paris.
Geikie, Sir Archibald: 1907, Journey through England; Annotated translation of Faujas de Saint Fond, Voyage en Angleterre.

Later Works with substantial reference to Thomas Beddoes

Cottle, A.: 1847, Reminiscences of Coleridge and Southey.
Thorpe, T. E.: 1896, Humphry Davy, Poet and Philosopher, Cassell.
Litchfield, R. B.: 1903, Tom Wedgwood 1771–1805, Duckworth and Co.
Hodgson, J. E.: James Sadler of Oxford, Aeronaut, Chemist, Engineer and Inventor, in Newcomen Society Transactions, Vol. VIII, 1927–8.
Weber, C. A.: 1935, Bristols Bedeutung für die Englische Romantik und die Deutsch-Englischen Beziehungen, Max Niemeyer, Halle (Saale).
Cartwright, F. F.: 1952, The English Pioneers of Anaesthesia.
King-Hele, D.: 1963, Erasmus Darwin, Macmillan.
Todd, A. C.: 1967, Beyond the Blaze, A Biography of Davies Gilbert [Giddy], D. Bradford Barton Ltd., Truro.
Emblen, D. L.: 1970, Peter Mark Roget, Longman Group Ltd.
Hartley, Sir Harold: 1972, Humphry Davy, E. P. Publishing Ltd., Wakefield.
Levere, T. H.: 1981, Poetry Realised in Nature, Samuel Taylor Coleridge and early nineteenth-century science, Cambridge University Press.

Additional Reading

Aspinall, A.: 1949, Politics and the Press 1780–1850, Harvester Press (1973).
Grant, Sir Alexander: 1884, The Story of Edinburgh University.
Watson, George: 1921, The Edgeworths and their Circle, in The Encyclopaedia and Dictionary of Education, I. Pitman.
Gunther, R. T.: 1923, Early Science in Oxford, Vol. I, Oxford Historical Society.

Hodgson, J. E.: 1924, History of Aeronautics in Great Britain, Oxford University Press, Oxford.

Garrison, F. H. and W. B. Saunders: 1929, History of Medicine, Philadelphia and London.

Partington, J. R.: 1930, A Text Book of Inorganic Chemistry, Macmillan and Co.

Gunther, R. T.: 1933, The Old Ashmolean, Oxford Historical Society.

LeFebvre, G. Trans. R. R. Palmer: 1947, The Coming of the French Revolution, Princeton Paperback Edition 1967, Princeton University Press, Princeton, New Jersey.

Ferguson, A. (ed.): 1972, Natural Philosophy Through the Eighteenth Century, First published 1948 as a Supplement to the Philosophical.

Goodwin, A.: 1953, The French Revolution, Hutchinson and Co.

Crosland, M. P.: 1962, Historical Studies in the Language of Chemistry, Dover Publications, Inc., New York.

Schofield, R. E.: 1963, The Lunar Society of Birmingham, Oxford.

Green, V. H. H.: 1964, Religion at Oxford and Cambridge.

Crane, V. W.: 1966, The Club of Honest Whigs, Friends of Science and Liberty, Vol. 23, Third Series, William and Mary Quarterly.

Stewart, W. A. Campbell and W. P. McCann: 1967, The Educational Innovators, 1750–1880.

Klingender, F. D. (ed. A. Elton): 1968, Art and the Industrial Revolution, Paladin, Granada Publishing Ltd., St. Albans.

Cone, C. B.: 1968, The English Jacobins, New York.

Musson, A. E. and E. Robinson: 1969, Science and Technology in the Industrial Revolution.

Ayling, S.: 1972, George the Third, Collins.

Trinder, B.: 1973, The Industrial History of Shropshire, Phillimore and Co. Ltd., Chichester.

Goodwin, A.: 1979, The Friends of Liberty, Hutchinson.

Smith, S.: 1979, A View from the Iron Bridge, Ironbridge Gorge Museum Trust.

Lawrence, C.: The Nervous System in the Scottish Enlightenment, Ch. I in Barnes, B. and S. Shapin (eds.): 1979, The Natural Order; Historical Studies of Scientific Culture.

Trinder, B. and J. Cox: 1980, Yeomen and Colliers in Telford Phillimore and Co. Ltd., Chichester.

Porter, R.: 1982, English Society in the Eighteenth Century, Pelican Books, Penguin Books Ltd., Harmondsworth.

Journal Articles

Jones, T. W.: 1925, Thomas Beddoes M.D., 'A Neglected Chemist', Science Progress, Vol. 19, Jan.–April 1925, pp. 628–39.

Parker, G.: 1928, 'The Discovery of the Anaesthetic Powers of Nitrous Oxide', The Lancet, 7 Jan. 1928, pp. 60–61.

Miller, A.: 1931, 'The Pneumatic Institution of Thomas Beddoes at Bristol', Annals of Medical History, New York.

Collier Abbott, C.: 1924, 'The Parents of T. L. Beddoes', Durham University Journal, June 1924, New series Vol. III, No. 3.

Todd, A. C.: 1952, 'T. L. Beddoes and his Guardian', Times Literary Supplement, Oct. 10, 1952.

Robinson, E.: 1955, 'The Reform of Science Teaching in Oxford', Annals of Science, II.

Todd, A. C.: 1957, 'Anna Maria, Mother of Thomas Lovell Beddoes', Studia Neophilologica, Vol. 29.

Anon. [Williams, T. I.] : Endeavour, July 1960.

Armytage, W. H. G.: 1960, 'Thomas Beddoes M.D. 1760–1808', British Medical Journal, 30 April 1960, pp. 1358–9.

Gibbs, F. W. and Smeaton, W. A.: 1961, 'Thomas Beddoes at Oxford', Ambix, 9 Feb. 1961.

Erdman, D. V.: 1965, 'Coleridge, Wordsworth and the Wedgwood Fund', Bulletin of the New York Public Library, Vol. 60, No. 9.

Gottleib, L. S.: 1965, 'Thomas Beddoes M.D. and the Pneumatic Institution at Clifton 1798–1801', Annals of Internal Medicine 63.5, pp. 530–33.

Cartwright, F. F.: 1967, 'The Association of Thomas Beddoes, M.D. with James Watt, F.R.S.', Notes and Records of the Royal Society of London, Vol. 22, Nos. 1 and 2.

Smith, W. D. A.: 1965–70, 'A History of Nitrous Oxide and Oxygen Anaesthesia, Part I: Joseph Priestley to Humphry Davy', Brit. J. Anaesth., Vol. 37, 1965, pp. 790–798; 'Part II: Davy's researches in relation to inhalation anaesthesia', ibid., Vol. 37, 1965, pp. 871–882; 'Part III: Parsons Shaw, Doctor Syntax and Nitrous Oxide', ibid., Vol 37, 1965, pp. 958–966; 'Part IV: Hickman and the "Introduction of certain gases into the lungs"', ibid., Vol. 38, 1966, pp. 58–72. 'Part IV$_A$: Further light on Hickman and his times', ibid., Vol. 42, 1970, pp. 347–353 and 445–458.

Key, T. E.: 1969, 'The Early Pneumatic Chemists and Physicians', Journal of Anaesthesiology – Journal of the American Society of American Society of Anaesthesiologists', Vol. 30, No. 4.

Levere, T. H.: 1977, 'Dr Thomas Beddoes and the establishment of his Pneumatic Institution', Notes and Records of the Royal Society of London, Vol. 32, 1977, pp. 41–49.

Stansfield, D. A.: 1979, 'Thomas Beddoes and education', History of Education Society Bulletin, No. 23, Spring 1979, pp. 7–14.

Levere, T. H.: 1981, 'Dr Thomas Beddoes at Oxford, Raidcal Politics in 1788–93, and the Fate of the Regius Chair in Chemistry', Ambix, Vol. 28, Part 2, July 1981.

Levere, T. H.: 1982, 'Thomas Beddoes, The Interaction of Pneumatic and Preventive Medicine with Chemistry', Interdisciplinary Science Reivew, Vol. 7, pp. 137–147.

INDEX OF NAMES

289

Voss, J. H. 115

Watt, Gregory 159
Watt, James 5, 26, 35, 62, 73, 174, 250
   and Pneumatic Institute 156–8
   breathing apparatus 152–4
Watt, James jnr 55, 157, 190–1, 232
Watt, Jessie 153, 156, 218
Watt, Robert (of Edinburgh) 3
Weber, C. A. 252
Webster, Dr Charles 32
Wedekind, G. 108
Wedgwood family 105
Wedgwood, John (1766–1844) 176
Wedgwood, Josiah I (1730–1795) 77,
   104, 155
Wedgwood, Josiah II (1769–1843) 94–
   95, 160, 167

Wedgwood, Thomas (1771–1805) 1, 94–
   95, 123, 139–140, 157, 159,
   167, 176, 189, 220, 231–232,
   242
Wells, H. 250
Whitehall, Mr R. 9, 223
Whitehall, Mrs 103, 122
Willick, A. F. M. 116
Willoughby, Capt. and Miss 241
Withering, Dr 28–9, 45, 73, 123, 150–1,
   155–6, 226
Woodhouse, Prof. 251
Wordsworth, D. 138, 140
Wordsworth, W. 80, 136, 138, 140, 144,
   173, 240, 252

Yonge, Dr 13, 79, 97, 121, 145
Yonge, Miss 122

INDEX OF SUBJECTS